SiGe and Si Strained-Layer Epitaxy for Silicon Heterostructure Devices

SiGe and Si Strained-Layer Epitaxy for Silicon Heterostructure Devices

Edited by
John D. Cressler

CRC Press
Taylor & Francis Group
Boca Raton London New York

CRC Press is an imprint of the
Taylor & Francis Group, an **informa** business

The material was previously published in *Silicon Heterostructure Handbook: Materials, Fabrication, Devices, Circuits and Applications of SiGe and Si Strained-Layer Epitaxy* © Taylor and Francis 2005.

CRC Press
Taylor & Francis Group
6000 Broken Sound Parkway NW, Suite 300
Boca Raton, FL 33487-2742

Printed in the United States of America on acid-free paper
10 9 8 7 6 5 4 3 2 1

International Standard Book Number-13: 978-1-4200-6685-2 (Hardcover)

Library of Congress Cataloging-in-Publication Data

SiGe and Si strained-layer epitaxy for silicon heterostructure devices / editor, John D. Cressler.
p. cm.
Includes bibliographical references and index.
ISBN 978-1-4200-6685-2 (alk. paper)
1. Bipolar transistors--Materials. 2. Heterostructures. 3.
Silicon--Electric properties. 4. Epitaxy. I. Cressler, John D.

TK7871.96.B55S53 2008
621.3815'28--dc22 2007030343

Visit the Taylor & Francis Web site at
http://www.taylorandfrancis.com

and the CRC Press Web site at
http://www.crcpress.com

For the tireless efforts
Of the many dedicated scientists and engineers
Who helped create this field and make it a success.
I tip my hat, and offer sincere thanks from all of us
Who have benefitted from your keen insights and imaginings.

And ...

For Maria:
My beautiful wife, best friend, and soul mate for these 25 years.
For Matthew John, Christina Elizabeth, and Joanna Marie:
God's awesome creations, and our precious gifts.
May your journey of discovery never end.

He Whose Heart Has Been Set
On The Love Of Learning And True Wisdom
And Has Exercised This Part of Himself,
That Man Must Without Fail Have Thoughts
That Are Immortal And Divine,
If He Lay Hold On Truth.

Plato

Εκείνος που έχει δώσει την
ψυχή του στην Αγάπη για Μάθηση
και Αληθινή Σοφία,
και έχει Ασκηθεί για τούτο,
Ένας τέτοιος μόνο Άνδρας μπορεί
το δίχως άλλο να κάνει σκέψεις
Αθάνατες και Θείες,
Εάν στηριχθεί στην Αλήθεια.

Πλάτωνας

Foreword

Progress in a given field of technology is both desired and expected to follow a stable and predictable long-term trajectory. Semilog plots of technology trends spanning decades in time and orders of magnitude in value abound. Perhaps the most famous exemplar of such a technology trajectory is the trend line associated with Moore's law, where technology density has doubled every 12 to 18 months for several decades. One must not, however, be lulled into extrapolating such predictability to other aspects of semiconductor technology, such as device performance, or even to the long-term prospects for the continuance of device density scaling itself. New physical phenomena assert themselves as one approaches the limits of a physical system, as when device layers approach atomic dimensions, and thus, no extrapolation goes on indefinitely.

Technology density and performance trends, though individually constant over many years, are the result of an enormously complex interaction between a series of decisions made as to the layout of a given device, the physics behind its operation, manufacturability considerations, and its extensibility into the future. This complexity poses a fundamental challenge to the device physics and engineering community, which must delve as far forward into the future as possible to understand when physical law precludes further progress down a given technology path. The early identification of such impending technological discontinuities, thus providing time to ameliorate their consequences, is in fact vital to the health of the semiconductor industry. Recently disrupted trends in CMOS microprocessor performance, where the "value" of processor-operating frequency was suddenly subordinated to that of integration, demonstrate the challenges remaining in accurately assessing the behavior of future technologies. However, current challenges faced in scaling deep submicron CMOS technology are far from unique in the history of semiconductors.

Bipolar junction transistor (BJT) technology, dominant in high-end computing applications during the mid-1980s, was being aggressively scaled to provide the requisite performance for future systems. By the virtue of bipolar transistors being vertical devices rather than lateral (as CMOS is), the length scale of bipolar transistors is set by the ability to control layer thicknesses rather than lateral dimensions. This allowed the definition of critical device dimensions, such as base width, to values far below the limits of optical lithography of the day. Although great strides in device performance had been made by 1985, with unity gain cutoff frequencies (f_T) in the range 20–30 GHz seemingly feasible, device scaling was approaching limits at which new physical phenomena became significant. Highly scaled silicon BJTs, having base widths below 1000 Å, demonstrated inordinately high reverse junction leakage. This was due to the onset of band-to-band tunneling between heavily doped emitter and base regions, rendering such devices unreliable. This and other observations presaged one of the seminal technology discontinuities of the past decade, silicon–germanium (SiGe) heterojunction bipolar transistor (HBT) technology being the direct consequence.

Begun as a program to develop bipolar technology with performance capabilities well beyond those possible via the continued scaling of conventional Si BJTs, SiGe HBT technology has found a wealth of applications beyond the realm of computing. A revolution in bipolar fabrication methodology, moving from device definition by implantation to device deposition and definition by epitaxy, accompanied by the exploitation of bandgap tailoring, took silicon-based bipolar transistor performance to levels never anticipated. It is now common to find SiGe HBTs with performance figures in excess of 300 GHz for both f_T and f_{max}, and circuits operable at frequencies in excess of 100 GHz.

A key observation is that none of this progress occurred in a vacuum, other than perhaps in the field of materials deposition. The creation of a generation of transistor technology having tenfold improved performance would of itself have produced far less ultimate value in the absence of an adequate eco-system to enable its effective creation and utilization. This text is meant to describe the eco-system that developed around SiGe technology as context for the extraordinary achievement its commercial rollout represented.

Early SiGe materials, of excellent quality in the context of fundamental physical studies, proved near useless in later device endeavors, forcing dramatic improvements in layer control and quality to then enable further development. Rapid device progress that followed drove silicon-based technology (recall that SiGe technology is still a silicon-based derivative) to unanticipated performance levels, demanding the development of new characterization and device modeling techniques. As materials work was further proven SiGe applications expanded to leverage newly available structural and chemical control.

Devices employing ever more sophisticated extensions of SiGe HBT bandgap tailoring have emerged, utilizing band offsets and the tailoring thereof to create SiGe-based HEMTs, tunneling devices, mobility-enhanced CMOS, optical detectors, and more to come. Progress in these diverse areas of device design is timely, as I have already noted the now asymptotic nature of performance gains to be had from continued classical device scaling, leading to a new industry focus on innovation rather than pure scaling. Devices now emerging in SiGe are not only to be valued for their performance, but rather their variety of functionality, where, for example, optically active components open up the prospect of the seamless integration of broadband communication functionality at the chip level.

Access to high-performance SiGe technology has spurred a rich diversity of exploratory and commercial circuit applications, many elaborated in this text. Communications applications have been most significantly impacted from a commercial perspective, leveraging the ability of SiGe technologies to produce extremely high-performance circuits while using back level, and thus far less costly, fabricators than alternative materials such as InP, GaAs, or in some instances advanced CMOS.

These achievements did not occur without tremendous effort on the part of many workers in the field, and the chapters in this volume represent examples of such contributions. In its transition from scientific curiosity to pervasive technology, SiGe-based device work has matured greatly, and I hope you find this text illuminating as to the path that maturation followed.

Bernard S. Meyerson
IBM Systems and Technology Group

Preface

While the idea of cleverly using silicon–germanium (SiGe) and silicon (Si) strained-layer epitaxy to practice bandgap engineering of semiconductor devices in the highly manufacturable Si material system is an old one, only in the past decade has this concept become a practical reality. The final success of creating novel Si heterostructure transistors with performance far superior to their Si-only homojunction cousins, while maintaining strict compatibility with the massive economy-of-scale of conventional Si integrated circuit manufacturing, proved challenging and represents the sustained efforts of literally thousands of physicists, electrical engineers, material scientists, chemists, and technicians across the world.

In the electronics domain, the fruit of that global effort is SiGe heterojunction bipolar transistor (SiGe HBT) BiCMOS technology, and strained Si/SiGe CMOS technology, both of which are at present in commercial manufacturing worldwide and are rapidly finding a number of important circuit and system applications. As with any new integrated circuit technology, the industry is still actively exploring device performance and scaling limits (at present well above 300 GHz in frequency response, and rising), new circuit applications and potential new markets, as well as a host of novel device and structural innovations. This commercial success in the electronics arena is also spawning successful forays into the optoelectronics and even nanoelectronics fields. The Si heterostructure field is both exciting and dynamic in its scope.

The implications of the Si heterostructure success story contained in this book are far-ranging and will be both lasting and influential in determining the future course of the electronics and optoelectronics infrastructure, fueling the miraculous communications explosion of the twenty-first century. While several excellent books on specific aspects of the Si heterostructures field currently exist (for example, on SiGe HBTs), this is the first reference book of its kind that "brings-it-all-together," effectively presenting a comprehensive perspective by providing very broad topical coverage ranging from materials, to fabrication, to devices (HBT, FET, optoelectronic, and nanostructure), to CAD, to circuits, to applications. Each chapter is written by a leading international expert, ensuring adequate depth of coverage, up-to-date research results, and a comprehensive list of seminal references. A novel aspect of this book is that it also contains "snap-shot" views of the industrial "state-of-the-art," for both devices and circuits, and is designed to provide the reader with a useful basis of comparison for the current status and future course of the global Si heterostructure industry.

This book is intended for a number of different audiences and venues. It should prove to be a useful resource as:

1. A hands-on reference for practicing engineers and scientists working on various aspects of Si heterostructure integrated circuit technology (both HBT, FET, and optoelectronic), including materials, fabrication, device physics, transistor optimization, measurement, compact modeling and device simulation, circuit design, and applications
2. A hands-on research resource for graduate students in electrical and computer engineering, physics, or materials science who require information on cutting-edge integrated circuit technologies

3. A textbook for use in graduate-level instruction in this field
4. A reference for technical managers and even technical support/technical sales personnel in the semiconductor industry

It is assumed that the reader has some modest background in semiconductor physics and semiconductor devices (at the advanced undergraduate level), but each chapter is self-contained in its treatment.

In this age of extreme activity, in which we are all seriously pressed for time and overworked, my success in getting such a large collection of rather famous people to commit their precious time to my vision for this project was immensely satisfying. I am happy to say that my authors made the process quite painless, and I am extremely grateful for their help. The list of contributors to this book actually reads like a global "who's who" of the silicon heterostructure field, and is impressive by any standard. I would like to formally thank each of my colleagues for their hard work and dedication to executing my vision of producing a lasting Si heterostructure "bible." In order of appearance, the "gurus" of our field include:

Bernd Tillack, IHP, Germany
Peter Zaumseil, IHP, Germany
Didier Dutartre, ST Microelectronics, France
F. Deléglise, ST Microelectronics, France
C. Fellous, ST Microelectronics, France
L. Rubaldo, ST Microelectronics, France
A. Talbot, ST Microelectronics, France
Michael Oehme, University of Stuttgart, Germany
Erich Kasper, University of Stuttgart, Germany
Thomas N. Adam, IBM Semiconductor Research and Development Center, USA
Anthony R. Peaker, University of Manchester, United Kingdom
V.P. Markevich, University of Manchester, United Kingdom
Armin Fischer, Innovations for High Performance Microelectronics (IHP), Germany
Judy L. Hoyt, Massachusetts Institute of Technology, USA
H. Jörg Osten, University of Hanover, Germany
C.K. Maiti, Indian Institute of Technology Kharagpur, India
S. Monfray, ST Microelectronics, France
Thomas Skotnicki, ST Microelectronics, France
S. Borel, CEA LETI, France
Michael Schröter, University of California at San Diego, USA
Ramana M. Malladi, IBM Microelectronics, USA

I would also like to thank my graduate students and post-docs, past and present, for their dedication and tireless work in this fascinating field. I rest on their shoulders. They include: David Richey, Alvin Joseph, Bill Ansley, Juan Roldán, Stacey Salmon, Lakshmi Vempati, Jeff Babcock, Suraj Mathew, Kartik Jayanaraynan, Greg Bradford, Usha Gogineni, Gaurab Banerjee, Shiming Zhang, Krish Shivaram, Dave Sheridan, Gang Zhang, Ying Li, Zhenrong Jin, Qingqing Liang, Ram Krithivasan, Yun Luo, Tianbing Chen, Enhai Zhao, Yuan Lu, Chendong Zhu, Jon Comeau, Jarle Johansen, Joel Andrews, Lance Kuo, Xiangtao Li, Bhaskar Banerjee, Curtis Grens, Akil Sutton, Adnan Ahmed, Becca Haugerud, Mustayeen Nayeem, Mustansir Pratapgarhwala, Guofu Niu, Emery Chen, Jongsoo Lee, and Gnana Prakash.

Finally, I am grateful to Tai Soda at Taylor & Francis for talking me into this project, and supporting me along the way. I would also like to thank the production team at Taylor & Francis for their able assistance (and patience!), especially Jessica Vakili.

The many nuances of the Si heterostructure field make for some fascinating subject matter, but this is no mere academic pursuit. In the grand scheme of things, the Si heterostructure industry is already reshaping the global communications infrastructure, which is in turn dramatically reshaping the way life on planet Earth will transpire in the twenty-first century and beyond. The world would do well to pay attention. It has been immensely satisfying to see both the dream of Si/SiGe bandgap engineering, and this book, come to fruition. I hope our efforts please you. Enjoy!

John D. Cressler
Editor

Editor

John D. Cressler received a B.S. in physics from the Georgia Institute of Technology (Georgia Tech), Atlanta, Georgia, in 1984, and an M.S. and Ph.D. in applied physics from Columbia University, New York, in 1987 and 1990. From 1984 to 1992 he was on the research staff at the IBM Thomas J. Watson Research Center in Yorktown Heights, New York, working on high-speed Si and SiGe bipolar devices and technology. In 1992 he left IBM Research to join the faculty at Auburn University, Auburn, Alabama, where he served until 2002. When he left Auburn University, he was Philpott–Westpoint Stevens Distinguished Professor of Electrical and Computer Engineering and director of the Alabama Microelectronics Science and Technology Center.

In 2002, Dr. Cressler joined the faculty at Georgia Tech, where he is currently Ken Byers Professor of Electrical and Computer Engineering. His research interests include SiGe devices and technology; Si-based RF/microwave/millimeter-wave mixed-signal devices and circuits; radiation effects; device-circuit interactions; noise and linearity; reliability physics; extreme environment electronics, 2-D/3-D device-level simulation; and compact circuit modeling. He has published more than 350 technical papers related to his research, and is author of the books *Silicon-Germanium Heterojunction Bipolar Transistors*, Artech House, 2003 (with Guofu Niu), and *Reinventing Teenagers: The Gentle Art of Instilling Character in Our Young People*, Xlibris, 2004 (a slightly different genre!).

Dr. Cressler was Associate Editor of the *IEEE Journal of Solid-State Circuits* (1998–2001), Guest Editor of the *IEEE Transactions on Nuclear Science* (2003–2006), and Associate Editor of the *IEEE Transactions on Electron Devices* (2005–present). He served on the technical program committees of the IEEE International Solid-State Circuits Conference (1992–1998, 1999–2001), the IEEE Bipolar/BiCMOS Circuits and Technology Meeting (1995–1999, 2005–present), the IEEE International Electron Devices Meeting (1996–1997), and the IEEE Nuclear and Space Radiation Effects Conference (1999–2000, 2002–2007). He currently serves on the executive steering committee for the IEEE Topical Meeting on Silicon Monolithic Integrated Circuits in RF Systems, as international program advisor for the IEEE European Workshop on Low-Temperature Electronics, on the technical program committee for the IEEE International SiGe Technology and Device Meeting, and as subcommittee chair of the 2004 Electrochemical Society Symposium of SiGe: Materials, Processing, and Devices. He was the Technical Program Chair of the 1998 IEEE International Solid-State Circuits Conference, the Conference Co-Chair of the 2004 IEEE Topical Meeting on Silicon Monolithic Integrated Circuits in RF Systems, and the Technical Program Chair of the 2007 IEEE Nuclear and Space Radiation Effects Conference.

Dr. Cressler was appointed an IEEE Electron Device Society Distinguished Lecturer in 1994, an IEEE Nuclear and Plasma Sciences Distinguished Lecturer in 2006, and was awarded the 1994 Office of Naval Research Young Investigator Award for his SiGe research program. He received the 1996 C. Holmes

MacDonald National Outstanding Teacher Award by Eta Kappa Nu, the 1996 Auburn University Alumni Engineering Council Research Award, the 1998 Auburn University Birdsong Merit Teaching Award, the 1999 Auburn University Alumni Undergraduate Teaching Excellence Award, an IEEE Third Millennium Medal in 2000, and the 2007 Georgia Tech Outstanding Faculty Leadership in the Development of Graduate Students Award. He is an IEEE Fellow.

On a more personal note, John's hobbies include hiking, gardening, bonsai, all things Italian, collecting (and drinking!) fine wines, cooking, history, and carving walking sticks, not necessarily in that order. He considers teaching to be his vocation. John has been married to Maria, his best friend and soul-mate, for 25 years, and is the proud father of three budding scholars: Matt, Christina, and Jo-Jo.

Dr. Cressler can be reached at School of Electrical and Computer Engineering, 777 Atlantic Drive, N.W., Georgia Institute of Technology, Atlanta, GA 30332-0250 U.S.A. or cressler@ece.gatech.edu http://users.ece.gatech.edu/~cressler/

Contents

1

The Big Picture

John D. Cressler
Georgia Institute of Technology

1.1 The Communications Revolution

We are at a unique juncture in the history of humankind, a juncture that amazingly we engineers and scientists have dreamed up and essentially created on our own. This pivotal event can be aptly termed the "Communications Revolution," and the twenty-first century, our century, will be the era of human history in which this revolution plays itself out.

This communications revolution can be functionally defined and characterized by the pervasive acquisition, manipulation, storage, transformation, and transmission of "information" on a global scale. This information, or more generally, knowledge, in its infinitely varied forms and levels of complexity, is gathered from our analog sensory world, transformed in very clever ways into logical "1"s and "0"s for ease of manipulation, storage, and transmission, and subsequently regenerated into analog sensory output for our use and appreciation. In 2005, this planetary communication of information is occurring at a truly mind-numbing rate, estimates of which are on the order of 80 Tera-bits/sec (10^{12}) of data transfer across the globe in 2005 solely in wired and wireless voice and data transmission, 24 hours a day, 7 days a week, and growing exponentially. The world is quite literally abuzz with information flow—communication.* It is for the birth of the Communications Revolution that we humans likely will be remembered for 1000 years hence. Given that this revolution is happening during the working careers of most of us, I find it a wonderful time to be alive, a fact of which I remind my students often.

Here is my point. No matter how one slices it, at the most fundamental level, it is semiconductor devices that are powering this communications revolution. Skeptical? Imagine for a moment that one could flip a switch and instantly remove all of the integrated circuits (ICs) from planet Earth. A moment's reflection will convince you that there is not a single field of human endeavor that would not come to a grinding halt, be it commerce, or agriculture, or education, or medicine, or entertainment. Life as we in the first world know it in 2005 would simply cease to exist. And yet, remarkably, the same result would not have been true 50 years ago; even 20 years ago. Given the fact that we humans have been on planet Earth in our present form for at least 1 million years, and within communities

*I have often joked with my students that it would be truly entertaining if the human retina was sensitive to longer wavelengths of electromagnetic radiation, such that we could "see" all the wireless communications signals constantly bathing the planet (say, in greens and blues!). It might change our feelings regarding our ubiquitous cell phones!

having entrenched cultural traditions for at least 15,000 years, this is truly a remarkable fact of history. A unique juncture indeed.

Okay, hold on tight. It is an easy case to make that the semiconductor silicon (Si) has single-handedly enabled this communications revolution.* I have previously extolled at length the remarkable virtues of this rather unglamorous looking silver-grey element [1], and I will not repeat that discussion here, but suffice it to say that Si represents an extremely unique material system that has, almost on its own, enabled the conception and evolving execution of this communications revolution. The most compelling attribute, by far, of Si lies in the economy-of-scale it facilitates, culminating in the modern IC fabrication facility, effectively enabling the production of gazillions of low-cost, very highly integrated, remarkably powerful ICs, each containing millions of transistors; ICs that can then be affordably placed into widgets of remarkably varied form and function.†

So what does this have to do with the book you hold in your hands? To feed the emerging infrastructure required to support this communications revolution, IC designers must work tirelessly to support increasingly higher data rates, at increasingly higher carrier frequencies, all in the design space of decreasing form factor, exponentially increasing functionality, and at ever-decreasing cost. And by the way, the world is going portable and wireless, using the same old wimpy batteries. Clearly, satisfying the near-insatiable appetite of the requisite communications infrastructure is no small task. Think of it as job security!

For long-term success, this quest for more powerful ICs must be conducted within the confines of conventional Si IC fabrication, so that the massive economy-of-scale of the global Si IC industry can be brought to bear. Therein lies the fundamental motivation for the field of Si heterostructures, and thus this book. Can one use clever nanoscale engineering techniques to custom-tailor the energy bandgap of fairly conventional Si-based transistors to: (a) improve their performance dramatically and thereby ease the circuit and system design constraints facing IC designers, while (b) performing this feat without throwing away all the compelling economy-of-scale virtues of Si manufacturing? The answer to this important question is a resounding "YES!" That said, getting there took time, vision, as well as dedication and hard work of literally thousands of scientists and engineers across the globe.

In the electronics domain, the fruit of that global effort is silicon–germanium heterojunction bipolar transistor (SiGe HBT) bipolar complementary metal oxide semiconductor (BiCMOS) technology, and is in commercial manufacturing worldwide and is rapidly finding a number of important circuit and system applications. In 2004, the SiGe ICs, by themselves, are expected to generate US$1 billion in revenue globally, with perhaps US$30 billion in downstream products. This US$1 billion figure is projected to rise to US$2.09 billion by 2006 [2], representing a growth rate of roughly 42% per year, a remarkable figure by any economic standard. The biggest single market driver remains the cellular industry, but applications in optical networking, hard disk drives for storage, and automotive collision-avoidance radar systems are expected to represent future high growth areas for SiGe. And yet, in the beginning of 1987, only 18 years ago, there was no such thing as a SiGe HBT. It had not been demonstrated as a viable concept. An amazing fact.

In parallel with the highly successful development of SiGe HBT technology, a wide class of "transport enhanced" field effect transistor topologies (e.g., strained Si CMOS) have been developed as a means to boost the performance of the CMOS side of Si IC coin, and such technologies have also recently begun

---

*The lone exception to this bold claim lies in the generation and detection of coherent light, which requires direct bandgap III–V semiconductor devices (e.g., GaAs of InP), and without which long-haul fiber communications systems would not be viable, at least for the moment.

†Consider: it has been estimated that in 2005 there are roughly 20,000,000,000,000,000,000 (2×10^{19}) transistors on planet Earth. While this sounds like a large number, let us compare it to some other large numbers: (1) the universe is roughly 4.2×10^{17}sec old (13.7 billion years), (2) there are about 1×10^{21} stars in the universe, and (3) the universe is about 4×10^{23} miles across (15 billion light-years)! Given the fact that all 2×10^{20} of these transistors have been produced since December 23, 1947 (following the invention of the point-contact transistor by Bardeen, Brattain, and Shockley), this is a truly remarkable feat of human ingenuity.

to enter the marketplace as enhancements to conventional core CMOS technologies. The commercial success enjoyed in the electronics arena has very naturally also spawned successful forays into the optoelectronics and even nanoelectronics fields, with potential for a host of important downstream applications.

The Si heterostructure field is both exciting and dynamic in its scope. The implications of the Si heterostructure success story contained in this book are far-ranging and will be both lasting and influential in determining the future course of the electronics and optoelectronics infrastructure, fueling the miraculous communications explosion of our twenty-first century. The many nuances of the Si heterostructure field make for some fascinating subject matter, but this is no mere academic pursuit. As I have argued, in the grand scheme of things, the Si heterostructure industry is already reshaping the global communications infrastructure, which is in turn dramatically reshaping the way life of planet Earth will transpire in the twenty-first century and beyond. The world would do well to pay close attention.

1.2 Bandgap Engineering in the Silicon Material System

As wonderful as Si is from a fabrication viewpoint, from a device or circuit designer's perspective, it is hardly the ideal semiconductor. The carrier mobility for both electrons and holes in Si is comparatively small compared to their III–V cousins, and the maximum velocity that these carriers can attain under high electric fields is limited to about 1×10^7 cm/sec under normal conditions, relatively "slow." Since the speed of a transistor ultimately depends on how fast the carriers can be transported through the device under sustainable operating voltages, Si can thus be regarded as a somewhat "meager" semiconductor. In addition, because Si is an indirect gap semiconductor, light emission is fairly inefficient, making active optical devices such as diode lasers impractical (at least for the present). Many of the III–V compound semiconductors (e.g., GaAs or InP), on the other hand, enjoy far higher mobilities and saturation velocities, and because of their direct gap nature, generally make efficient optical generation and detection devices. In addition, III–V devices, by virtue of the way they are grown, can be compositionally altered for a specific need or application (e.g., to tune the light output of a diode laser to a specific wavelength). This atomic-level custom tailoring of a semiconductor is called bandgap engineering, and yields a large performance advantage for III–V technologies over Si [3]. Unfortunately, these benefits commonly associated with III–V semiconductors pale in comparison to the practical deficiencies associated with making highly integrated, low-cost ICs from these materials. There is no robust thermally grown oxide for GaAs or InP, for instance, and wafers are smaller with much higher defect densities, are more prone to breakage, and are poorer heat conductors (the list could go on). These deficiencies translate into generally lower levels of integration, more difficult fabrication, lower yield, and ultimately higher cost. In truth, of course, III–V materials such as GaAs and InP fill important niche markets today (e.g., GaAs metal semiconductor field effect transistor (MESFETs) and HBTs for cell phone power amplifiers, AlGaAs- or InP-based lasers, efficient long wavelength photodetectors, etc.), and will for the foreseeable future, but III–V semiconductor technologies will never become mainstream in the infrastructure of the communications revolution if Si-based technologies can do the job.

While Si ICs are well suited to high-transistor-count, high-volume microprocessors and memory applications, RF, microwave, and even millimeter-wave (mm-wave) electronic circuit applications, which by definition operate at significantly higher frequencies, generally place much more restrictive performance demands on the transistor building blocks. In this regime, the poorer intrinsic speed of Si devices becomes problematic. That is, even if Si ICs are cheap, they must deliver the required device and circuit performance to produce a competitive system at a given frequency. If not, the higher-priced but faster III–V technologies will dominate (as they indeed have until very recently in the RF and microwave markets).

The fundamental question then becomes simple and eminently practical: is it possible to improve the performance of Si transistors enough to be competitive with III–V devices for high-performance applications, while preserving the enormous yield, cost, and manufacturing advantages associated with conventional Si fabrication? The answer is clearly "yes," and this book addresses the many nuances associated with using SiGe and Si-strained layer epitaxy to practice bandgap engineering in the Si material system, a process culminating in, among other things, the SiGe HBT and strained Si CMOS, as well as a variety of other interesting electronic and optoelectronic devices built from these materials. This totality can be termed the "Si heterostructures" field.

1.3 Terminology and Definitions

A few notes on modern usage and pronunciation in this field are in order (really!). It is technically correct to refer to silicon–germanium alloys according to their chemical composition, $Si_{1-x}Ge_x$, where x is the Ge mole fraction. Following standard usage, such alloys are generally referred to as "SiGe" alloys. Note, however, that it is common in the material science community to also refer to such materials as "Ge:Si" alloys.

A SiGe film that is carbon doped (e.g., less than 0.20% C) in an attempt to suppress subsequent boron out-diffusion (e.g., in HBTs) is properly referred to as a SiGe:C alloy, or simply SiGeC (pronounced "silicon germanium carbon," *not* "silicon germanium carbide"). This class of SiGe alloys should be viewed as optimized SiGe alloys, and are distinct from SiGe films with a much higher C content (e.g., 2% to 3% C) that might be used, for instance, to lattice-match SiGeC alloys to Si.

Believe it or not, this field also has its own set of slang pronunciations. The colloquial usage of the pronunciation \'sig-ee\ to refer to "silicon–germanium" (begun at IBM in the late 1990s) has come into vogue (heck, it may make it to the dictionary soon!), and has even entered the mainstream IC engineers's slang; pervasively.*

In the electronics domain, it is important to be able to distinguish between the various SiGe technologies as they evolve, both for CMOS (strained Si) and bipolar (SiGe HBT). Relevant questions in this context include: Is company X's SiGe technology more advanced than company Y's SiGe technology? For physical as well as historical reasons, one almost universally defines CMOS technology (Si, strained Si, or SiGe), a lateral transport device, by the drawn lithographic gate length (the CMOS technology "node"), regardless of the resultant intrinsic device performance. Thus, a "90-nm" CMOS node has a drawn gate length of roughly 90 nm. For bipolar devices (i.e., the SiGe HBT), however, this is not so straightforward, since it is a vertical transport device whose speed is not nearly as closely linked to lithographic dimensions.

In the case of the SiGe HBT it is useful to distinguish between different technology generations according to their resultant *ac* performance (e.g., peak common-emitter, unity gain cutoff frequency (f_T), which is (a) easily measured and unambiguously compared technology to technology, and yet is (b) a very strong function of the transistor vertical doping and Ge profile and hence nicely reflects the degree of sophistication in device structural design, overall thermal cycle, epi growth, etc.) [1]. The peak f_T generally nicely reflects the "aggressiveness," if you will, of the transistor scaling which has been applied to a given SiGe technology. A higher level of comparative sophistication can be attained by also invoking the maximum oscillation frequency (f_{max}), a parameter which is well correlated to both intrinsic profile and device parasitics, and hence a bit higher on the ladder of device performance metrics, and thus more representative of actual large-scale circuit performance. The difficulty in this case is that f_{max} is far more ambiguous than f_T, in the sense that it can be inferred from various gain definitions (e.g., U vs. MAG), and in practice power gain data are often far less ideal in its behavior over frequency, more sensitive to accurate deembedding, and ripe with extraction "issues."

We thus term a SiGe technology having a SiGe HBT with a peak f_T in the range of 50 GHz as "first generation;" that with a peak f_T in the range of 100 GHz as "second generation;" that with a peak f_T in the range of 200 GHz as "third generation;" and that with a peak f_T in the range of 300 GHz as "fourth generation." These are loose definitions to be sure, but nonetheless useful for comparison purposes.

*I remain a stalwart holdout against this snowballing trend and stubbornly cling to the longer but far more satisfying "silicon–germanium."

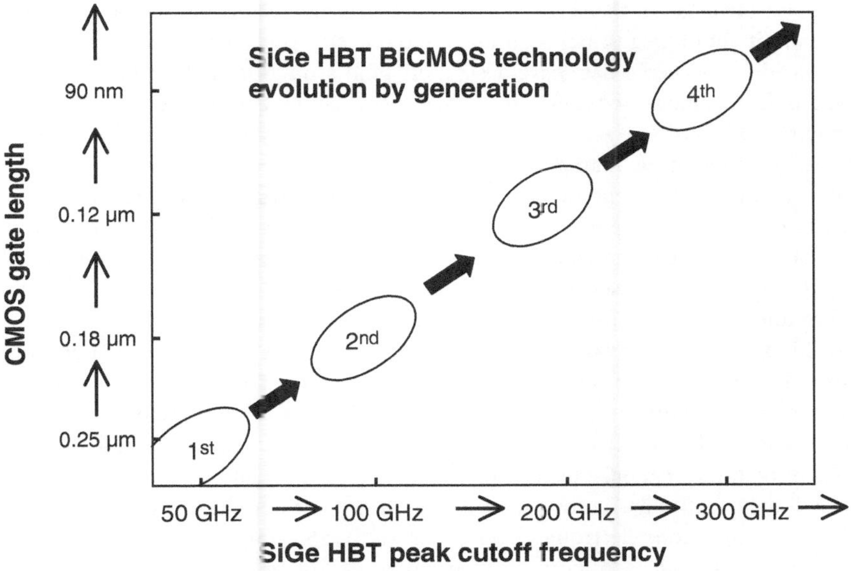

FIGURE 1.1 Evolution of SiGe HBT BiCMOS technology generations, as measured by the peak cutoff frequency of the SiGe HBT, and the CMOS gate length.

A complicating factor in SiGe technology terminology results from the fact that most, if not all, commercial SiGe HBT technologies today also contain standard Si CMOS devices (i.e., SiGe HBT BiCMOS technology) to realize high levels of integration and functionality on a single die (e.g., single-chip radios complete with RF front-end, data converters, and DSP). One can then speak of a given generation of SiGe HBT BiCMOS technology as the most appropriate intersection of both the SiGe HBT peak f_T and the CMOS technology node (Figure 1.1). For example, for several commercially important SiGe HBT technologies available via foundry services, we have:

- IBM SiGe 5HP—50 GHz peak f_T SiGe HBT + 0.35 μm Si CMOS (first generation)
- IBM SiGe 7HP—120 GHz peak f_T SiGe HBT + 0.18 μm Si CMOS (second generation)
- IBM SiGe 8HP—200 GHz peak f_T SiGe HBT + 0.13 μm Si CMOS (third generation)
- Jazz SiGe 60—60 GHz peak f_T SiGe HBT + 0.35 μm Si CMOS (first generation)
- Jazz SiGe 120—150 GHz peak f_T SiGe HBT + 0.18 μm Si CMOS (second generation)
- IHP SiGe SGC25B—120 GHz peak f_T SiGe HBT + 0.25 μm Si CMOS (second generation)

All SiGe HBT BiCMOS technologies can thus be roughly classified in this manner. It should also be understood that multiple transistor design points typically exist in such BiCMOS technologies (multiple breakdown voltages for the SiGe HBT and multiple threshold or breakdown voltages for the CMOS), and hence the reference to a given technology generation implicitly refers to the most aggressively scaled device within that specific technology platform.

1.4 The Application Space

It goes without saying in our field of semiconductor IC technology that no matter how clever or cool a new idea appears at first glance, its long-term impact will ultimately be judged by its marketplace "legs" (sad, but true). That is, was the idea good for a few journal papers and an award or two, or did someone actually build something and sell some useful derivative products from it? The sad reality is that the semiconductor field (and we are by no means exceptional) is rife with examples of cool new devices that never made it past the pages of the IEDM digest! The ultimate test, then, is one of stamina. And sweat. Did the idea make it out of the research laboratory and into the hands of the manufacturing lines? Did it pass the qualification-checkered flag, have design kits built around it, and get delivered to real circuit designers who built ICs, fabricated them, and tested them? Ultimately, were the derivative ICs inserted into real systems—widgets—to garner leverage in this or that system metric, and hence make the products more appealing in the marketplace?

Given the extremely wide scope of the semiconductor infrastructure fueling the communications revolution, and the sheer volume of widget possibilities, electronic to photonic to optoelectronic, it is useful here to briefly explore the intended application space of Si heterostructure technologies as we peer out into the future. Clearly I possess no crystal ball, but nevertheless some interesting and likely lasting themes are beginning to emerge from the fog.

SiGe HBT BiCMOS is the obvious ground-breaker of the Si heterostructures application space in terms of moving the ideas of our field into viable products for the marketplace. The field is young, but the signs are very encouraging. As can be seen in Figure 1.2, there are at present count 25 + SiGe HBT industrial fabrication facilities on line in 2005 around the world, and growing steadily. This trend points to an obvious recognition that SiGe technology will play an important role in the emerging electronics infrastructure of the twenty-first century. Indeed, as I often point out, the fact that virtually every major player in the communications electronics field either: (a) has SiGe up and running in-house, or (b) is using someone else's SiGe fab as foundry for their designers, is a remarkable fact, and very encouraging in the grand scheme of things. As indicated above, projections put SiGe ICs at a US$2.0 billion level by 2006, small by percentage perhaps compared to the near trillion dollar global electronics market, but growing rapidly.

The intended application target? That obviously depends on the company, but the simple answer is, gulp, a little bit of everything! As depicted in Figure 1.3 and Figure 1.4, the global communications landscape is exceptionally diverse, ranging from low-frequency wireless (2.4 GHz cellular) to the fastest high-speed wireline systems (10 and 40 Gbit/sec synchronous optical network (SONET)). Core CMOS technologies are increasingly being pushed into the lower frequency wireless space, but the compelling drive to higher carrier frequencies over time will increasingly favor SiGe technologies.

At present, SiGe ICs are making inroads into: the cellular industry for handsets [global system for mobile communications—GSM, code division multiple access (CDMA), wideband CDMA (W-CDMA), etc.], even for power amplifiers; various wireless local area networks (WLAN) building blocks, from components to fully integrated systems ranging from 2.4 to 60 GHz and up; ultrawide band (UWB) components; global positioning systems (GPS); wireless base stations; a variety of wireline networking products, from 2.5 to 40 Gbit/sec (and higher); data converters (D/A and A/D); high-speed memories; a variety of instrumentation electronics; read-channel memory storage products; core analog functions (op amps, etc.); high-speed digital circuits of various flavors; radiation detector

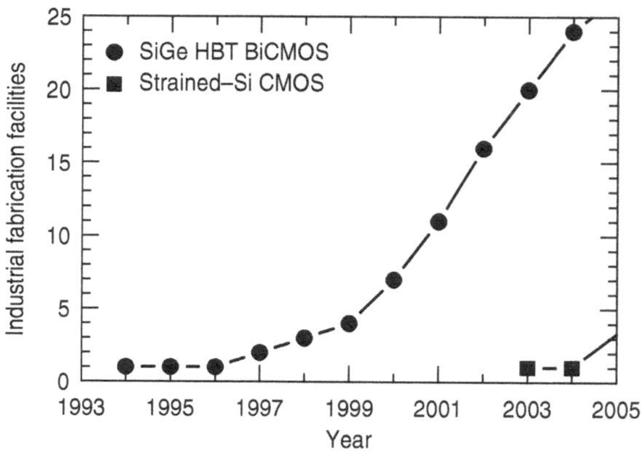

FIGURE 1.2 Number of industrial SiGe and strained Si fabrication facilities.

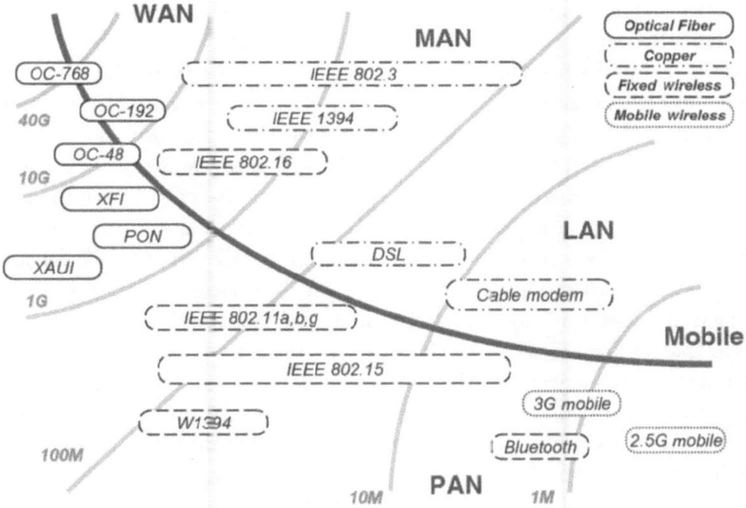

FIGURE 1.3 The global communications landscape, broken down by the various communications standards, and spanning the range of: wireless to wireline; fixed to mobile; copper to fiber; low data rate to broadband; and local area to wide area networks. WAN is *wide area network*, MAN is *metropolitan area network*, the so-called "last mile" access network, LAN is *local area network*, and PAN is *personal area network*, the emerging in-home network. (Used with the permission of Kyutae Lim.)

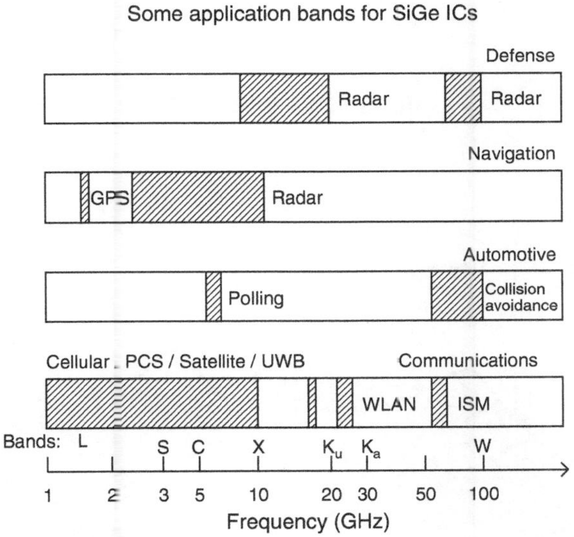

FIGURE 1.4 Some application frequency bands for SiGe integrated circuits.

electronics; radar systems (from 3 to 77 GHz and up); a variety space-based electronics components; and various niche extreme environment components (e.g., cryogenic (77 K) hybrid superconductor–semiconductor systems). The list is long and exceptionally varied—this is encouraging. Clearly, however, some of these components of "everything" are more important than others, and this will take time to shake out.

The strength of the BiCMOS twist to SiGe ICs cannot be overemphasized. Having both the high-speed SiGe HBT together on-chip with aggressively scaled CMOS allows one great flexibility in system design, the depths of which is just beginning to be plumbed. While debates still rage with respect to the most cost-effective partitioning at the chip and package level (system-on-a-chip versus system-in-a-package, etc.), clearly increased integration is viewed as a good thing in most camps (it is just a question of how much), and SiGe HBT BiCMOS is well positioned to address such needs across a broad market sector.

The envisioned high-growth areas for SiGe ICs over the new few years include: the cellular industry, optical networking, disk drives, and radar systems. In addition, potential high-payoff market areas span the emerging mm-wave space (e.g., the 60 GHz ISM band WLAN) for short range, but very high data rate (Gbit/sec) wireless systems. A SiGe 60 GHz single-chip/package transceiver (see Figure 1.5 for IBM's vision of such a beast) could prove to be the "killer app" for the emerging broadband multimedia market. Laughable? No. The building blocks for such systems have already been demonstrated using third-generation SiGe technology [4], and fully integrated transceivers are under development.

The rest of the potential market opportunities within the Si heterostructures field can be leveraged by successes in the SiGe IC field, both directly and indirectly. On the strained Si CMOS front, there are existent proofs now that strained Si is likely to become a mainstream component of conventional CMOS scaling at the 90-nm node and beyond (witness the early success of Intel's 90-nm logic technology built around uniaxially strained Si CMOS; other companies are close behind). Strained Si would seem to represent yet another clever technology twist that CMOS device technologists are pulling from their bag of tricks to keep the industry on a Moore's law growth path. This was not an obvious development (to me anyway) even a couple of years back. A wide variety of "transport enhanced" Si-heterostructure-based FETs have been demonstrated (SiGe-channel FETs, Si-based high electron mobility transistors (HEMTs), as well as both uniaxially and biaxially strained FETs, etc). Most of these devices, however, require complex substrate engineering that would have seemed to preclude giga-scale integration level needs for microprocessor-level integration. Apparently not so. The notion of using Si heterostructures (either

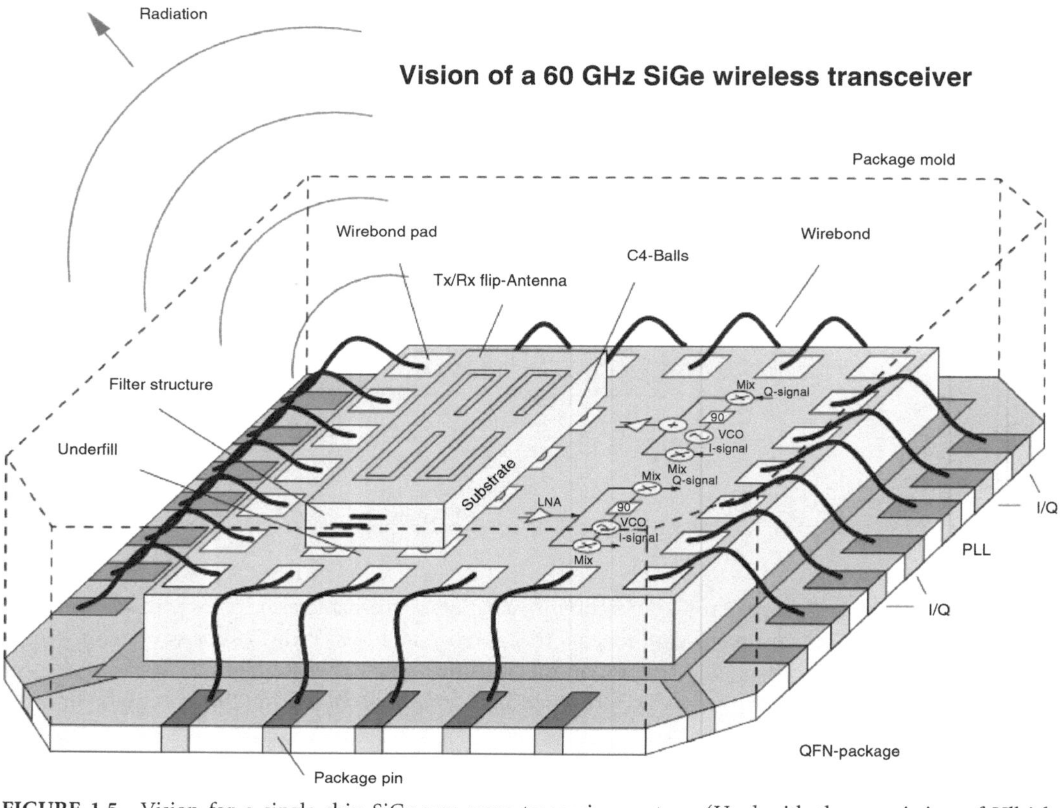

FIGURE 1.5 Vision for a single-chip SiGe mm-wave transceiver system. (Used with the permission of Ullrich Pfeiffer.)

uniaxial or biaxial strain or both) to boost conventional CMOS performance appears to be an appealing path for the future, a natural merging point I suspect for SiGe strained layers found in SiGe HBT BiCMOS (which to date contains only conventional Si CMOS) and strained Si CMOS.

From the optoelectronics camp, things are clearly far less evolved, but no less interesting. A number of functional optoelectronic devices have been demonstrated in research laboratories. Near-term successes in the short wavelength detector arena and light emitting diodes (LEDs) are beginning to be realized. The achievement of successful coherent light emission in the Si heterostructure system (e.g., via quantum cascade techniques perhaps) would appear to be the "killer app" in this arena, and research in this area is in progress. More work is needed.

1.5 Performance Limits and Future Directions

We begin with device performance limits. Just how fast will SiGe HBTs be 5 years from now? Transistor-level performance in SiGe HBTs continues to rise at a truly dizzying pace, and each major conference seems to bear witness to a new performance record (Figure 1.6). Both first- and second-generation SiGe HBT BiCMOS technology is widely available in 2005 (who would have thought even 3 years ago that fully integrated 100+ GHz Si-based devices would be "routine" on 200 mm wafers?), and even at the 200 GHz (third-generation) performance level, six companies (at last count) have achieved initial technology demonstrations, including IBM (Chapter 7), Jazz (Chapter 8), IHP (Chapter 11), ST Microelectronics (Chapter 12), Hitachi (Chapter 9), and Infineon (Chapter 10). (see *Fabrication of SiGe HBT BiCMOS Technology* for these chapters.) Several are now either available in manufacturing, or are very close (e.g., [5]). At press time, the most impressive new stake-in-the-ground is the report (June 2004) of the newly optimized "SiGe 9T" technology, which simultaneously achieves 302 GHz peak f_T and 306 GHz peak f_{max}, a clear record for any Si-based transistor, from IBM (Figure 1.7) [6]. This level of ac performance was achieved at a BV_{CEO} of 1.6 V, a BV_{CBO} of 5.5 V, and a current gain of 660. Noise measurements on these devices yielded NF_{min}/G_{assoc} of 0.45 dB/14 dB and 1.4 dB/8 dB at 10 and 25 GHz, respectively. Measurements of earlier (unoptimized) fourth-generation IBM SiGe HBTs have yielded record values of 375 GHz peak f_T [7] at 300 K and above 500 GHz peak f_T at 85 K. Simulations suggest that THz-level (1000 GHz) intrinsic transistor performance is not a laughable proposition in SiGe HBTs (Chapter 16, see *Silicon Heterostructure Devices*). This fact still amazes even me, the eternal optimist of SiGe performance! I, for one, firmly believe that we will see SiGe HBTs above-500 GHz peak

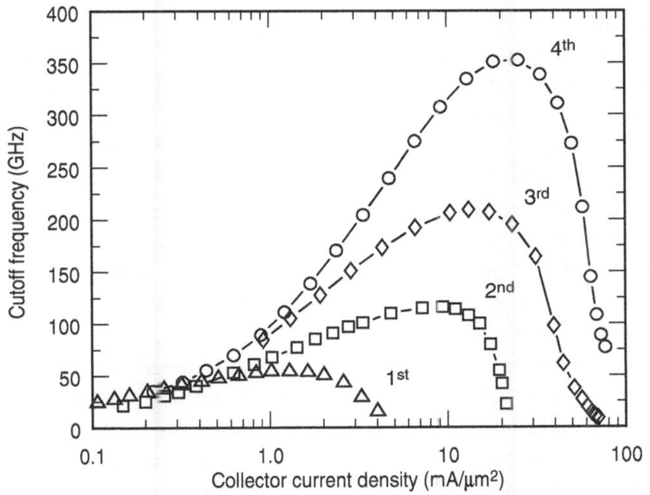

FIGURE 1.6 Measured cutoff frequency as a function of bias current density for four different SiGe HBT technology generations.

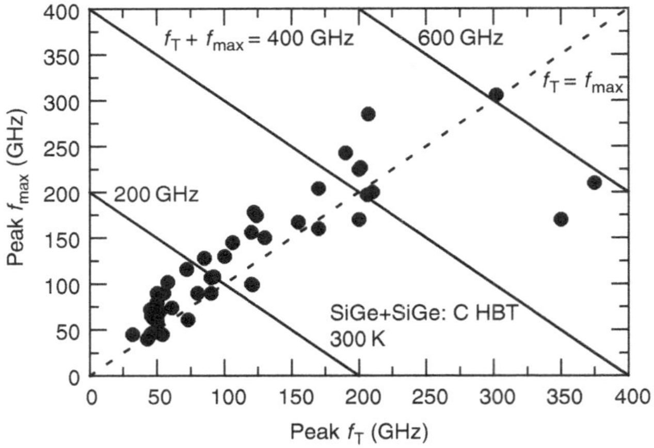

FIGURE 1.7 Measured maximum oscillation frequency versus cutoff frequency for a variety of generations of SiGe HBT BiCMOS technology shown in Figure 1.1.

f_T and f_{max} fully integrated with nanometer-scale (90 nm and below) Si CMOS (possibly strained Si CMOS) within the next 3 to 5 years.

One might logically ask, particularly within the confines of the above discussion on ultimate market relevance, why one would even attempt to build 500 GHz SiGe HBTs, other than to win a best-paper award, or to trumpet that "because-it's-there" Mount Everest mentality we engineers and scientists love so dearly. This said, if the future "killer app" turns out to be single-chip mm-wave transceiver systems with on-board DSP for broadband multimedia, radar, etc., then the ability of highly scaled, highly integrated, very high performance SiGe HBTs to dramatically enlarge the circuit/system design space of the requisite mm-wave building blocks may well prove to be a fruitful (and marketable) path.

Other interesting themes are emerging in the SiGe HBT BiCMOS technology space. One is the very recent emergence of complementary SiGe (C-SiGe) HBT processes (npn + pnp SiGe HBTs). While very early pnp SiGe HBT prototypes were demonstrated in the early 1990s, only in the last 2 years or so have fully complementary SiGe processes been developed, the most mature of which to date is the IHP SGC25C process, which has 200 GHz npn SiGe HBTs and 80 GHz pnp SiGe HBTs (Chapter 11, see *Fabrication of SiGe HBT BiCMOS Technology*). Having very high-speed pnp SiGe HBTs on-board presents a fascinating array of design opportunities aimed particularly at the analog/mixed-signal circuit space. In fact, an additional emerging trend in the SiGe field, particularly for companies with historical pure analog circuit roots, is to target lower peak f_T, but higher breakdown voltages, while simultaneously optimizing the device for core analog applications (e.g., op amps, line drivers, data converters, etc.), designs which might, for instance, target better noise performance, and higher current gain-Early voltage product than mainstream SiGe technologies. One might even choose to park that SiGe HBT platform on top of thick film SOI for better isolation properties (Chapter 13, see *Fabrication of SiGe HBT BiCMOS Technology*). Another interesting option is the migration of high-speed vertical SiGe HBTs with very thin film CMOS-compatible SOI (Chapter 5, see *Fabrication of SiGe HBT BiCMOS Technology*). This technology path would clearly favor the eventual integration of SiGe HBTs with strained Si CMOS, all on SOI, a seemingly natural migratory path.

If one accepts the tenet that integration is a good thing from a system-level perspective, the Holy Grail in the Si heterostructure field would, in the end, appear to be the integration of SiGe HBTs for RF through mm-wave circuitry (e.g., single-chip mm-wave transceivers complete with on-chip antennae), strained Si CMOS for all DSP and memory functionality, both perhaps on SOI, Si-based light emitters, SiGe HBT modulator electronics, and detectors for such light sources, together with on-chip waveguides to steer the light, realized all on one Si wafer to produce a "Si-based optoelectronic superchip" [8], that could do-it-all. These diverse blocks would be optional plug-in modules around a core SiGe

HBT + strained Si CMOS IC technology platform, perhaps with flip-chip (or other) packaging techniques to join different sub-die to the main superchip (e.g., for a Si-based detector or laser).

I know, I know. It is not obvious that even if each of these blocks could be realized, that it would make economic sense to do so for real systems. I have no quarrel with that. I think such a Si-based superchip is a useful paradigm, however, to bind together all of the clever objects we wish to ultimately build with Si heterostructures, from electronic to photonic, and maintain the vision of the one overarching constraint that guides us as we look forward—keep whatever you do compatible with high-volume manufacturing in Si fabrication facilities if you want to shape the path of the ensuing communications revolution. This Si-based superchip clearly remains a dream at present. A realizable dream? And if realizable, commercially viable? Who knows? Only time will tell. But it is fun to think about.

As you peruse this book you hold in your hands, which spans the whole Si heterostructure research and development space, from materials, to devices, to circuit and system applications, I think you will be amazed at both the vision, cleverness, and smashing successes of the many scientists and engineers who make up our field. Do not count us out! We are the new architects of an oh-so-very-interesting future.

References

1. JD Cressler and G Niu. *Silicon–Germanium Heterojunction Bipolar Transistors.* Boston, MA: Artech House, 2003.
2. "SiGe devices market to hit $2 billion in 2006," article featured on CompoundSemicoductor.net, http://compoundsemiconductor.net/articles/news/8/3/22/1.
3. F Capasso. Band-gap engineering: from physics and materials to new semiconductor devices. *Science*, 235:172–176, 1987.
4. S Reynolds, B Floyd, U Pfeiffer, and T. Zwick. 60 GHz transciever circuits in SiGe bipolar technology. Technical Digest of the IEEE International Solid-State Circuits Conference, San Francisco, 2004, pp 442–443.
5. AJ Joseph, D Coolbaugh, D Harame, G Freeman, S Subbanna, M Doherty, J Dunn, C Dickey, D Greenberg, R Groves, M Meghelli, A Rylyakov, M Sorna, O Schreiber, D Herman, and T Tanji. 0.13 μm 210 GHz f_T SiGe HBTs—expanding the horizons of SiGe BiCMOS. Technical Digest of the IEEE International Solid-State Circuits Conference, San Francisco, 2002, pp 180–182.
6. J-S Rieh, D Greenberg, M Khater, KT Schonenberg, J-J Jeng, F Pagette, T Adam, A Chinthakindi, J Florkey, B Jagannathan, J Johnson, R Krishnasamy, D Sanderson, C Schnabel, P Smith, A Stricker, S Sweeney, K Vaed, T Yanagisawa, D Ahlgren, K Stein, and G Freeman. SiGe HBTs for millimeter-wave applications with simultaneously optimized f_T and f_{max}. Proceedings of the IEEE Radio Frequency Integrated Circuits (RFIC) Symposium, Fort Worth, 2004, pp 395–398.
7. JS Rieh, B Jagannathan, H Chen, KT Schonenberg, D Angell, A Chinthakindi, J Florkey, F Golan, D Greenberg, S-J Jeng, M Khater, F Pagette, C Schnabel, P Smith, A Stricker, K Vaed, R Volant, D Ahlgren, G Freeman, K Stein, and S Subbanna. SiGe HBTs with cutoff frequency of 350 GHz. Technical Digest of the IEEE International Electron Devices Meeting, San Francisco, 2002, pp 771–774.
8. R Soref. Silicon-based photonic devices. Technical Digest of the IEEE International Solid-State Circuits Conference, 1995, pp 66–67.

2

A Brief History of the Field

John D. Cressler
Georgia Institute of Technology

In the historical record of any field of human endeavor, being "first" is everything. It is often said that "hindsight is 20–20," and it is tempting in many cases to ascribe this or that pivotal event as "obvious" or "easy" once the answer is known. Anyone intimately involved in a creative enterprise knows, however, that it is never easy being first, and often requires more than a little luck and maneuvering. Thus the triumphs of human creativity, the "firsts," should be appropriately celebrated. Still, later chroniclers often gloss over, and then eventually ignore, important (and sometimes very interesting) twists and turns, starts and stops, of the winners as well as the second and third place finishers, who in the end may in fact have influenced the paths of the winners, sometimes dramatically. The history of our field, for instance, is replete with interesting competitive battles, unusual personalities and egos, no small amount of luck, and various other fascinating historical nuances.

There is no concise history of our field available, and while the present chapter is not intended to be either exhaustive or definitive, it represents my firm conviction that the history of any field is both instructive and important for those who follow in the footsteps of the pioneers. Hopefully this brief history does not contain too many oversights or errors, and is offered as a step in the right direction for a history of pivotal events that helped shape the Si heterostructures field.

2.1 Si–SiGe Strained Layer Epitaxy

The field of Si-based heterostructures solidly rests on the shoulders of materials scientists and crystal growers, those purveyors of the semiconductor "black arts" associated with the deposition of pristine films of nanoscale dimensionality onto enormous Si wafers with near infinite precision. What may seem routine today was not always so. The Si heterostructure story necessarily begins with materials, and circuit designers would do well to remember that much of what they take for granted in transistor performance owes a great debt to the smelters of the crystalline world. Table 2.1 summarizes the key steps in the development of SiGe–Si strained layer epitaxy.

Given that Ge was the earliest and predominant semiconductor pursued by the Bell Laboratories transistor team, with a focus on the more difficult to purify Si to come slightly later, it is perhaps not surprising that the first study of SiGe alloys, albeit unstrained bulk alloys, occurred as early as 1958 [1]. It was recognized around 1960 [2] that semiconductor epitaxy* would enable more robust and controllable transistor fabrication. Once the move to Si-based processing occurred, the field of Si epitaxy was

*The word "epitaxy" (or just "epi") is derived from the Greek word *epi*, meaning "upon" or "over."

TABLE 2.1 Milestones in the Development of SiGe–Si Strained Layer Epitaxy

Historical Event	Year	Ref.
First investigation of the bandgap of unstrained SiGe alloys	1958	[1]
First epitaxially grown layer to be used in a transistor	1960	[2]
First investigation of high-temperature Si epitaxy	1963	[3]
Concept of critical thickness for epitaxial strained layers	1963	[4]
Energy minimization approach for critical thickness	1963	[5]
Force-balance approach for critical thickness	1974	[6]
First growth of SiGe strained layers	1975	[7]
First growth of SiGe epitaxy by MBE	1984	[8]
First stability calculations of SiGe strained layers	1985	[9]
First measurements of energy bandgap in SiGe strained layers	1985	[10,11]
First growth of Si epitaxy by LRP-CVD	1985	[12]
First 2D electron gas in the SiGe system	1985	[13]
First growth of Si epitaxy by UHV/CVD	1986	[14]
First measurements of band alignments in SiGe–Si	1986	[15]
First growth of SiGe epitaxy by UHV/CVD	1988	[16]
First step-graded relaxed SiGe substrate	1988	[16]
First growth of SiGe epitaxy by LRP-CVD	1989	[17]
First growth of Si epitaxy by AP-CVD	1989	[18]
First 2D hole gas in the SiGe system	1989	[19]
First growth of SiGe epitaxy by AP-CVD	1991	[20]
First majority hole mobility measurements in SiGe	1991	[21]
First minority electron mobility measurements in SiGe	1992	[22]
First growth of lattice-matched SiGeC alloys	1992	[23]
First growth of SiGe layers with carbon doping	1994	[24]
First stability calculations to include a Si cap layer	2000	[25]

launched, the first serious investigation of which was reported in 1963 [3]. Early Si epitaxy was exclusively conducted under high-temperature processing conditions, in the range of 1100°C, a temperature required to obtain a chemically pure and pristine growth interface on the Si host substrate for the soon-to-be-grown crystalline Si epi. High-temperature Si epi has been routinely used in basically this same form for over 40 years now, and represents a mature fabrication technique that is still widely practiced for many types of Si devices (e.g., high-speed bipolar transistors and various power devices).

Device engineers have long recognized the benefits of marrying the many virtues of Si as a host material for manufacturing electronic devices, with the bandgap engineering principles routinely practiced in the III–V system. Ultimately this requires a means by which one can perform epitaxial deposition of thin Si layers on large Si substrates, for both p- and n-type doping of arbitrary abruptness, with very high precision, across large wafers, and doping control at high dynamic range. Only a moment's reflection is required to appreciate that this means the deposition of the Si epi must occur at very low growth temperatures, say 500°C to 600°C (not "low" per se, but low compared to the requisite temperatures needed for solid-state diffusion of dopants in Si). Such a low-temperature Si epi would then facilitate the effective marriage of Si and Ge, two chemically compatible elements with differing bandgaps, and enable the doping of such layers with high precision, just what is needed for device realizations. Clearly the key to Si-based bandgap engineering, Si-heterostructures, our field, is the realization of device quality, low-temperature Si epi (and hence SiGe epi), grown pseudomorphically* on large Si host substrates. Conquering this task proved to be remarkably elusive and time consuming.

In the III–V semiconductor world, where very low processing temperatures are much easier to attain, and hence more common than for Si, the deposition of multiple semiconductors on top of one another proved quite feasible (e.g., GaAs on InP), as needed to practice bandgap engineering, for instance,

*The word "pseudo" is derived from the Greek word *pseudēs*, meaning "false," and the word "morphic" is derived from the Greek word *morphē*, meaning "form." Hence, pseudomorphic literally means *false-form*.

resulting in complex material composites having differing lattice constants in intimate physical contact. To accommodate the differing lattice constants while maintaining the crystallinity of the underlying films, strain is necessarily induced in the composite film, and the notion of a film "critical thickness," beyond which strain relaxation occurs via fundamental thermodynamic driving forces, was defined as early as 1963 [4], as were the energy minimization techniques needed for calculating such critical thicknesses [5]. Alternative "force-balance" techniques for addressing the so-called stability issues in strained layer epitaxy came from the III–V world in 1974, and were applied to SiGe strained layer epitaxy in 1985 [9]. Interestingly, however, research continues today on stability in complicated (e.g., compositionally graded) SiGe films, and only very recently have reasonably complete theories been offered which seem to match well with experiment [25].

The first reported growth of SiGe strained layers was in 1975 in Germany [7], but the field did not begin to seriously heat up until the early 1980s, when several teams pioneered the application of molecular beam epitaxy (MBE) to facilitate materials studies of device-quality strained SiGe on Si in 1984 [8]. Optical studies on these films resulted in encouraging findings concerning the beneficial effects of strain on the band-edge properties of SiGe [10,11], paving the way for serious contemplation of devices built from such materials. Parallel paths toward other low-temperature Si epi growth techniques centered on the ubiquitous chemical vapor deposition (CVD) approach were simultaneously pursued, culminating in the so-called limited-reaction-processing CVD (LRP-CVD) technique (Si epi in 1985 [12], and SiGe epi in 1989 [17]), the ultrahigh-vacuum CVD (UHV/CVD) technique (Si epi in 1986 [14] and SiGe epi in 1988 [16]), and various atmospheric pressure CVD (AP-CVD) techniques (e.g., Si epi in 1989 [18], and SiGe epi in 1991 [20]). These latter two techniques, in particular, survive to this day, and are widely used in the SiGe heterojunction bipolar transistor (HBT) industry.

Device-quality SiGe–Si films enabled a host of important discoveries to occur, which have important bearing on device derivatives, including the demonstration of both two-dimensional electron and hole gases [13,19], and the fortuitous observation that step-graded SiGe buffer layers could be used to produce device-quality strained Si on SiGe, with its consequent conduction band offsets [16]. This latter discovery proved important in the development of SiGe–Si heterostructure-based FETs. Both majority and minority carrier mobility measurements occurred in the early 1990s [21,22], although reliable data, particularly involving minority carriers, remain sparse in the literature. Also in the early 1990s, experiments using high C content as a means to relieve strain in SiGe and potentially broaden the bandgap engineering space by lattice-matching SiGe:C materials to Si substrates (a path that has to date not borne much fruit, unfortunately), while others began studying efficacy of C-doping of SiGe, a result that ultimately culminated in the wide use today of C-doping for dopant diffusion suppression in SiGe:C HBTs [23,24].

The Si–SiGe materials field continues to evolve. Commercial single wafer (AP-CVD) and batch wafer (UHV/CVD) Si–SiGe epi growth tools compatible with 200 mm (and soon 300 mm) Si wafers exist in literally dozens of industrial fabrication facilities around the world, and SiGe growth can almost be considered routine today in the ease in which it can be integrated into CMOS-compatible fabrication processes. It was clearly of paramount importance in the ultimate success of our field that some of the "black magic" associated with robust SiGe film growth be removed, and this, thankfully, is the case in 2005.

2.2 SiGe HBTs

Transistor action was first demonstrated by Bardeen and Brattain in late December of 1947 using a point contact device [26]. Given all that has transpired since, culminating in the Communications Revolution, which defines our modern world (refer to the discussion in Chapter 1), this pivotal event surely ranks as one of the most significant in the course of human history—bold words, but nevertheless true. This demonstration of a solid-state device exhibiting the key property of amplification (power gain) is also unique in the historical record for the precision with which we can locate it in time—December 23,

1947, at about 5 p.m. Not to be outdone, Shockley rapidly developed a theoretical basis for explaining how this clever object worked, and went on to demonstrate the first true bipolar junction transistor (BJT) in 1951 [27]. The first BJT was made, ironically in the present context, from Ge. The first silicon BJT was made by Teal in 1954 using grown junction techniques. The first diffused silicon BJT was demonstrated in 1956 [28], and the first epitaxially grown silicon BJT was reported in 1960, see Ref. [2].

The concept of the HBT is surprisingly an old one, dating in fact to the fundamental BJT patents filed by Shockley in 1948 [29]. Given that the first bipolar transistor was built from Ge, and III–V semiconductors were not yet on the scene, it seems clear that Shockley envisioned the combination of Si (wide bandgap emitter) and Ge (narrow bandgap base) to form a SiGe HBT. The basic formulation and operational theory of the HBT, for both the traditional wide bandgap emitter plus narrow bandgap base approach found in most III–V HBTs, as well as the drift-base (graded) approach used in SiGe HBTs today, was pioneered by Kroemer, and was largely in place by 1957 [30–32]. It is ironic that Kroemer in fact worked hard early on to realize a SiGe HBT, without success, ultimately pushing him toward the III–V material systems for his heterostructure studies, a path that proved in the end to be quite fruitful for him, since he shared the Nobel Prize in physics in 2000 for his work in (III–V) bandgap engineering for electronic and photonic applications [33]. While III–V HBT (e.g., AlGaAs–GaAs) demonstrations began appearing in the 1970s, driven largely by the needs for active microwave components in the defense industry, reducing the SiGe HBT to practical reality took 30 years after the basic theory was in place due to material growth limitations. As pointed out [34] the semiconductor device field is quite unique in the scope of human history because "science" (theoretical understanding) preceded the "art" (engineering and subsequent technological advancement). Once device-quality SiGe films were finally achieved in the mid-1980s, however, progress was quite rapid. Table 2.2 summarizes the key steps in the evolution of SiGe HBTs.

The first functional SiGe HBT was demonstrated by an IBM team in December 1987 at the IEDM [35]. The pioneering result showed a SiGe HBT with functional, albeit leaky, dc characteristics; but it was a SiGe HBT, it worked (barely), and it was the first.* It is an often overlooked historical point, however, that at least four independent groups were simultaneously racing to demonstrate the first functional SiGe HBT, all using the MBE growth technique: the IBM team [35], a Japanese team [62], a Bell Laboratories team [63], and a Linköping University team [64]. The IBM team is fairly credited with the victory, since it presented (and published) its results in early December of 1987 at the IEDM (it would have been submitted to the conference for review in the summer 1987) [35]. Even for the published journal articles, the IBM team was the first to submit its paper for review (on November 17, 1987) [65]. All four papers appeared in print in the spring of 1988. Other groups soon followed with more SiGe HBT demonstrations.

The first SiGe HBT demonstrated using (the ultimately more manufacturable) CVD growth technique followed shortly thereafter, in 1989, first using LRP-CVD [17], and then with UHV/CVD [36]. Worldwide attention became squarely focused on SiGe technology, however, in June 1990 at the IEEE VLSI Technology Symposium with the demonstration of a non-self-aligned UHV/CVD SiGe HBT with a peak cutoff frequency of 75 GHz [37,38]. At that time, this SiGe HBT result was roughly twice the performance of state-of-the-art Si BJTs, and clearly demonstrated the future performance potential of the technology (doubling of transistor performance is a rare enough event that it does not escape significant attention!). Eyebrows were raised, and work to develop SiGe HBTs for practical circuit applications began in earnest in a large number of industrial and university laboratories around the world.†

The feasibility of implementing pnp SiGe HBTs was also demonstrated in June 1990 [40]. In December 1990, the simplest digital circuit, an emitter-coupled-logic (ECL) ring oscillator, using self-

*An interesting historical perspective of early SiGe HBT development at IBM is contained in Ref. [61].

†A variety of zero-Dt, mesa-isolated, III–V-like high-speed SiGe HBTs were reported in the early 1990s (e.g., Ref. [66]), but we focus here on fully integrated, CMOS-compatible SiGe HBT technologies, because they are inherently more manufacturable, and hence they are the only ones left standing today, for obvious reasons.

TABLE 2.2 Milestones in the Development of SiGe HBTs

Historical Event	Year	Ref.
First demonstration of transistor action	1947	[26]
Basic HBT concept	1948	[29]
First demonstration of a bipolar junction transistor	1951	[27]
First demonstration of a silicon bipolar transistor	1956	[28]
Drift-base HBT concept	1954	[30]
Fundamental HBT theory	1957	[31,32]
First epitaxial silicon transistors	1960	[2]
First SiGe HBT	1987	[35]
First ideal SiGe HBT grown by CVD	1989	[17]
First SiGe HBT grown by UHV/CVD	1989	[36]
First high-performance SiGe HBT	1990	[37,38]
First self-aligned SiGe HBT	1990	[39]
First SiGe HBT ECL ring oscillator	1990	[39]
First pnp SiGe HBT	1990	[40]
First operation of SiGe HBTs at cryogenic temperatures	1990	[41]
First SiGe HBT BiCMOS technology	1992	[42]
First LSI SiGe HBT integrated circuit	1993	[43]
First SiGe HBT with peak f_T above 100 GHz	1993	[44,45]
First SiGe HBT technology in 200-mm manufacturing	1994	[46]
First SiGe HBT technology optimized for 77 K	1994	[47]
First radiation tolerance investigation of SiGe HBTs	1995	[48]
First report of low-frequency noise in SiGe HBTs	1995	[49]
First SiGe:C HBT	1996	[50]
First high-power SiGe HBTs	1996	[51,52]
First sub-10 psec SiGe HBT ECL circuits	1997	[53]
First high-performance SiGe:C HBT technology	1999	[54]
First SiGe HBT with peak f_T above 200 GHz	2001	[55]
First SiGe HBT with peak f_T above 300 GHz	2002	[56]
First complementary (npn + pnp) SiGe HBT technology	2003	[57]
First C-SiGe technology with npn and pnp f_T above 100 GHz	2003	[58]
First vertical SiGe HBT on thin film (CMOS compatible) SOI	2003	[59]
First SiGe HBT with both f_T and f_{max} above 300 GHz	2004	[60]

aligned, fully integrated SiGe HBTs was produced [39]. The first SiGe BiCMOS technology (SiGe HBT + Si CMOS) was reported in December 1992 [42]. Theoretical predictions of the inherent ability of SiGe HBTs to operate successfully at cryogenic temperatures (in contrast to Si BJTs) were first confirmed in 1990 [41], and SiGe HBT profiles optimized for the liquid nitrogen temperature environment (77 K) were reported in 1994 [48]. The first LSI SiGe HBT circuit (a 1.2 Gsample/sec 12-bit digital-to-analog converter—DAC) was demonstrated in December 1993 [43]. The first SiGe HBTs with frequency response greater than 100 GHz were described in December 1993 by two independent teams [44,45], and the first SiGe HBT technology entered commercial production on 200-mm wafers in December 1994 [46].

The first report of the effects of ionizing radiation on advanced SiGe HBTs was made in 1995 [48]. Due to the natural tolerance of epitaxial-base bipolar structures to conventional radiation-induced damage mechanisms without any additional radiation-hardening process changes, SiGe HBTs are potentially very important for space-based and planetary communication systems applications, spawning an important new sub-discipline for SiGe technology. The first demonstration that epitaxial SiGe strained layers do not degrade the superior low-frequency noise performance of bipolar transistors occurred in 1995, opening the way for very low-phase noise frequency sources [49].

Carbon-doping of epitaxial SiGe layers as a means to effectively suppress boron out-diffusion during fabrication has rapidly become the preferred approach for commercial SiGe technologies, particularly those above first-generation performance levels. Carbon-doping of SiGe HBTs has its own interesting history, dating back to the serendipitous discovery [50] in 1996 that incorporating small amounts of C into a SiGe epi layer strongly retards (by an order of magnitude) the diffusion of the boron (B) base layer during subsequent thermal cycles. Given that maintaining a thin base profile during fabrication is perhaps the most challenging aspect of building a manufacturable SiGe technology, it is somewhat surprising that it took so long for the general adoption of C-doping as a key technology element. I think it is fair to say that most SiGe practitioners at that time viewed C-doping with more than a small amount of skepticism, given that C can act as a deep trap in Si, and C contamination is generally avoided at all costs in Si epi processes, particularly for minority carrier devices such as the HBT. At the time of the discovery of C-doping of SiGe in 1996, most companies were focused on simply bringing up a SiGe process and qualifying it, relegating the potential use of C to the back burner. In fairness, most felt that C-doping was not necessary to achieve first-generation SiGe HBT performance levels. The lone visionary group to solidly embrace C-doping of SiGe HBTs at the onset was the IHP team in Germany, whose pioneering work eventually paid off and began to convince the skeptics of the merits of C-doping. The minimum required C concentration for effective out-diffusion suppression of B was empirically established to be in the vicinity of 0.1% to 0.2% C (i.e., around $1 \times 10^{20}\,cm^{-3}$). Early on, much debate ensued on the physical mechanism of *how* C impedes the B diffusion process, but general agreement for the most part now exists and is discussed in Chapter 11. The first high-performance, fully integrated SiGe:C HBT technology was reported in 1999 [54].

The first "high-power" SiGe HBTs (S band, with multiwatt output power) were reported in 1996 using thick collector doping profiles [51,52]. The 10-psec ECL circuit performance barrier was broken in 1997 [53]. The 200-GHz peak f_T performance barrier was broken in November 2001 for a non-self-aligned device [55], and for a self-aligned device in February 2002 [67]. By 2004, a total of six industrial laboratories had achieved 200 GHz performance levels. A SiGe HBT technology with a peak f_T of 350 GHz (375 GHz values were reported in the IEDM presentation) was presented in December 2002 [56], and this 375 GHz f_T value remains a record for room temperature operation (it is above 500 GHz at cryogenic temperatures), and an optimized version with both f_T and f_{max} above 300 GHz was achieved in June 2004 [60]. This combined level of 300+ GHz for both f_T and f_{max} remains a solid record for any Si-based semiconductor device.

Other recent and interesting developments in the SiGe HBT field include the first report of a complementary (npn + pnp) SiGe HBT (C-SiGe) technology in 2003 [57], rapidly followed by a C–SiGe technology with f_T for both the npn and pnp SiGe HBTs above 100 GHz [58]. In addition, a novel vertical npn SiGe HBT has been implemented in thin-film (120 nm) CMOS-compatible SOI [59]. Besides further transistor performance enhancements, other logical developments to anticipate in this field include the integration of SiGe HBTs with strained-Si CMOS for a true all-Si-heterostructure technology.

Not surprisingly, research and development activity involving SiGe HBTs, circuits built from these devices, and various SiGe HBT technologies, in both industry and at universities worldwide, has grown very rapidly since the first demonstration of a functional SiGe HBT in 1987, only 18 years in the past.

2.3 SiGe–Strained Si FETs and Other SiGe Devices

The basic idea of using an electric field to modify the surface properties of materials, and hence construct a "field-effect" device, is remarkably old (1926 and 1935), predating even the quest for a solid-state amplifier [68]. Given the sweeping dominance of CMOS technology in the grand scheme of the electronics industry today, it is ironic that the practical demonstration of the BJT preceded that of the MOSFET by 9 years. This time lag from idea to realization was largely a matter of dealing with the many perils associated with obtaining decent dielectric materials in the Si system—doubly ironic given that Si has such a huge natural advantage over all other semiconductors in this regard. Bread-and-butter notions of ionic contamination, de-ionized water, fixed oxide charge, surface state passivation, and clean-room techniques in semiconductor fabrication had to be learned the hard way. Once device-quality SiO_2 was obtained in the late 1950s, and a robust gate dielectric could thus be fabricated, it was not long until the first functional MOSFET was demonstrated in 1960 [69]. The seemingly trivial (remember, however, that hindsight is 20–20!) connection of n-channel and p-channel MOSFETs to form low-power CMOS in 1963 [70] paved the way (eventually) to the high-volume, low-cost, highly integrated microprocessor, and the enormous variety of computational engines that exist today as a result.

Like their cousin, the SiGe HBT, SiGe–strained Si FETs did not get off the ground until the means for accomplishing the low-temperature growth of Si epitaxy could be realized. Once that occurred in the mid-1980s the field literally exploded. Table 2.3 summarizes the milestones in the evolution of SiGe–strained Si FETs, as well as a veritable menagerie of other electronic and optoelectronic components built from SiGe–strained Si epitaxy.

It was discovered as early as 1971 that direct oxidation of SiGe was a bad idea for building gate dielectrics [71]. Given that gate oxide quality, low-temperature deposited oxides, did not exist in the mid-1980s, the earliest FET demonstrations were modulation-doped, Schottky-gated, FETs, and both n-channel and p-channel SiGe MODFETs were pioneered as early as 1986 using MBE-grown material [72,73]. Before the SiGe MOSFET field got into high gear in the 1990s, a variety of other novel device demonstrations occurred, including: the first SiGe superlattice photodetector [74], the first SiGe Schottky barrier diodes (SBD) in 1988 [75], the first SiGe hole-transport resonant tunneling diode (RTD) in 1988 [76], and the first SiGe bipolar inversion channel FET (BiCFET) in 1989, a now-extinct dinosaur [77]. Meanwhile, early studies using SiGe in conventional CMOS gate stacks to minimize dopant depletion effects and tailor work functions, a fairly common practice in CMOS today, occurred in 1990 [78], and the first SiGe waveguides on Si substrates were produced in 1990 [79].

The first functional SiGe channel pMOSFET was published in 1991, and shortly thereafter, a wide variety of other approaches aimed at obtaining the best SiGe pMOSFETs (see, for instance, Refs. [93–95]). The first electron-transport RTD was demonstrated in 1991 [81], and the first LED in SiGe

TABLE 2.3 Milestones in the Development of SiGe–Strained Si FETs and Other Devices

Historical Event	Year	Ref.
Field effect device concept	1926	[68]
First Si MOSFET	1960	[69]
First Si CMOS	1963	[70]
First oxidation study of SiGe	1971	[71]
First SiGe nMODFET	1986	[72]
First SiGe pMODFET	1986	[73]
First SiGe photodetector	1986	[74]
First SiGe SBD	1988	[75]
First SiGe hole RTD	1988	[76]
First SiGe BiCFET	1989	[77]
First SiGe gate CMOS technology	1990	[78]
First SiGe waveguide	1990	[79]
First SiGe pMOSFET	1991	[80]
First SiGe electron RTD	1991	[81]
First SiGe LED	1991	[82]
First SiGe solar cell	1992	[83]
First a-SiGe phototransistor	1993	[84]
First SiGe pMOSFET on SOI	1993	[85]
First strained Si pMOSFET	1993	[86]
First strained Si nMOSFET	1994	[87]
First SiGe:C pMOSFET	1996	[88]
First SiGe pFET on SOS	1997	[89]
First submicron strained Si MOSFET	1998	[90]
First vertical SiGe pFET	1998	[91]
First strained Si CMOS technology	2002	[92]

also in 1991 (a busy year for our field). In 1992, the first a-SiGe solar cell was discussed [83], and in 1993, the first high-gain a-SiGe phototransistor [84]. The first SiGe pMOSFETs using alternate substrate materials were demonstrated, first in SOI in 1993 [85], and then on sapphire in 1997 [88], the first SiGe:C channel pMOSFET was demonstrated in 1996 [89], and the first vertical SiGe FET was published in 1998 [92].

Because of the desire to use Si-based bandgap engineering to improve not only the p-channel MOSFET, but also the n-channel MOSFET, research in the early- to mid-1990s in the FET field began to focus on strained Si MOSFETs on relaxed SiGe layers, with its consequent improvement in both electron and hole transport properties. This work culminated in the first strained Si pMOSFET in 1993 [87], and the first stained Si nMOSFET in 1994 [88], and remains an intensely active research field today. Key to the eventual success of strained Si CMOS approaches was that significant mobility enhancement could be achieved in both nFETs and pFETs down to very short (sub-micron) gate lengths, and this was first demonstrated in 1998 [90]. Strained Si CMOS at the 90-nm node and below is rapidly becoming mainstream for most serious CMOS companies, and the first commercial 90 nm strained Si CMOS technology platform was demonstrated by Intel in 2002 [91]. At last count, there were upwards of a half-dozen companies (e.g., Texas Instruments and IBM) also rapidly pushing toward 90 nm (and below) strained Si CMOS technologies, utilizing a variety of straining techniques, and thus it would appear that strained Si CMOS will be a mainstream IC technology in the near future, joining SiGe HBT BiCMOS technology. This is clearly outstanding news for our field. The merger of SiGe HBTs with strained Si CMOS would be a near-term logical extension.

References

1. R Braunstein, AR Moore, and F Herman. Intrinsic optical absorption in germanium–silicon alloys. *Physical Review B* 32:1405–1408, 1958.
2. HC Theuerer, JJ Kleimack, HH Loar, and H Christensen. Epitaxial diffused transistors. *Proceedings of the IRE* 48:1642–1643, 1960.
3. BA Joyce and RR Bradley. Epitaxial growth of silicon from the pyrolysis of monosilane on silicon substrates. *Journal of the Electrochemical Society* 110:1235–1240, 1963.
4. JH van der Merwe. Crystal interfaces. Part I. Semi-infinite crystals. *Journal of Applied Physics* 34:117–125, 1963.
5. JH van der Merwe. Crystal interfaces. Part II. Finite overgrowths. *Journal of Applied Physics* 34:123–127, 1963.
6. JW Matthews and AE Blakeslee. Defects in epitaxial multilayers: I. Misfit dislocations in layers. *Journal of Crystal Growth* 27:118–125, 1974.
7. E Kasper, HJ Herzog, and H Kibbel. A one-dimensional SiGe superlattice grown by UHV epitaxy. *Journal of Applied Physics* 8:1541–1548, 1975.
8. JC Bean, TT Sheng, LC Feldman, AT Fiory, and RT Lynch. Pseudomorphic growth of Ge_xSi_{1-x} on silicon by molecular beam epitaxy. *Applied Physics Letters* 44:102–104, 1984.
9. R People and JC Bean. Calculation of critical layer thickness versus lattice mismatch for Ge_xSi_{1-x}/Si strained layer heterostructures. *Applied Physics Letters* 47:322–324, 1985.
10. R People. Indirect bandgap of coherently strained $Si_{1-x}Ge_x$ bulk alloys on $\langle 0\ 0\ 1 \rangle$ silicon substrates. *Physical Review B* 32:1405–1408, 1985.
11. DV Lang, R People, JC Bean, and AM Sergent. Measurement of the bandgap of Ge_xSi_{1-x}/Si strained-layer heterostructures. *Applied Physics Letters* 47:1333–1335, 1985.
12. JF Gibbons, CM Gronet, and KE Williams. Limited reaction processing: silicon epitaxy. *Applied Physics Letters* 47:721–723, 1985.
13. G Abstreiter, H Brugger, T Wolf, H Joke, and HJ Kerzog. Strain-induced two-dimensional electron gas in selectively doped Si/Si_xGe_{1-x} superlattices. *Physical Review* 54:2441–2444, 1985.
14. BS Meyerson. Low-temperature silicon epitaxy by ultrahigh vacuum/chemical vapor deposition. *Applied Physics Letters* 48:797–799, 1986.

15. R People and JC Bean. Band alignments of coherently strained Ge_xSi_{1-x}/Si heterostructures on ⟨0 0 1⟩ Ge_ySi_{1-y} substrates. *Applied Physics Letters* 48:538–540, 1986.
16. BS Meyerson, KJ Uram, and FK LeGoues. Cooperative phenomena is silicon/germanium low temperature epitaxy. *Applied Physics Letters* 53:2555–2557, 1988.
17. CA King, JL Hoyt, CM Gronet, JF Gibbons, MP Scott, and J Turner. Si/Si_{1-x}/Ge_x heterojunction bipolar transistors produced by limited reaction processing. *IEEE Electron Device Letters* 10:52–54, 1989.
18. TO Sedgwick, M Berkenbilt, and TS Kuan. Low-temperature selective epitaxial growth of silicon at atmospheric pressure. *Applied Physics Letters* 54:2689–2691, 1989.
19. PJ Wang, FF Fang, BS Meyerson, J Mocera, and B Parker. Two-dimensional hole gas in Si/$Si_{0.85}Ge_{0.15}$ modulation doped heterostructures. *Applied Physics Letters* 54:2701–2703, 1989.
20. P Agnello, TO Sedgwick, MS Goorsky, J Ott, TS Kuan, and G Scilla. Selective growth of silicon–germanium alloys by atmospheric-pressure chemical vapor deposition at low temperatures. *Applied Physics Letters* 59:1479–1481, 1991.
21. T Manku and A Nathan. Lattice mobility of holes in strained and unstrained $Si_{1-x}Ge_x$ alloys. *IEEE Electron Device Letters* 12:704–706, 1991.
22. T Manku and A Nathan. Electron drift mobility model for devices based on unstrained and coherently strained $Si_{1-x}Ge_x$ grown on ⟨0 0 1⟩ silicon subtrate. *IEEE Transactions on Electron Devices* 39:2082–2089, 1992.
23. K Erbel, SS Iyer, S Zollner, JC Tsang, and FK LeGoues. Growth and strain compensation effects in the ternary $Si_{1-x-y}Ge_xC_y$ alloy system. *Applied Physics Letters* 60:3033–3035, 1992.
24. HJ Osten, E Bugiel, and P Zaumseil. Growth of inverse tetragonal distorted SiGe layer on Si(0 0 1) by adding small amounts of carbon. *Applied Physics Letters* 64:3440–3442, 1994.
25. A Fischer, H-J Osten, and H Richter. An equilibrium model for buried SiGe strained layers. *Solid-State Electronics* 44:869–873, 2000.
26. J Bardeen and WH Brattain. The transistor, a semi-conductor triode. *Physical Review* 71:230–231, 1947.
27. W Shockley, M Sparks, and GK Teal. p–n junction transistors. *Physical Review* 83:151–162, 1951.
28. M Tanenbaum and DE Thomas. Diffused emitter and base silicon transistors. *Bell System Technical Journal* 35:23–34, 1956.
29. See, for instance, W Shockley. U.S. Patents 2,502,488, 2,524,035, and 2,569,347.
30. H Kroemer. Zur theorie des diffusions und des drifttransistors. Part III. *Archiv der Elektrischen Ubertragungstechnik* 8:499–504, 1954.
31. H Kroemer. Quasielectric and quasimagnetic fields in nonuniform semiconductors. *RCA Review* 18:332–342, 1957.
32. H Kroemer. Theory of a wide-gap emitter for transistors. *Proceedings of the IRE* 45:1535–1537, 1957.
33. B Brar, GJ Sullivan, and PM Asbeck. Herb's bipolar transistors. *IEEE Transactions on Electron Devices* 48:2473–2476, 2001.
34. RM Warner. Microelectronics: Its unusual origin and personality. *IEEE Transactions on Electron Devices* 48:2457–2467, 2001.
35. SS Iyer, GL Patton, SL Delage, S Tiwari, and J.M.C. Stork. Silicon–germanium base heterojunction bipolar transistors by molecular beam epitaxy. Technical Digest of the IEEE International Electron Devices Meeting, San Francisco, 1987, pp. 874–876.
36. GL Patton, DL Harame, JMC Stork, BS Meyerson, GJ Scilla, and E Ganin. Graded-SiGe-base, poly-emitter heterojunction bipolar transistors. *IEEE Electron Device Letters* 10:534–536, 1989.
37. GL Patton, JH Comfort, BS Meyerson, EF Crabbé, E de Frésart, JMC Stork, JY-C Sun, DL Harame, and J Burghartz. 63-75 GHz f_T SiGe-base heterojunction-bipolar technology. Technical Digest IEEE Symposium on VLSI Technology, Honolulu, 1990, pp. 49–50.
38. GL Patton, JH Comfort, BS Meyerson, EF Crabbé, GJ Scilla, E de Frésart, JMC Stork, JY-C Sun, DL Harame, and J Burghartz. 75 GHz f_T SiGe base heterojunction bipolar transistors. *IEEE Electron Device Letters* 11:171–173, 1990.

39. JH Comfort, GL Patton, JD Cressler, W Lee, EF Crabbé, BS Meyerson, JY-C Sun, JMC Stork, P-F Lu, JN Burghartz, J Warnock, K Kenkins, K-Y Toh, M D'Agostino, and G Scilla. Profile leverage in a self-aligned epitaxial Si or SiGe-base bipolar technology. Technical Digest IEEE International Electron Devices Meeting, Washington, 1990, pp. 21–24.
40. DL Harame, JMC Stork, BS Meyerson, EF Crabbé, GL Patton, GJ Scilla, E de Frésart, AA Bright, C Stanis, AC Megdanis, MP Manny, EJ Petrillo, M Dimeo, RC Mclntosh, and KK Chan. SiGe-base PNP transistors fabrication with n-type UHV/CVD LTE in a "NO DT" process. Technical Digest IEEE Symposium on VLSI Technology, Honolulu, 1990, pp. 47–48.
41. EF Crabbeé, GL Patton, JMC Stork, BS Meyerson, and JY-C Sun. Low temperature operation of Si and SiGe bipolar transistors. Technical Digest IEEE International Electron Devices Meeting, Washington, 1990, pp. 17–20.
42. DL Harame, EF Crabbé, JD Cressler, JH Comfort, JY-C Sun, SR Stiffler, E Kobeda, JN Burghartz, MM Gilbert, J Malinowski, and AJ Dally. A high-performance epitaxial SiGe-base ECL BiCMOS technology. Technical Digest IEEE International Electron Devices Meeting, Washington, 1992, pp. 19–22.
43. DL Harame, JMC Stork, BS Meyerson, KY-J Hsu, J Cotte, KA Jenkins, JD Cressler, P Restle, EF Crabbé, S Subbanna, TE Tice, BW Scharf, and JA Yasaitis. Optimization of SiGe HBT technology for high speed analog and mixed-signal applications. Technical Digest IEEE International Electron Devices Meeting, San Francisco, 1993, pp. 71–74.
44. E Kasper, A Gruhle, and H Kibbel. High speed SiGe-HBT with very low base sheet resistivity. Techncial Digest IEEE International Electron Devices Meeting, San Francisco, 1993, pp. 79–81.
45. EF Crabbé, BS Meyerson, JMC Stork, and DL Harame. Vertical profile optimization of very high frequency epitaxial Si- and SiGe-base bipolar transistors. Technical Digest IEEE International Electron Devices Meeting, Washington, 1993, pp. 83–86.
46. DL Harame, K Schonenberg, M Gilbert, D Nguyen-Ngoc, J Malinowski, S-J Jeng, BS Meyerson, JD Cressler, R Groves, G Berg, K Tallman, K Stein, G Hueckel, C Kermarrec, T Tice, G Fitzgibbons, K Walter, D Colavito, T Houghton, N Greco, T Kebede, B Cunningham, S Subbanna, JH Comfort, and EF Crabbé. A 200 mm SiGe-HBT technology for wireless and mixed-signal applications. Technical Digest IEEE International Electron Devices Meeting, Washington, 1994, pp. 437–440.
47. JD Cressler, EF Crabbé, JH Comfort, JY-C Sun, and JMC Stork. An epitaxial emitter cap SiGe-base bipolar technology for liquid nitrogen temperature operation. *IEEE Electron Device Letters* 15:472–474, 1994.
48. JA Babcock, JD Cressler, LS Vempati, SD Clark, RC Jaeger, and DL Harame. Ionizing radiation tolerance of high performance SiGe HBTs grown by UHV/CVD. *IEEE Transactions on Nuclear Science* 42:1558–1566, 1995.
49. LS Vempati, JD Cressler, RC Jaeger, and DL Harame. Low-frequency noise in UHV/CVD Si- and SiGe-base bipolar transistors. Proceedings of the IEEE Bipolar/BiCMOS Circuits and Technology Meeting, Minnneapolis, 1995, pp. 129–132.
50. L Lanzerotti, A St Amour, CW Liu, JC Sturm, JK Watanabe, and ND Theodore. $Si/Si_{1-x-y}Ge_xC_y/Si$ heterojunction bipolar transistors. *IEEE Electron Device Letters* 17:334–337, 1996.
51. A Schüppen, S Gerlach, H Dietrich, D Wandrei, U Seiler, and U König. 1-W SiGe power HBTs for mobile communications. *IEEE Microwave and Guided Wave Letters* 6:341–343, 1996.
52. PA Potyraj, KJ Petrosky, KD Hobart, FJ Kub, and PE Thompson. A 230-Watt S-band SiGe heterojunction junction bipolar transistor. *IEEE Transactions on Microwave Theory and Techniques* 44:2392–2397, 1996.
53. K Washio, E Ohue, K Oda, M Tanabe, H Shimamoto, and T Onai. A selective-epitaxial SiGe HBT with SMI electrodes featuring 9.3-ps ECL-Gate Delay. Technical Digest IEEE International Electron Devices Meeting, San Francisco, 1997, pp. 795–798.
54. HJ Osten, D Knoll, B Heinemann, H Rücker, and B Tillack. Carbon doped SiGe heterojunction bipolar transistors for high frequency applications. Proceedings of the IEEE Bipolar/BiCMOS Circuits and Technology Meeting, Minneapolis, 1999, pp. 109–116.

55. SJ Jeng, B Jagannathan, J-S Rieh, J Johnson, KT Schonenberg, D Greenberg, A Stricker, H Chen, M Khater, D Ahlgren, G Freeman, K Stein, and S Subbanna. A 210-GHz f_T SiGe HBT with non-self-aligned structure. *IEEE Electron Device Letters* 22:542–544, 2001.
56. JS Rieh, B Jagannathan, H Chen, KT Schonenberg, D Angell, A Chinthakindi, J Florkey, F Golan, D Greenberg, S-J Jeng, M Khater, F Pagette, C Schnabel, P Smith, A Stricker, K Vaed, R Volant, D Ahlgren, G Freeman, K Stein, and S Subbanna. SiGe HBTs with cut-off frequency of 350 GHz. Technical Digest of the IEEE International Electron Devices Meeting, San Francisco, 2002, pp. 771–774.
57. B El-Kareh, S Balster, W Leitz, P Steinmann, H Yasuda, M Corsi, K Dawoodi, C Dirnecker, P Foglietti, A Haeusler, P Menz, M Ramin, T Scharnagl, M Schiekofer, M Schober, U Schulz, L Swanson, D Tatman, M. Waitschull, JW Weijtmans, and C Willis. A 5V complementary SiGe BiCMOS technology for high-speed precision analog circuits. Proceedings of the IEEE Bipolar/BiCMOS Circuits and Technology Meeting, Toulouse, 2003, pp. 211–214.
58. B Heinemann, R Barth, D Bolze, J Drews, P Formanek, O Fursenko, M Glante, K Glowatzki, A Gregor, U Haak, W Höppner, D Knoll, R Kurps, S Marschmeyer, S Orlowski, H Rücker, P Schley, D Schmidt, R Scholz, W Winkler, and Y Yamamoto. A complementary BiCMOS technology with high speed npn and pnp SiGe:C HBTs. Technical Digest of the IEEE International Electron Devices Meeting, Washington, 2003, pp. 117–120.
59. J Cai, M Kumar, M Steigerwalt, H Ko, K Schonenberg, K Stein, H Chen, K Jenkins, Q Ouyang, P Oldiges, and T Ning. Vertical SiGe-base bipolar transistors on CMOS-compatible SOI substrate. Proceedings of the IEEE Bipolar/BiCMOS Circuits and Technology Meeting, Toulouse, 2003, pp. 215–218.
60. J-S Rieh, D Greenberg, M Khater, KT Schonenberg, J-J Jeng, F Pagette, T Adam, A Chinthakindi, J Florkey, B Jagannathan, J Johnson, R Krishnasamy, D Sanderson, C Schnabel, P Smith, A Stricker, S Sweeney, K Vaed, T Yanagisawa, D Ahlgren, K Stein, and G Freeman. SiGe HBTs for millimeter-wave applications with simultaneously optimized f_T and f_{max}. Proceedings of the IEEE Radio Frequency Integrated Circuits (RFIC) Symposium, Fort Worth, 2004, pp. 395–398.
61. DL Harame and BS Meyerson. The early history of IBM's SiGe mixed signal technology. *IEEE Transactions on Electron Devices* 48:2555–2567, 2001.
62. T Tatsumi, H Hirayama, and N Aizaki. Si/$Ge_{0.3}Si_{0.7}$ heterojunction bipolar transistor made with Si molecular beam epitaxy. *Applied Physics Letters* 52:895–897, 1988.
63. H Temkin, JC Bean, A Antreasyan, and R Leibenguth. Ge_xSi_{1-x} strained-layer heterostructure bipolar transistors. *Applied Physics Letters* 52:1089–1091, 1988.
64. D-X Xu, G-D Shen, M Willander, W-X Ni, and GV Hansson. n-Si/p-Si_{1-x}/n-Si double-heterojunction bipolar transistors. *Applied Physics Letters* 52:2239–2241, 1988.
65. GL Patton, SS Iyer, SL Delage, S Tiwari, and JMC Stork. Silicon–germanium-base heterojunction bipolar transistors by molecular beam epitaxy. *IEEE Electron Device Letters* 9:165–167, 1988.
66. A Gruhle, H Kibbel, U König, U Erben, and E Kasper. MBE-Grown Si/SiGe HBTs with high β, f_T, and f_{max}. *IEEE Electron Device Letters* 13:206–208, 1992.
67. AJ Joseph, D Coolbaugh, D Harame, G Freeman, S Subbanna, M Doherty, J Dunn, C Dickey, D Greenberg, R Groves, M Meghelli, A Rylyakov, M Sorna, O Schreiber, D Herman, and T Tanji. 0.13 μm 210 GHz f_T SiGe HBTs—expanding the horizons of SiGe BiCMOS. Technical Digest IEEE International Solid-State Circuits Conference, San Francisco, 2002, pp. 180–182.
68. H. Lilienfeld Patent, 1926; O. Heil, British patent number 439,457, 1935.
69. D Khang and MM Atalla. Silicon–silicon dioxide field induced surface devices. Solid State Research Conference, Pittsburgh, 1960.
70. FM Wanlass and CT Sah. Nanowatt logic using field-effect metal-oxide-semiconductor triodes (MOSTs). IEEE International Solid-State Circuits Conference, Philadelphia, 1963, pp. 32–33.
71. P Balk. Surface properties of oxidized germanium-doped silicon. *Journal of the Electrochemical Society* 118:494–495, 1971.
72. H Daembkes, H-J Herzog, H Jorke, H. Kibbel, and E Kasper. The n-channel SiGe/Si modulation doped field-effect transistor. *IEEE Transactions on Electron Devices* 33:633–638, 1986.

73. TP Pearsall and JC Bean. Enhancement and depletion-mode p-channel Ge_xSi_{1-x} modulation-doped field effect transistor. *IEEE Electron Device Letters* 7:308–310, 1986.
74. H Temkin, TP Pearsall, JC Bean, RA Logan, and S. Luryi. Ge_xSi_{1-x} strained-layer superlattice waveguide photodetectors operating near 1.3 μm. *Applied Physics Letters* 48:963–965, 1986.
75. RD Thompson, KN Tu, J Angillelo, S Delage, and SS Iyer. Interfacial reaction between Ni and MBE grown SiGe alloys. *Journal of the Electrochemical Society* 135:3161–3163, 1988.
76. HC Liu, D Landheer, M Buchmann, and DC Houghton. Resonant tunneling diode in the $Si_{1-x}Ge_x$ system. *Applied Physics Letters* 52:1809–1811, 1988.
77. RC Taft, JD Plummer, and SS Iyer. Demonstration of a p-channel BiCFET in the Ge_xSi_{1-x}/Si system. *IEEE Electron Device Letters* 10:14–16, 1989.
78. TJ King, JR Pfriester, JD Scott, JP McVittie, and KC Saraswat. A polycrystalline SiGe gate CMOS technology. Technical Digest of the IEEE International Electron Devices Meeting, Washington, 1990, pp. 253–256.
79. RA Soref, F Namavar, and JP Lorenzo. Optical waveguiding in a single-crystal layer of germanium–silicon grown on silicon. *Optics Letters* 15:270–272, 1990.
80. DK Nayak, JCS Woo, JS Park, KL Wang, and KP MacWilliams. Enhancement-mode quantum-well Ge_xSi_{1-x} PMOS. *IEEE Electron Device Letters* 12:154–156, 1991.
81. K Ismail, BS Meyerson, and PJ Wang. Electron resonant tunneling in Si/SiGe double barrier diodes. *Applied Physics Letters* 59:973–975, 1991.
82. DC Houghton, JP Noel, and NL Rowell. Electroluminescence and photoluminesence from SiGe alloys grown on (1 0 0) silicon by MBE. *Materials Science and Engineering B* 9:237–244, 1991.
83. DS Chen, JP Conde, V Chu, S Aljishi, JZ Liu, and S Wagner. Amorphous silicon–germanium thin-film photodetector array. *IEEE Electron Device Letters* 13:5–7, 1992.
84. S-B Hwang, YK Fang, K-H Chen, C-R Liu, J-D Hwang, and M-H Chou. An a-Si:H/a-Si, Ge:H bulk barrier phototransistor with a-SiC:H barrier enhancement layer for high-gain IR optical detector. *IEEE Transactions on Electron Devices* 40:721–726, 1993.
85. DK Nayak, JCS Woo, GK Yabiku, KP MacWilliams, JS Park, and KL Wang. High mobility GeSi PMOS on SIMOX. *IEEE Electron Device Letters* 14:520–522, 1993.
86. DK Nayak, JCS Woo, JS Park, KL Wang, and KP MacWilliams. High-mobility p-channel metal-oxide semiconductor field-effect transistor on strained Si. *Applied Physics Letters* 62:2853–2855, 1993.
87. J Welser, JL Hoyt, and JF Gibbons. Electron mobility enhancement in strained-Si n-type metal-oxide semiconductor field-effect transistors. *IEEE Electron Device Letters* 15:100–102, 1994.
88. SK Ray, S John, S Oswal, and SK Banerjee. Novel SiGeC channel heterojunction pMOSFET. Technical Digest of the IEEE International Electron Devices Meeting, Washington, 1996, pp. 261–264.
89. SJ Mathew, WE Ansley, WB Dubbelday, JD Cressler, JA Ott, JO Chu, PM Mooney, KL Vavanagh, BS Meyerson, and I Lagnado. Effect of Ge profile on the frequency response of a SiGe pFET on sapphire technology. Technical Digest of the IEEE Device Research Conference, Boulder, 1997, pp. 130–131.
90. K Rim, JL Hoyt, and JF Gibbons. Transconductance enhancement in deep submicron strained-Si n-MOSFETs. Technical Digest of the IEEE International Electron Devices Meeting, Washington, 1998, pp. 707–710.
91. KC Liu, SK Ray, SK Oswal, and SK Banerjee. $Si_{1-x}Ge_x$/Si vertical pMOSFET fabricated by Ge ion implantation. *IEEE Electron Device Letters* 19:13–15, 1998.
92. S Thompson, N. Anand, M Armstrong, C Auth, B Arcot, M Alavi, P Bai, J Bielefeld, R Bigwood, J Brandenburg, M Buehler, S Cea, V Chikarmane, C Choi, R Frankovic, T Ghani, G Glass, W Han, T Hoffmann, M Hussein, P Jacob, A Jain, C Jan, S Joshi, C Kenyon, J Klaus, S Klopcic, J Luce, Z Ma, B McIntyre, K Mistry, A Murthy, P Nguyen, H Pearson, T Sandford, R Schweinfurth, R Shaheed, S Sivakumar, M Taylor, B Tufts, C Wallace, P Wang, C Weber, and M Bohr. A 90 nm logic technology featuring 50 nm strained silicon channel transistors, 7 layers of Cu interconnects, low k ILD, and 1 μm^2 SRAM Cell. Technical Digest of the IEEE International Electron Devices Meeting, Washington, 2002, pp. 61–64.

93. VP Kesan, S Subbanna, PJ Restle, MJ Tejwani, JM Aitken, SS Iyer, and JA Ott. High performance 0.25 μm p-MOSFETs with silicon–germanium channels for 300 K and 77 K operation. Technical Digest of the IEEE International Electron Devices Meeting, San Francisco, 1991, pp. 25–28, 1991.
94. S Verdonckt-Vanderbroek, E Crabbé, BS Meyerson, DL Harame, PJ Restle, JMC Stork, AC Megdanis, CL Stanis, AA Bright, GMW Kroesen, and AC Warren. High-mobility modulation-doped, graded SiGe-channel p-MOSFETs. *IEEE Electron Device Letters* 12:447–449, 1991.
95. S Verdonckt-Vanderbroek, E Crabbé, BS Meyerson, DL Harame, PJ Restle, JMC Stork, and JB Johnson. SiGe-channel heterojunction p-MOSFETs. *IEEE Transactions on Electron Devices* 41:90–102, 1994.

3

Overview: SiGe and Si Strained-Layer Epitaxy

John D. Cressler
Georgia Institute of Technology

The field of silicon heterostructures necessarily begins with materials, and the crystal growers of our field have learned through much hard work how to practice modern miracles in their growth of near defect-free, nanoscale films of Si and SiGe strained-layer epitaxy, which are compatible with conventional high-volume silicon integrated circuit manufacturing.

This book tells the materials side of the story, and details the many advances in the Si–SiGe strained-layer epitaxy for device applications. Chapter 4, "Strained SiGe and Si Epitaxy," by B. Tillack of IHP, reviews the underlying materials science of Si–SiGe epitaxy, while Chapters 5 to 7 discuss modern SiGe epitaxial growth techniques: RTCVD in Chapter 5, "SiGe:C Eiptaxy by RTCVD," by D. Dutartre of ST Microelectronics, MBE in Chapter 6, "MBE Growth Techniques," by M. Oehme and E. Kaspar of the University of Stuttgart, and UHV/CVD in Chapter 7, "UHV/CVD Growth Techniques," by T. Adam of IBM. The complexity of epi defects and the dopant diffusion characteristics in such films are discussed by A. Peaker of the University of Manchester in Chapter 8, "Defects and Diffusion in Strained SiGe and Si," and the most recent (and robust) stability theory is covered in Chapter 9, "Stability Constraints in SiGe Epitaxy," by A. Fischer of IHP. The electrical transport properties of SiGe, strained Si, and Si–C alloys are detailed by J. Hoyt of MIT in Chapter 10, "Electronic Properties of Strained Si–SiGe and $Si_{1-y}C_y$ alloys." The basic mechanisms underlying the now-pervasive use of C-doping in SiGe HBTs as a boron-doping diffusion inhibitor are reviewed in Chapter 11, "Carbon Doping of SiGe," by J. Osten of the University of Hanover, and Chapter 12, "Contact Metallization on SiGe," by C. Maiti of IIT, covers ohmic and Schottky contacts to SiGe and strained Si. Finally, Chapter 13, "Selective Etching Techniques for SiGe–Si," by S. Monfray of ST Microelectronics, discusses the use of SiGe for selective etching in various emerging MEMS applications.

In addition to this material, and the numerous references contained in each chapter, a number of review articles and books on SiGe–strained Si materials exist, including Refs. [1–6].

References

1. R People. Physics and applications of Ge_xSi_{1-x}/Si strained layer heterostructures. *IEEE Journal of Quantum Electronics* 22:1696–1710, 1986.

2. JC Bean. Silicon-based semiconductor heterostructures: column IV bandgap engineering. *Proceedings of the IEEE* 80:571–587, 1992.
3. B Meyerson. UHV/CVD growth of Si and SiGe alloys: chemistry, physics, and device applications. *Proceedings of the IEEE* 80:1592–1608, 1992.
4. SC Jain. *Germanium–Silicon Strained Layers and Heterostructures.* New York, NY: Academic Press, 1994.
5. E Kaspar. *Properties of Strained and Relaxed Silicon Germanium.* London: INSPEC, EMIS Datareviews Series No. 12, 1995.
6. CK Maiti, NB Chakrabarti, and SK Ray. *Strained Silicon Heterostructures: Materials and Devices.* London: The Institute of Electrical Engineers, 2001.

4

Strained SiGe and Si Epitaxy

Bernd Tillack and
Peter Zaumseil
IHP

4.1 Introduction

By adding Ge to Si new properties of the material can be created, which offer applications in a wide range of electronic devices. In this way the capability of Si electronics is widened. The SiGe heterojunction bipolar transistor (SiGe HBT), which makes use of SiGe bandgap engineering and strain, has reached industrial level (see Refs. [1,2]) with increasing market potential. For future CMOS technologies SiGe and Si strained layers are becoming increasingly important, for instance, for higher mobility channel material [3,4]. After the first report on growth of epitaxial SiGe layers dating back to 1962 [5], the pioneering work of Kasper (see, e.g., Ref. [6]) and Meyerson (see, e.g., Ref. [7]) became important milestones for the development of SiGe strained-layer epitaxy. After demonstrating the stability and high-volume production capability of SiGe strained-layer epitaxy for HBT base deposition using ultrahigh vacuum CVD (UHV CVD) [8] and low-pressure CVD (LP CVD) [9] the low-temperature epitaxial deposition of SiGe by CVD was ready to be used in manufacturing. Especially, the demonstration of the ability to grow device-quality SiGe layers without using UHV deposition techniques [10–14] has greatly influenced the development of strained-layer deposition process technology and tools.

Adding C to SiGe has extended the capability of the material and has been another important step in the success story of SiGe:C strained-layer epitaxy for HBT application [15,16]. C can significantly suppress B diffusion without negative impact on device parameters [16,17]. The first BiCMOS technology using SiGe:C HBTs was demonstrated by IHP in 1999 [18]. As an example, Figure 4.1 shows a transmission electron microscopy (TEM) cross section of an HBT with SiGe:C base layer deposited by CVD.

In this chapter, we will review the basic considerations regarding SiGe heteroepitaxy (Section 4.2) and strain relaxation. Moreover, characterization of SiGe layers (Section 4.3) and process aspects of growth of strained SiGe and SiGe:C on Si (Section 4.4) will be discussed.

4.2 Heteroepitaxy of SiGe and SiGe:C on Si

Silicon and germanium have the same crystallographic structure. Both materials can be alloyed as $Si_{1-x}Ge_x$ with any value of $0 \leq x \leq 1$. The lattice constant of Ge is 4.18% larger than that of Si, and

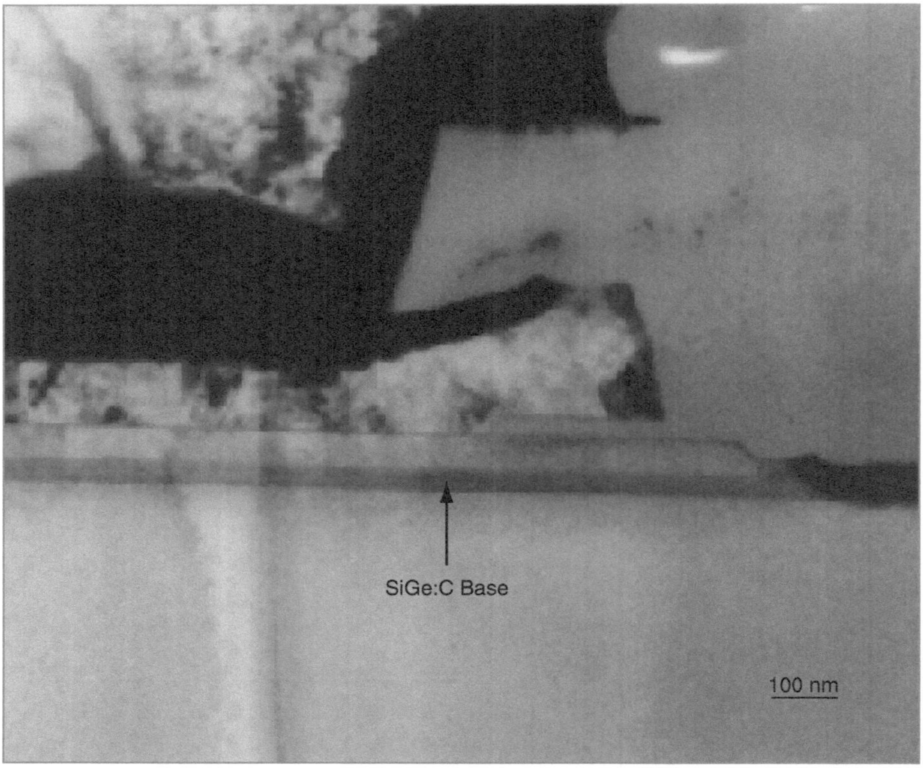

FIGURE 4.1 TEM cross section of HBT with pseudomorphic SiGe:C on Si (IHP's first SiGe:C HBT generation).

for a $Si_{1-x}Ge_x$ alloy it does not exactly follow Vegard's law. The relative change of the lattice constant is given by [19]

$$\varepsilon = \frac{a_{SiGe} - a_{Si}}{a_{Si}} = 0.00501x^2 + 0.03675x.$$

Growing a $Si_{1-x}Ge_x$ layer with $x > 0$ on a Si substrate means that the layer is under compressive stress. A perfect epitaxial growth of such a strained heteroepitaxial layer is only possible as long as its thickness does not exceed a critical thickness of stability [20]. Above this value, the strain is relaxed through the formation of misfit dislocations.

The dislocation-free (pseudomorphic) SiGe layer on a 001 Si substrate surface shows a tetragonally distorted unit cell (see also Figure 4.2) with in-plane ($a_{\|}$) and perpendicular ($a_{\perp}$) lattice constants given by

$$a_{\|} = a_{Si}, \quad a_{\perp} = a_{Si}(1 + k\varepsilon),$$

with

$$k = 1 + \frac{2C_{12}}{C_{11}} \cong 1.75.$$

In this case, the degree of relaxation is zero. For a fully relaxed layer the lattice constants $a_{\|}$ and $a_{\perp}$ are equal:

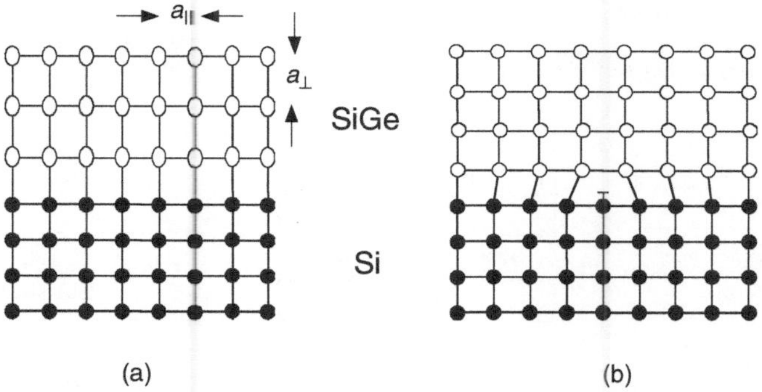

FIGURE 4.2 Structural scheme of pseudomorphic (a) and fully relaxed (b) SiGe layer grown on Si substrate.

$$a_{\parallel} = a_{\perp} = a_{Si}(1 + \varepsilon).$$

For a partly relaxed layer the lattice constants depend on both the Ge content and the degree of relaxation [21]. The consequence is that always $a_{\parallel}$ and $a_{\perp}$ must be measured to determine the Ge content (and the degree of relaxation) as long as it is not definitely clear that a SiGe layer is either pseudomorphic or fully relaxed.

Experimentally it was found that using low deposition temperatures (550°C and lower) it was possible to deposit pseudomorphic SiGe layers with thicknesses exceeding the critical thickness value [22,23]. In these cases films are metastable. Nevertheless, by capping metastable SiGe films with Si, stabilization could be obtained resulting in SiGe–Si stacks that withstand thermal treatment during device processing (see Chapter 9, "Stability Constraints in SiGe Epitaxy").

The strain situation is completely different for $Si_{1-y}C_y$ layers. Carbon atoms are much smaller than Si atoms, and in consequence the lattice constant of $Si_{1-y}C_y$ is smaller than that of silicon. Here, a deviation from Vegard's law was also found [24]:

$$\varepsilon = \frac{a_{SiC} - a_{Si}}{a_{Si}} = 0.10504y^2 - 0.44909y.$$

A $Si_{1-x-y}Ge_xC_y$ layer can be treated in first approximation as a mixture of a $Si_{1-x}Ge_x$ and a $Si_{1-y}C_y$ layer. The compressive strain of the SiGe can be (partly) compensated by the tensile strain of the added C, which will be demonstrated in detail in Section 4.3.

Different modes were found during heteroepitaxy depending on the strain in the heteroepitaxial film and the growth conditions (mainly growth temperature) (Figure 4.3). Which growth mode for the epitaxial deposition is taking place is determined by the free energy of the interfaces and the lattice mismatch of the heteroepitaxial system. For most of the applications layer-by-layer growth (two-dimensional growth) is desired. In this case, the pseudomorphic films are obtained if the strain in the film does not exceed the critical thickness limitation or if the films are metastable, and partly relaxed films are obtained if the strain is relaxed by misfit dislocation formation [20,25]. The mode can change from two-dimensional to three-dimensional (island growth) during growth if the strain increases or for high deposition temperatures (Stranski–Krastanow growth). Dislocation-free Stranski–Krastanow growth in particular has been investigated (e.g., Ref. [26]) because of its capability for optical applications.

For selective heteroepitaxial growth it was found that the dislocation density depends on the area of the deposited films [27]. The dislocation density decreases with decreasing area. Even dislocation-free films with thicknesses above the critical thickness are possible for small areas.

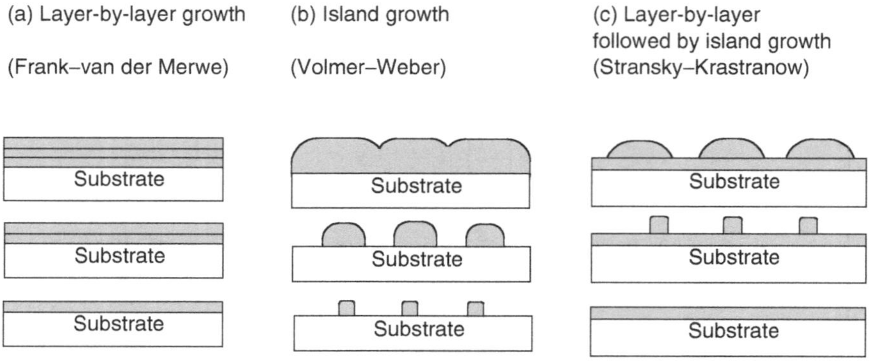

FIGURE 4.3 Growth modes for heteroepitaxial systems. (From JW Matthews and AE Blakeslee. *J Cryst Growth* 32:265–273, 1976; JH Van der Merwe. *J Appl Phys* 34:123–127, 1963; L Vescan, W Jäger, C Dieker, K Schmidt, A Hartmann, and H Lüth. MRS Symposium Proceedings, Vol. 263, Mechanism of Heteroepitaxial Growth Symposium, 1992, pp. 23–28. With permission.)

4.3 Characterization of Strained SiGe and Si Layers

One of the main structural features of SiGe or SiGe:C layers is the difference in its lattice constant relative to the silicon substrate, which was already mentioned in Section 4.2. This difference in lattice constant, or strain, which correlates in the case of a pseudomorphically grown SiGe layer directly to the Ge content, offers the possibility of an easy characterization by x-ray diffractometry (XRD), where the lattice constant is transferred to a measurable diffraction angle via Bragg's law

$$2d \sin \Theta = n\lambda.$$

In the following, we will discuss the application of XRD to characterize SiGe and SiGe:C structures. Later on, we will compare the results obtained by XRD with those obtained by other techniques, and give an outlook to further developments. Here, we restrict our discussion to pseudomorphic structures; fully or partly relaxed structures will be discussed elsewhere.

The typical XRD arrangement consists of the x-ray source, a monochromator or collimator, the sample, and the detector. In the simple case, the collimator consists of a perfect Si crystal of the same orientation as the sample (Figure 4.4). Modern diffractometers often use the so-called Bartel's monochromators and additional mirrors as collimator to make the arrangement more flexible. For some applications, an additional analyzer crystal is used in front of the detector. A rocking curve is measured by rotating the sample around a substrate diffraction peak $\Delta\Theta$ (typically of the netplanes parallel to the surface) and correcting the detector position in such a way that the diffracted beam enters the detector window at the same position always ($\Theta/2\Theta$ scan). How the diffractometer arrangement influences the measured rocking curve of a SiGe structure was demonstrated in Ref. [28].

Figure 4.5 shows as an example the CuK_α-400-diffraction of a 117 nm thick $Si_{0.8}Ge_{0.2}$ layer with a 56 nm thick Si cap layer on top. For such a relatively simple structure, the Ge content can be directly obtained from the angular distance between Si and SiGe peak. The width of the SiGe peak is a direct measure of the SiGe layer thickness. But, the comparison with the calculated curve without the Si cap layer shows that it is difficult to estimate the thickness of the cap layer from the diffraction pattern directly. The situation becomes even worse for more complicated structures, for example, structures with graded SiGe layers (see below). The determination of the depth profile of such structures is only possible by creating a reasonable layer model, simulation of the diffraction curve of this model, and fitting the simulated curve to the experimental one by modification of free parameters.

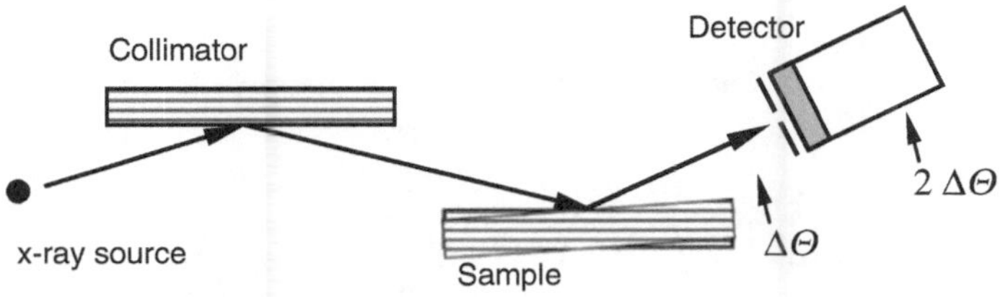

FIGURE 4.4 Scheme of a double crystal diffractometer arrangement.

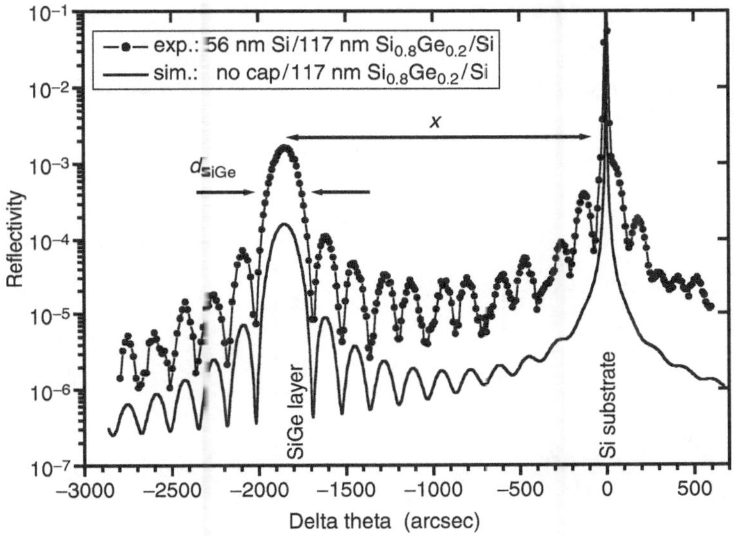

FIGURE 4.5 Experimental diffraction curve of a SiGe layer with a Si cap layer on top and a simulated curve without the cap layer; CuK_α radiation, 400 reflection. The simulated curve is shifted by one order of magnitude relative to the experimental one.

For $Si_{1-y}C_y$ layers, the diffraction curve would be similar but with the $Si_{1-y}C_y$ peak on the right-hand side of the Si substrate peak (high angle side).

A $Si_{1-x-y}Ge_xC_y$ layer can be treated in first approximation as a mixture of a $Si_{1-x}Ge_x$ and a $Si_{1-y}C_y$ layer. The compressive strain of the SiGe can be (partly) compensated by the tensile strain of the SiC. Figure 4.6 shows an XRD measurement with calculated rocking curves of a 100-nm thick SiGeC layer with 20% Ge. The SiGeC peak shift as a function of the C content indicates strain compensation by adding C to SiGe.

Since XRD measures primarily the strain of the layer relative to the substrate, and for $Si_{1-x-y}Ge_xC_y$ this strain is the sum of two components, it is impossible to determine the Ge and the C content without independent information about one of the components. Usually, to get the C content of a SiGeC layer we deposit a SiGe layer with the same Ge deposition parameters for comparison. Then the Ge content is measured at the SiGe sample, and assuming the same for the SiGeC sample the C content can be obtained.

To characterize an HBT structure under conditions of a routine process control, the following requirements must be fulfilled: the depth resolution should be in the order of about 1 nm; the accuracy of the absolute Ge content should be better than 0.5%; it must be nondestructive, fast, and reliable. This is a real challenge for XRD, especially when the Ge content is not constant over the layer thickness as in typical HBT structures.

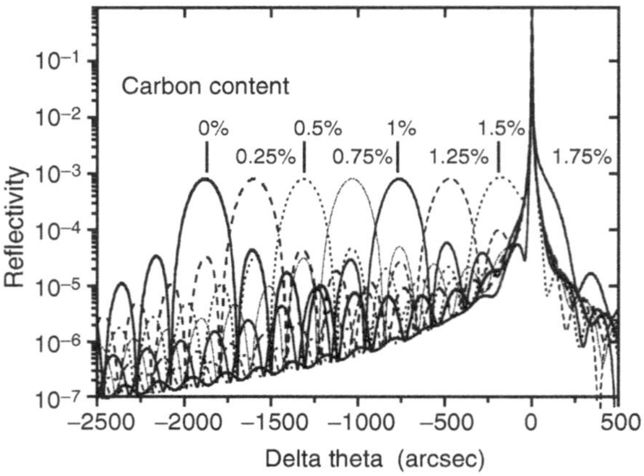

FIGURE 4.6 Calculated rocking curves of a 100 nm thick SiGeC layer with 20% Ge and increasing carbon content; CuK_{α} radiation, 400 reflection.

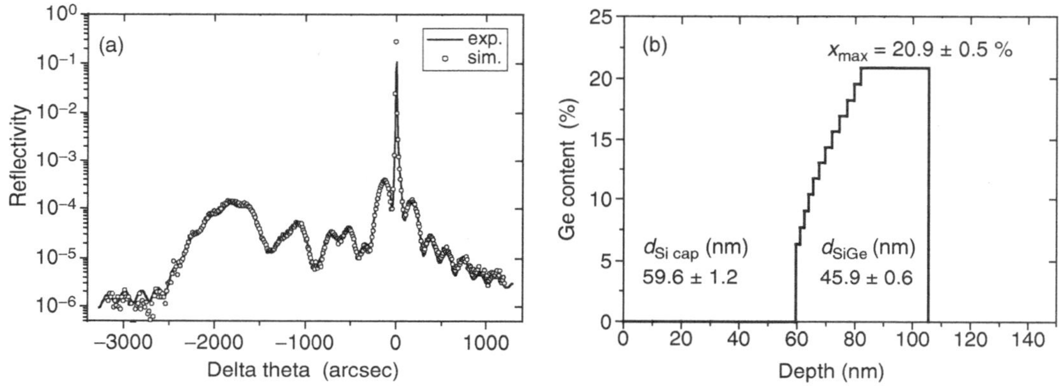

FIGURE 4.7 XRD rocking curve (a) of a SiGe HBT structure. The full line represents the best-fitted simulated curve. The Ge depth profile is shown in (b).

Figure 4.7 shows as an example the characterization of an HBT structure. Due to the gradient part of the Ge profile, the rocking curve (a) shows less details compared to the simple layer structure (Figure 4.5). Here, it is absolutely necessary to simulate curves with a suited model and to fit this in a trial-and-error procedure to the experimental one, since it is practically impossible to get any direct information from the rocking curve. The depth profile of Ge content that gave the best fit of the rocking curve is shown in Figure 4.7b. The gradient part of the Ge profile is divided into 11 lamellae of constant strain. Since the thickness of each lamella is a free parameter in the fitting process, the shape of this profile part need not be linear. It should fit to the real shape within the sensitivity limits of this technique.

The accuracy depends on the stability of the fitting procedure, supposing the model used describes the real situation in a proper way [29]. This is mainly influenced by the statistical intensity fluctuations (noise) in the RC range far away from the Si substrate peak. Following this, the accuracy is given by the intensity of the x-ray source used or the measuring time. For similar structures to that shown in Figure 4.7, an error in the layer thickness of 0.4 to 1.0 nm and of less than 0.4% of the maximum Ge content is typically achievable with an intensity of about 500 kcps in the incident beam and a measuring time of less than 1 h.

TABLE 4.1 Summary of Main Features of Investigated Techniques for HBT Stack Characterization

Technique	Destructive	Area	Accuracy of Measurement			IF Roughness
			$d_{Si\ cap}$ (nm)	d_{SiGe} (nm)	x_{max}(%)	
XRD	No	>0.5 × 0.5 mm^2	1.0	0.6	0.5	(−)
XRR	No	(cm)	0.5	0.5	3–6	++
SE	No	14 × 28 μm^2	1.0	1.0	1	−
AES	Yes	0.1 × 0.1 μm^2	—	5.0	1–3	−
SIMS	Yes	60 × 60 μm^2	3.0	3.0	3–5	−
TEM	Yes	(μm)	1.0	1.0	—	(+)

Besides XRD, there are many other analytical techniques available to study SiGe or SiGe:C HBT structures. In Ref. [30], the capabilities of six different techniques are discussed; three non-destructive methods: XRD, x-ray reflectometry (XRR), and spectroscopic ellipsometry (SE); and three destructive methods: Auger electron spectroscopy (AES), secondary ion mass spectroscopy (SIMS), and TEM. The main result was that every technique has its advantages and disadvantages, and they all can be used successfully either in process development, failure analysis, or in-line monitoring of the epitaxy process. Within the error limits of each technique, the HBT parameters obtained agreed quite well. Table 4.1 summarizes the main features of these techniques.

XRD plays an outstanding role, since this technique was used to calibrate other techniques, such as SE, AES, and SIMS, with suitable simple SiGe layer structures. The main disadvantage of XRD for an in-line routine application is its limited lateral resolution. A minimum spot size for laboratory devices of about 0.5 × 0.5 mm^2 is sufficient to measure on monitoring areas of the same size [31] but far too large for measurements on real device structures.

The alternative for future use is SE. This method allows measurements in micrometer areas; it is fast and well established in microelectronics technology. The procedure [32] includes the creation of databases for the refractive index dispersion of all components of HBT stacks using simple one-layer structures with thickness and composition calibrated by XRD. Then these databases (e.g., SiGe:C optical constants versus Ge-content) can be applied for thickness and composition determination of graded HBTs with different shapes of profiles. The achievable accuracy in layer thickness and Ge content determination is comparable to XRD.

4.4 Growth of Strained SiGe on Si

The critical thickness limitation and the thermal stability (metastable layers) of strained SiGe films on Si cause severe limitations for the integration into Si process technology. The deposition of pseudomorphic SiGe layers itself requires low-temperature process technique. For the integration into CMOS or BiCMOS technologies the impact of the thermal budget of the deposition process on existing structures as well as the interaction of thermal treatment of the processing with the deposited SiGe layer has to be considered. The low temperature requirement for the deposition techniques has different aspects:

1. For low-temperature deposition the moisture and oxygen level in the reaction chamber is more critical compared to high temperature.
2. The low-temperature deposition process is controlled by the kinetics, which means that process conditions, especially process temperature and partial pressures of source gases, are essentially impacting the parameters of the deposited films like deposition rate, incorporation of Ge, C, and dopants. To grow films according to manufacturing requirements with uniform thickness, composition, and dopant distribution, the temperature has to be controlled during the process and across the wafer very accurately.

3. For the integration of SiGe into CMOS or BiCMOS low thermal budget deposition is required. The thermal budget of SiGe heteroepitaxy using CVD techniques is mainly determined by the H_2 prebake before deposition to clean the substrate surface at temperatures typically between 950°C and 1000°C.

Different low-temperature deposition techniques have been developed for strained SiGe epitaxy on Si. Today, heteroepitaxy by CVD is in production with proven stability and manufacturability. In the following part, key aspects, requirements, and challenges for SiGe CVD epitaxy will be discussed.

For the deposition of epitaxial SiGe layers with low defect densities a clean substrate surface with low contamination level is essential. Cleaning of the substrate is achieved by combination of *ex situ* wet chemical treatment and *in situ* H_2 prebake in the reaction chamber before epitaxial deposition [33,34]. The cleaning effect of the prebake improves with increasing temperature. However, it has to be compromised with the demand of a minimum thermal budget necessary for process integration into CMOS and BiCMOS. The kind of surface passivation after wet chemical treatment impacts the minimum H_2 prebake temperature necessary for effective contamination removal. An oxide passivation was achieved by a final SC-2 treatment at standard conditions ($HCl + H_2O_2 + H_2O$ at 75°C to 85°C). The chemically grown oxide during SC-2 has to be removed during the prebake, resulting in higher bake temperatures of about 890°C. By applying hydrogen passivation generated by final DHF dip followed by a DI water rinse in combination with an optimized bake regime (maximum temperature for about 3 sec, with a ramp rate of higher than 70°C/sec) the bake temperature could be lowered to temperatures below 800°C [33]. Figure 4.8 demonstrates, by means of SIMS, the effect of the prebake temperature on the O and C concentrations at the interface between substrate and epitaxial layer for oxygen-passivated surface (Figure 4.8a) and hydrogen-passivated surface (Figure 4.8b). The impact of defects caused by insufficient H_2 prebake during the growth of the epitaxial Si–SiGe–Si layer stack on HBT leakage currents was shown in Ref. [34].

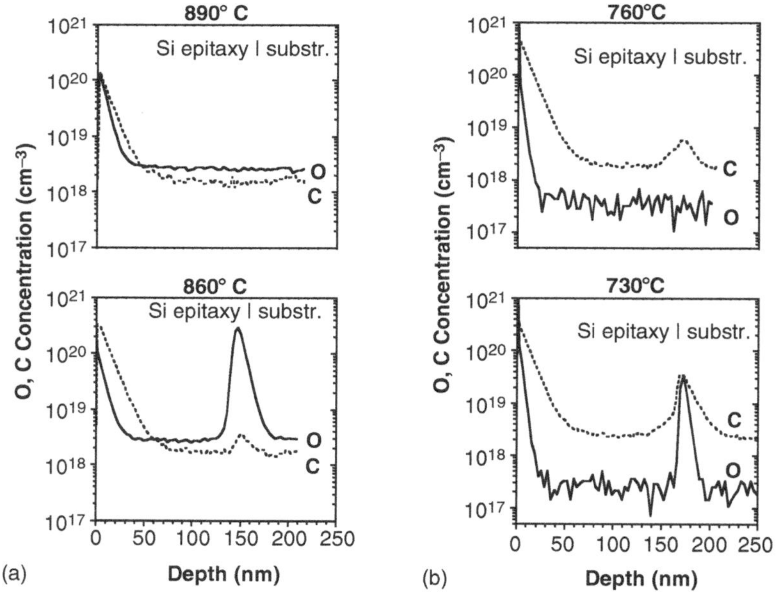

FIGURE 4.8 SIMS profiles illustrating the oxygen and carbon removal at the epitaxy substrate interface for (a) oxygen passivated (standard RCA cleaning), and (b) hydrogen passivated surface (DHF treatment) with corresponding hydrogen prebake temperatures.

After cleaning of the Si surface by H_2 prebake the SiGe (or SiGe:C) heteroepitaxy is performed at temperatures typically between 500°C and 700°C depending on the target layer parameters, especially the Ge content. In this temperature range the growth is kinetically controlled. Figure 4.9 shows the growth rate as a function of the reciprocal temperature for different GeH_4 partial pressures at constant SiH_4 partial pressure. GeH_4 and SiH_4 are the sources for the SiGe layer growth. Hydrogen is used as carrier gas. The activation energies determined from the plots in Figure 4.9 are decreasing with increasing GeH_4 partial pressure (increasing Ge concentration). For the lowest GeH_4 partial pressure, the activation energy is 1.9 eV, which is close to the value obtained for pure Si deposition and discussed as the activated energy of the desorption of H from the Si surface. Adding GeH_4 is supporting the desorption of H, resulting in lower activation energies. Deposition pressures between low-pressure conditions (typical 1 to 2 Torr) to atmospheric pressure have been used. The most common tools are working in the reduced pressure range of about 100 Torr for the deposition.

The Ge concentration in the SiGe layer is controlled by the GeH_4 partial pressure for constant SiH_4 partial pressure and temperature (Figure 4.10). The incorporation of Ge is impacted by the growth

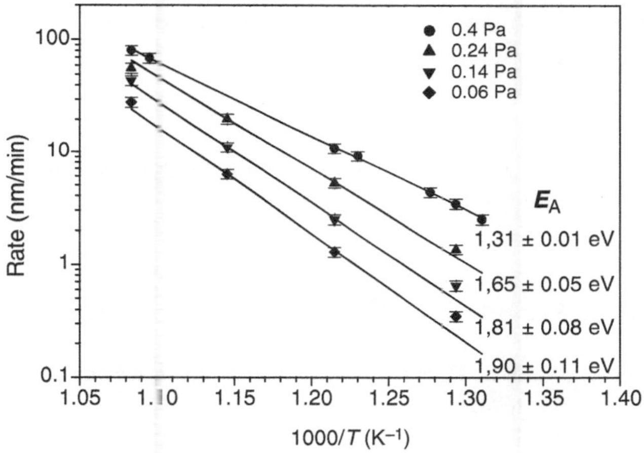

FIGURE 4.9 SiGe deposition rate as function of the reciprocal temperature for different GeH_4 partial pressures, constant SiH_4 pressure and H_2 as carrier gas, and resulting activation energies.

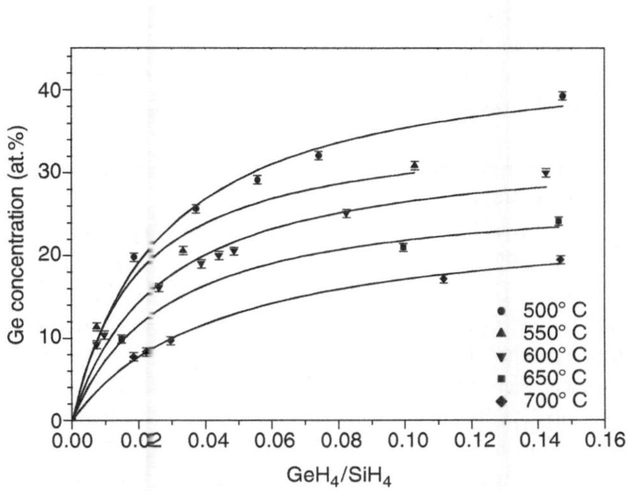

FIGURE 4.10 Ge concentration in SiGe as a function of the GeH_4 to SiH_4 ratio for different growth temperatures.

temperature. The Ge content in the SiGe layer decreases with increasing temperature for constant GeH_4 and SiH_4 partial pressure. Therefore, for higher Ge content lower growth temperatures have to be applied. Moreover, for SiGe layers with high Ge content grown at high temperature, island growth and relaxation are more likely.

In the case of SiGe:C epitaxy the growth temperature is impacting the incorporation of C into SiGe. Lower growth temperatures are beneficial for incorporation of C on substitutional sites. At high growth temperature (and high C content), C tends to be incorporated interstitially degrading the crystalline and electrical properties of the layers.

4.5 Summary

Today SiGe-strained epitaxy is meeting manufacturing requirements and it is a proven process in microelectronics technology for bipolar (HBT) and CMOS (strained Si) applications. There is still a great potential of SiGe, Si, and Ge layers for future devices and technologies. Despite the fact that strained-layer epitaxy is managed very well using commercially available deposition tools there is room for further improvement, for example, in low-temperature processing and increase in throughput. A very interesting topic for further development of epitaxy is the atomic layer processing approach for atomic level control of doping and deposition [35–37].

Acknowledgment

The authors would like to thank the IHP technology team for support and for the preparation of the SiGe–SiGe:C layers and HBT processing.

References

1. JD Cressler and G Niu. *Silicon–Germanium Heterojunction Bipolar Transistors.* Boston, MA: Artech House, 2003.
2. R Singh, DL Harame, and MM Oprysko. *Silicon Germanium Technology, Modeling, and Design.* Piscataway, NJ: IEEE Press, 2004.
3. S Verdonckt-Vanderbroek, F Crabbe, BS Meyerson, DL Harame, PJ Restle, JMC Stork, and JB Johnson. SiGe channel heterojunction p-MOSFET's. *IEEE Trans Electron Dev* 41:90–101, 1994.
4. T Ghani, M Armstrong, C Auth, M Bost, P Charvat, G Glass, T Hoffmann, K Johnson, C Kenyon, J Klaus, B McIntyre, K Mistry, A Murthy, J Sandford, M Siberstein, S Sivakumar, P Smith, K Zawadzki, S Thompson, and M Bohr. A 90 nm high volume manufacturing logic technology featuring novel 45 nm gate length strained silicon CMOS transistors. Technical Digest of the IEEE International Electron Devices Meeting, Washington, 2003, pp. 978–980.
5. KJ Miller and MJ Grieco. Epitaxial silicon-germanium alloy films on silicon substrates. *J Electrochem Soc* 109:70–71, 1962.
6. E Kasper and HJ Herzog. Elastic strain and misfit dislocation density in $Si_{0.92}Ge_{0.08}$ films on silicon substrates. *Thin Solid Films* 44:357–370, 1977.
7. BS Meyerson. UHV/CVD growth of Si and Si:Ge alloys: Chemistry, physics, and device applications. *Proc IEEE* 80:1592–1608, 1992.
8. DC Ahlgren, M Gilbert, D Greenberg, SJ Jeng, J Malikowski, D Nguyen-Ngoc, K Schonenberg, K Stein, R Groves, K Walter, G Hueckel, D Colavito, G Freeman, D Sunderland, DL Harame, and B Meyerson. Manufacturability demonstration of an integrated SiGe HBT technology for the analog and wireless marketplace. Technical Digest of the IEEE International Electron Devices Meeting, San Francisco, 1996, pp. 859–862.
9. B Tillack, D Bolze, G Fischer, G Kissinger, D Knoll, G Ritter, P Schley, and D Wolansky. SiGe heteroepitaxy for high frequency circuits. MRS Symposium Proceedings, Vol. 525, Rapid Thermal and Integrated Processing VII, 1998, pp. 379–384.

10. G Ritter, B Tillack, and D Knoll. Successful preparation of high frequency HBT by integrated RTCVD processes. MRS Symposium Proceedings, Vol. 387, Rapid Thermal and Integrated Processing IV, 1995, pp. 341–346.
11. D Dutartre, P Warren, I Berbezier, and P Perret. Low temperature silicon and $Si_{1-x}Ge_x$ epitaxy by rapid thermal chemical vapour deposition using hydrides. *Thin Solid Films* 222:52–56, 1992.
12. JC Sturm, PV Schwartz, EJ Prinz, and H Manoharan, Growth of $Si_{1-x}Ge_x$ by rapid thermal chemical vapor deposition and application to heterojunction bipolar transisitors. *J Vac Sci Technol B* 9:2011–2016, 1991.
13. WB de Boer and DJ Meyer. Low temperature chemical vapor deposition of epitaxial Si and SiGe layers at atmospheric pressure. *Appl Phys Lett* 58:1286–1288, 1991.
14. CA King, JL Hoyt, DB Noble, CM Gronet, JF Gibbons, MP Scott, SS Laderman, TI Kamins, and J Turner. Epitaxial growth of $Si_{1-x}Ge_x$/Si heterostructures by limited reaction processing for minority carrier device applications. MRS Symposium Proceedings, Vol. 146, Rapid Thermal Annealing/Chemical Vapor Deposition and Integrated Processing Symposium, 1989, pp. 71–82.
15. B Heinemann, D Knoll, G Fischer, D Krüger, G Lippert, HJ Osten, H Rücker, W Röpke, P Schley, and B Tillack. Control of steep boron profiles in Si/SiGe heterojunction bipolar transistors. ESSDERC: Proceedings of the 27th European Solid-State Device Research Conference, Stuttgart, 1997, pp. 544–547.
16. HJ Osten, G Lippert, D Knoll, R Barth, B Heinemann, H Rücker, and P Schley. The effect of carbon incorporation on SiGe heterobipolar transistor performance and process margin. Technical Digest of the IEEE International Electron Devices Meeting, Washington, 1997, pp. 803–806.
17. D Knoll, B Heinemann, HJ Osten, KE Ehwald, B Tillack, P Schley, R Barth, M Matthes, KS Park, Y Kim, and W Winkler. Si/SiGe:C heterojunction bipolar transistors in an epi-free well, single-polysilicon technology. Technical Digest of the IEEE International Electron Devices Meeting, San Francisco, 1998, pp. 703–706.
18. HJ Osten, D Knoll, B Heinemann, H Rücker, and B Tillack. Carbon doped SiGe heterojunction bipolar transistors for high frequency applications. Proceedings of the Bipolar/BiCMOS Circuits and Technology Meeting, Minneapolis, 1999, pp. 109–116.
19. JP Dismukes, L Ekstrom, and RJ Paff. Lattice parameter and density in germanium–silicon alloys. *J Phys Chem* 68:3021–3027, 1964.
20. JW Matthews and AE Blakeslee. Defects in epitaxial multilayers. III. Preparation of almost perfect layers. *J Cryst Growth* 32:265–273, 1976.
21. P Zaumseil. A fast x-ray method to determine Ge content and relaxation of partly relaxed $Si_{1-x}Ge_x$ layers on silicon substrates. *Phys Stat Sol (a)* 141:155–161, 1994.
22. R People, and J Bean. Erratum: calculation of critical layer thickness versus lattice mismatch for Ge_xSi_{1-x}/Si strained-layer heterostructures. *Appl Phys Lett* 49:229, 1986.
23. E Kasper, A Schuh, G Bauer, B Holländer, and H Kibbel. Test of Vegard's Law in thin epitaxial SiGe layers. *J Cryst Growth* 157:68–72, 1995.
24. F Berti, D De Salvador, AV Drigo, F Romanato, J Stangl, S Zerlauth, F Schäffler, and G Bauer, Lattice parameter in $Si_{1-y}C_y$ epilayers: deviation from Vegard's rule. *Appl Phys Lett* 72:1602–1604, 1998.
25. JH Van der Merwe. Crystal interfaces. Part II. Finite overgrowth. *J Appl Phys* 34:123–127, 1963.
26. L Vescan, W Jäger, C Dieker, K Schmidt, A Hartmann, and H Lüth, Formation of heterogeneous thickness modulations during epitaxial growth of LPCVD-$Si_{1-x}Ge_x$/Si quantum well structures. MRS Symposium Proceedings, Vol. 263, Mechanism of Heteroepitaxial Growth Symposium, 1992, pp. 23–28.
27. T Stoica and L Vescan. Misfit dislocations in finite lateral size $Si_{1-x}Ge_x$ films grown by selective epitaxy. *J Cryst Growth* 131:32–40, 1993.
28. P Zaumseil. A comparison of different multiple-crystal diffractometer arrangements to measure the reflection curve of SiGe layers on Si substrates. *Cryst Res Technol* 31:529–537, 1996.
29. P Zaumseil. High resolution determination of the Ge depth profile in SiGe heterobipolar transistor structures by x-ray diffractometry. *Phys Stat Sol (a)* 165:195–204, 1998.

30. P Zaumseil, D Krüger, R Kurps, O Fursenko, and P Formanek. Precise measurement of Ge depth profiles in SiGe HBT's—a comparison of different methods. *Solid State Phenomena* 95–96:473–482, 2004.
31. P Zaumseil, TA Lafford, and M Taylor. Inline characterization of SiGe structures on 8-inch Si wafers using the Bede QC200 x-ray diffractometer. *J Phys D Appl Phys* 34:A52–A56, 2001.
32. O Fursenko, J Bauer, P Zaumseil, D Krüger, A Goryachko, Y Yamamoto, K Köpke, and B. Tillack. Spectroscopic ellipsometry for in-line process control of SiGe:C HBT. Proceedings of the Second International SiGe Technology and Device Meeting, ISTDM 2004, Frankfurt(Oder), Germany, 2004, pp. 53–54.
33. D Wolansky, B Tillack, K Blum, KD Bolze, KD Glowatzki, K Köpke, D Krüger, R Kurps, G Ritter, and P Schley. Low temperature clean for Si/SiGe epitaxy for CMOS integration of HBTs. *Electrochem Soc Proc* 98-1:812–821, 1998.
34. D Wolansky, GG Fischer, D Knoll, D Bolze, B Tillack, P Schley, and Y Yamamoto. Impact of defects on the leakage currents of Si/SiGe/Si heterojunction bipolar transistors. *Solid State Phenomena* 95–96:249–254, 2004.
35. J Murota, M Sakuraba, and B Tillack. Atomically controlled technology for future Si-based devices. *Solid State Phenomena* 95–96:607–616, 2004.
36. B Tillack, Y Yamamoto, D Knoll, B Heinemann, P Schley, B Senapati, and D Krüger. High performance SiGe:C HBTs using atomic layer base doping. *Appl Surf Sci* 224:55–58, 2004.
37. B Tillack, B Heinemann, and D Knoll. Atomic layer doping of SiGe—fundamentals and device applications. *Thin Solid Films* 369:189–194, 2000.

5

Si–SiGe(C) Epitaxy by RTCVD

Didier Dutartre, F. Deléglise, C. Fellous, L. Rubaldo, and A. Talbot

ST Microelectronics

5.1 Introduction

At this time, about 15 to 20 years after a real breakthrough in $Si_{1-x}Ge_x$ (SiGe) growth, it is of interest to reflect upon developments and progress made in this field. On the one hand, Si-based alloys, namely Si_{1-x-y} Ge_xC_y (SiGeC), are well known to be key materials for extending the capabilities of the silicon technology that is very dominant in electronics. These alloys, indeed, are fully compatible with this technology and have various characteristics (electronic, chemical, mechanical, and optical) that can be used for a number of proven and potential applications as detailed throughout this book.

On the other hand, the development of SiGe growth techniques has been highly competitive, it is probably unique that the deposition of a material was simultaneously studied over the complete pressure domain available, from 10^{-8} Torr by molecular beam epitaxy to 760 Torr by chemical vapor deposition (CVD), and using a variety of CVD techniques like ultrahigh vacuum (UHV), very low pressure, low pressure, reduced pressure, atmospheric and plasma-enhanced CVD, chemical beam epitaxy, etc. However, rapid thermal chemical vapor deposition (RTCVD), first invented and developed in home-made tools or in prototypes, made rapid and impressive progresses, especially in low-temperature SiGe epitaxy (epi), and was rapidly introduced in industrial tools. Finally, this technique took the leadership for SiGe epi. Today, one can say that RTCVD and SiGe(C) epitaxy have been married for the better: RTCVD has been demonstrated to be a very effective technique for growing SiGe epitaxial layers and SiGe a powerful booster for RTCVD.

Thus, blanket epitaxies of SiGe on full-sheet silicon wafers, and after SiGeC ones, were rapidly demonstrated in pioneer RTCVD studies. However, epitaxial depositions that are required today may be much more complex, and have in most cases to be integrated in advanced technologies: i.e., on wafers patterned with very fine structures and with drastic thermal budget limitations.

This chapter focuses on the current developments of SiGe and SiGeC epitaxies by RTCVD and their integration in complex technologies. Section 5.2 presents the RTCVD technique in terms of equipments and process capabilities for SiGe epi. This technique is also compared to the other important epi techniques and our conclusion at the advantage of RTCVD is in accordance with the success of this technique. Section 5.3 details the important points of the Si-based alloy epitaxy: surface preparation, low-temperature epi (LTE), germanium and carbon incorporation, and selective epi. LTE is usually obtained with hydrides like silane- or chlorides-like dichlorosilane (for selective epi). Both systems are considered and surface reactions that are known to play a major role are reviewed. On the other hand, the main features of strained SiGe epi and of carbon incorporation are reported and discussed. Section 5.4 is devoted to the integration issues. Among the huge number of possible points, we chose those that are actually met by the process engineer who has to manage epi for the creation of devices in modern technologies. Note that some of these points, pattern-induced defectivity as an example, are not so frequently reported in literature. Finally, Section 5.5 illustrates the RTCVD capabilities giving a few examples of current applications developed at STMicroelectronics. As a number of applications will be detailed in following chapters, the selection is very limited: epitaxial base of heterojunction bipolar transistor, epi for gate-all-around MOS and epi on ultrathin silicon on insulator films. Examples were chosen in order to give a certain panorama of applications and of epi issues.

5.2 Rapid Thermal Chemical Vapor Deposition

Background

CVD consists in the deposition of a solid film on a substrate by the reaction of vapor-phase reactants (precursors). The substrate temperature provides the energy to activate the chemical reactions. As illustrated in Figure 5.1, the sequential steps of this process are the following:

- Transport of precursor gas into reactor by forced convection
- Diffusion to surface and adsorption of precursor molecules on surface
- Surface reactions (decomposition and recombination) and incorporation into solid film
- Desorption of by-product molecules and diffusion into the gas phase
- Evacuation of gaseous by-products from reactor

RTCVD is defined as a CVD technique capable of a rapid switching of the process temperature. This technique is very interesting as the temperature agility allows the thermal budget to be minimized and different films to be grown using different "adapted" temperatures. Gibbons et al. [1] first reported such a technique, and referred it as "limited reaction processing" (LRP); a stable gas flow was established with the wafer at low temperature and deposition was switched on and off by rapidly heating and cooling the

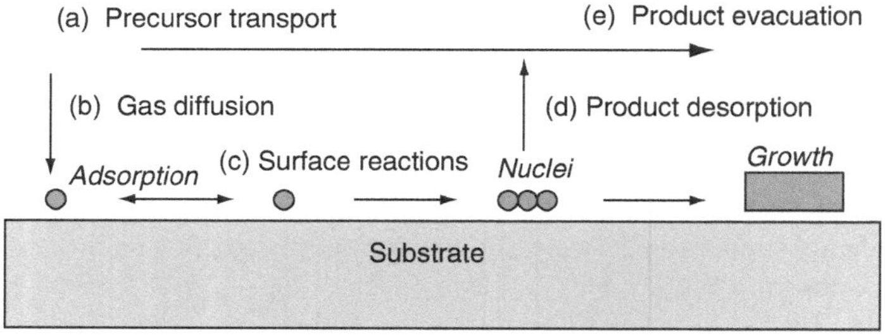

FIGURE 5.1 Generic steps of a CVD process: (a) reactant transport, (b) reactant diffusion, (c) surface reactions, (d) by-product desorption, and (e) by-product evacuation.

wafer. This technique was demonstrated to be very effective for some processes, Si epitaxy as an example, but it also presents important drawbacks especially in terms of process control. Thus, the majority of investigators preferred to use gas flow switching rather than temperature to control growth, and then the technique has been more often referred as RTCVD.

On the other hand, RTCVD has been found to significantly relax the stringent temperature or cleanliness conditions for epitaxial growth; and within a few years, a considerable amount of research was devoted to the application of this technique to the silicon and SiGe epi. The present contribution will be restricted to these applications.

Equipments

Beyond the pioneer work, various reactors were developed and important refinements were introduced: infrared pyrometer for temperature control and loadlock for moisture contamination reduction [2,3]. Basically, the minimal common features of these systems were: single-wafer susceptor-less configuration, lamp heating, cold walls, and low pressure (a few Torr). Rapidly, using relatively simple tools, very impressive material demonstrations were made in the domain of Si–SiGe epi: high structural quality epi (proved by excitonic emission) [4,5], functional HBT structures [6], ultra-abrupt dopant profiles [7], etc.

However, in the majority of RTCVD reactors, as a consequence of poor temperature control and gas flow design, the deposition uniformity was not good enough for industrial applications. Around 1990, for the first time a new production epitaxy reactor, the "Epsilon One" from ASM company [8], included some above-mentioned features for RTCVD: single-wafer, lamp heating, loadlock, etc., but it also used a susceptor and substrate rotation for better temperature control and uniformity improvement. More recently, "Applied Materials" also introduced the "Centura-HTF" reactor, which is a similar system. Despite the temperature ramps are slowed down by the presence of a rotating susceptor, in the author's opinion these tools have been considered as RTCVD epi reactors.

The schematics of these modern vapor-phase epitaxy (VPE) systems typically consists of four modules: the process module, the transfer module, the gas control box, and automation. These systems are fully automatic and 25 or 50 wafers can be processed either using a given recipe, in production mode, or various recipes, for research and development (R&D); a variety of process parameters and hardware configurations can also be controlled. The gas distribution system is designed with ultrapurity standards and allows precise gas injection onto the wafer with short (<1 sec) switching transients. The transfer module, preferably at reduced pressure, isolates the process chamber from any atmospheric contamination at each loading or unloading of wafers. These systems process a single wafer at a time in a low-profile, horizontal deposition chamber that is shown schematically in Figure 5.2. The chamber walls are made of fused quartz, and air-cooling allows the quartz to remain at moderate temperatures, about half the wafer temperature (in Celsius). Any stainless steel part is outside of the chamber or protected by quartz liners, and is water-cooled,

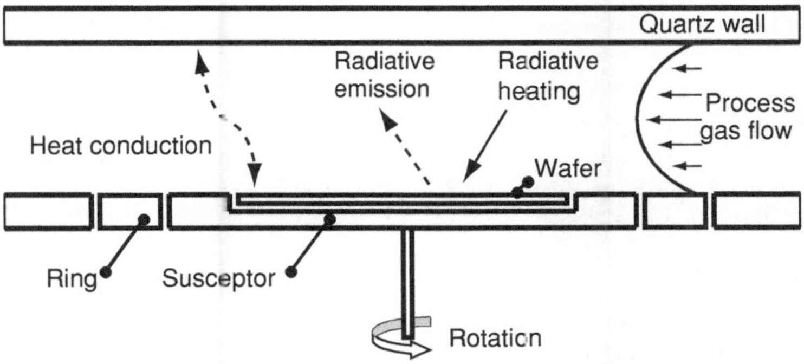

FIGURE 5.2 Schematic cross section of a typical RTCVD reactor. The process gases make a single pass over the rotating wafer and susceptor assembly that are heated from top and bottom by tungsten halogen lamps.

this way any metal corrosion by chloride is minimized. For process, the wafer is placed on a SiC-coated graphite susceptor and both parts are radiantly heated from both sides by external tungsten halogen lamps. The top and bottom heating allow the temperature difference between the wafer and the susceptor to be minimized, then improving control and uniformity of temperature (see "Low-Temperature Epi"). In the same way, the radiative heating can be adjusted along the wafer radius to tune the radial uniformity. For control, temperature is measured either by a pyrometer (Applied Materials) or a thermocouple (ASM). A process gas enters from the right of the figure and makes a single pass over the wafer. Finally, the chamber is evacuated by combination of a high-capacity Roots blower with a mechanical dry pump, and processes can be run in a large pressure range (1 to 760 Torr).

Process Capabilities

Let us start with temperature agility, the first characteristic of RTCVD. In susceptor-less lamp-heated systems, the time constant, related to the thermal inertia of wafer, like in rapid thermal annealing tools, is in the range 1 to 4 sec. That roughly corresponds to the time of ramp up and down between a process temperature and a temperature significantly lower (a temperature at which the kinetics of important physical mechanisms has decreased by a decade). Unfortunately, the systems with the highest level of development do have a susceptor. In this case, the time constant is less aggressive, between 10 and 30 sec as a function of system and process. This characteristic is nevertheless useful to facilitate and make some processes more effective, including epi. A rapid hydrogen bake is possible, 30 sec at 900°C for example, and the epi temperature can be adapted to the epi deposition or even to the different parts of a complex stack. As seen below, silicon epi requires higher temperatures, as compared to SiGe epi. On the other hand, the temperature control, maybe the most critical point as it is not as perfect as desired, will be discussed in "Low-Temperature Epi."

Considering the chamber design, we believe that the low-profile horizontal geometry is important. In most of the experimental conditions, it induces first a laminar flow beneficial in terms of uniformity and particles, and second a relatively high chemical efficiency as precursor molecules have to pass close to the wafer. On the other hand, the susceptor rotation, with a significant speed (10 to 50 rpm), is useful to smooth temperature variations and to level the flow and depletion of reactants. In addition, as the two industrial tools mentioned present the capability of reduced pressure, the residence time of precursors in chamber can be then reduced by an important (10 to 100) factor. This reduced time, together with a limited hot volume because of the cold walls, will eliminate or minimize the possibility of homogeneous gas-phase nucleation. Thus, it appears that RTCVD is a technique well adapted for the utilization of very reactive precursor gases.

Typical processes and some process details will be given in the next section. However, the major advantages of this important technique can be summarized as follows:

- Temperature agility
- Cold walls (at least significantly colder than the substrate)
- External heater
- Cleanliness of gases
- Process flexibility (temperature and pressure range)
- Chemistry flexibility (hydrides and chlorides)
- Ability to work in both deposition regimes (surface or mass transport-limited kinetics)
- *In situ* chamber cleaning (usually with HCl at high temperature)
- Mature and industrial technique

as well as the main drawbacks:

- Complex temperature uniformity and temperature control
- Complexity of CVD growth kinetics
- Productivity of low-temperature and growth rate processes

TABLE 5.1 Performances of Current Epitaxy Techniques for Si and SiGe(C) Applications

Process Capabilities	Technique					
	PECVD (single)	MBE (single)	CBE (single)	UHVCVD (batch)	LPCVD (batch)	RTCVD (single)
Epi domain						
Thin films	Y	Y	Y	Y	Y	Y
Thick films	Y	N	N	N	N	Y
Temperature domain						
High *T*	N	N?	N?	N	N	Y
Low *T*	Y	Y	Y	Y	Y	Y
Ultralow *T*	Y[a]	Y[a]	N	N	N	N
Process domain						
Surface-limited	?	Y?	Y?	Y	Y	Y
Mass transport	Y	Y	Y	N	N	Y
Selective epi (against Si_3N_4)	N	N	Y?	N	N?	Y
Crystal quality	Y	Y	Y	Y	Y	Y
Low defectivity: particle-induced	Y	Y?	Y[b]	Y?	Y?	Y
High chemical purity	Y	N?[c]	Y	Y	Y	Y
Process flexibility (*P*, *T*, etc.)	Some	Some	Some	Some	Some	Y
Memory effect elimination	Y	N	Y?	N?	N?	Y
In situ chamber etch	Y	N	N	N	N?	Y
Chemistry flexibility	Y	Y[d]	Y	N	N	Y
Temperature agility (*T* change)	N	N	Y?	N[e]	N[e]	Y
Repeatability	Y	N	Y	Y	Y	Y
Uniformity (all elements)	Y	N?	Y[f]	?	?	Y
Productivity (>6 w/h)	Y	N	N?	Y	Y	Y
Industrial tool (8/12 in.)	Y	N	N	N	Y	Y

?, questionable.
[a] Questionable crystal quality.
[b] Y if H_2 bake possible.
[c] Metal contamination often observed [9].
[d] In terms of atoms or evaporated radicals.
[e] Difficult.
[f] Not proved for any element.

Pros and Cons

In order to complete the picture, RTCVD is compared to the other existing techniques in Table 5.1. In comparing these pros and cons to those of other techniques, the balance is clearly in favor of RTCVD. That certainly explains the extensive development and success of these equipments. Today, they are sold in large numbers for conventional epitaxies in manufacturing plants, as well as for advanced epitaxies and R&D at major semiconductor suppliers and institutes.

5.3 Epitaxy Processes

Today, epitaxy technology is facing new challenges, which come from the continuous progress of silicon technology, namely smaller geometries and new epi applications in devices (epitaxial base of HBTs as an example). In addition, as technology and devices, especially CMOS transistors, approach their theoretical limits (optical lithography limit tunnel current through the gate oxide, etc.), great efforts are made in order to improve or exchange device architecture. For these applications, epitaxial deposition of new materials like Si-based alloys is very desirable, and in most cases it has to be run in the course of device fabrication.

Most of the well-established epi processes in production today are carried out by CVD with a silicon precursor diluted in hydrogen; they are run in the very front end of technology at high temperatures (1000 to 1100°C). However, as soon as a substantial amount of the device is present on the wafer and strained or metastable materials are concerned, conventional epitaxy processes with high thermal budgets can no longer be used.

This section presents and discusses the important issues that are specific to Si and SiGeC processes required by new epi applications and new devices in Si technology.

Surface Preparation

Epitaxy essence is that the growing material forms a monocrystal that prolongs the substrate lattice. That is to say that the presence of an amorphous film on the surface cannot be tolerated; the typical example is the native oxide that naturally exists on silicon substrates. In the same way, to achieve high-quality epi, it is essential to remove any particles and contamination from the substrate surface prior deposition initiation. Contaminants would prevent surface migration of silicon atoms and form precipitates, generating lattice defects.

Thus, the perfection of the epitaxial growth on silicon substrates relies critically upon the surface preparation, and all the following conditions are required:

- No oxide on the surface
- A good crystal quality (etching or implant may have been used in previous operations)
- No precipitates of dopant or other impurities
- No surface contamination or particles

In conventional epi processes carried out at high temperature, the silicon surface preparation relies on both a wet *ex situ* clean and an *in situ* hydrogen bake. The *ex situ* clean generally consists of an RCA cleaning procedure [10] that eliminates particles, native oxide, organics, metal and carbon, and re-passivates the reactive silicon surface with a thin (0.6 nm) layer of suboxide that is hydrophilic, stable, not highly reactive and relatively easily removed in situ. The substrate is then submitted to a high-temperature hydrogen bake (above 1000°C) to remove the RCA-regrown oxide before epi deposition. It is admitted that this oxide is reduced via the two following reactions:

$$SiO_2(\text{solid}) + Si(\text{solid}) \rightarrow 2SiO(\text{gas}) \tag{5.1}$$

$$SiO_2(\text{solid}) + H_2(\text{gas}) \rightarrow SiO(\text{gas}) + H_2O(\text{gas}) \tag{5.2}$$

Since reaction (5.1) is very effective each time a silicon–oxide interface is in direct contact with vapor and the RCA oxide is somewhat porous, the author supposes that it is the most effective in the case of pre-epi bake. The hydrogen bake that is also capable to anneal eventual crystal imperfections, to dissolve or diffuse eventual high dopant concentrations or contamination, and to dissolve or evaporate most of small particles is very beneficial to the epi. Thus, this combination of *ex situ* and *in situ* clean-ups, usually carried out on "full-sheet" silicon wafer, is very effective in producing high-quality epi.

In the domain of LTE, there are two very different situations. In a first situation, epitaxy has to be run at low temperature because metastable films (strained, SiGeC, etc.) or sharp dopant profiles have to be created, but the substrate is resistant enough to high temperatures. This corresponds to processes placed at the very front end of the technology, or used for R&D structures basically deposited on full-sheet wafers. In this case, a conventional surface preparation is used and very low defect densities (a few defects per wafer) are achieved. In a second situation, which is more frequent, as sensitive structures are already present, the substrate cannot withstand high temperatures. In this case, the strategy of surface preparation has to be changed. Indeed, at moderate temperatures (below 900°C), reactions (5.1) and (5.2) responsible for the oxide reduction are not effective enough to remove the superficial oxide. It is admitted that a 6-Å thick chemical oxide as prepared by RCA cleaning requires a hydrogen bake above 1000°C to be fully removed and to leave a perfectly clean surface suited for high-quality epi.

The best solution would be to use etch processes, carried out directly in the epitaxy chamber or in an clustered annex, that would remove effectively the chemical oxide at low temperatures. Various treatments have been proposed. Low-energy plasmas (hydrogen, argon, etc.) have been investigated. Encouraging results have been obtained but these solutions were not developed at an industrial level [11]. However, as the interest for low thermal budgets becomes stronger and stronger these treatments may come back on the scene. Vapor-phase HF cleaning has also been developed and more or less integrated in epi tools. However, this chemistry presents some difficulties: the oxide etch rate from anhydrous HF-based process is unstable and the process control difficult, and maybe more tricky is the fact that hydrocarbons on the oxide surface block the adsorption of the species that are required for etching [12]. All these new chemistries, carried out in prototype reactors, are not yet common.

Consequently, most researchers today adopt a strategy based on the more conventional "HF-last" clean. The HF bath removes the oxide from the wafer surface and passivates the silicon surface with atomic hydrogen. The UHV/CVD growth process reported by Meyerson is based on such a HF-last clean [13]. Originally, the wafers were etched in diluted HF just prior to loading, without any water rinse. However, this procedure presents two major difficulties: (i) transport and manipulation of the wafers coming from HF acid without any rinse are very critical and (ii) removal of the residual HF is very difficult when hydrophilic patterns are present on the wafer surface. Because of these difficulties, a possible and more widely used surface preparation procedure is: RCA clean + HF last + water rinse + IPA dry + loading in the epi tool + *in situ* hydrogen bake (800°C to 900°C for about 1 min). Of course, a wet clean using an "*in situ*" rinse, made by displacing an extremely diluted HF solution with DI water in the same tank, is recommended because it eliminates the transfer of hydrophobic wafers from the HF to the DI-water bath. On the other hand, clustering this precleaning with the epitaxy tool would also be preferable [14].

In the author's opinion, this type of surface preparation allows high-quality epitaxies to be grown: typically, no oxygen or carbon is visible by SIMS at the interface (detection limit around 1×10^{12} atoms/cm^2), and defect densities are in the range of 0.1 to 10 defects/cm^2 (light point defect $> 0.16\,\mu$m), depending on precise experimental conditions. Such low values measured on silicon full-sheets are also expected on patterned wafers. On the other hand, some operations, such as dry-etch or implant, can cause the silicon at the surface and in the subsurface region to be highly defective and not compatible with a high-quality epitaxial growth. In these cases, the processes of these operations require modifications in order to respect the crystalline quality of the silicon (for example, by adding a soft-etch step to an etch recipe), or a sacrificial oxide, typically 10 to 20 nm thick, can be grown (and removed) for damage removal before epitaxy.

Low-Temperature Epi

Silicon epi can be deposited using $SiCl_4$ (sil tet), $SiHCl_3$ (TCS), SiH_2Cl_2 (DCS), SiH_4, Si_2H_6, or even Si_3H_8 as precursors. Growth rate (GR) depends on several parameters: reactor geometry, temperature, gas source, flow rate, deposition pressure, and concentrations. According to the gas source, for similar conditions it is admitted that the deposition kinetics increases when Cl atoms are reduced and Si atoms increased in the molecule, namely from $SiCl_4$ to Si_3H_8. However, as the reactivity increases, the stability decreases, and these molecules are increasingly subject to thermal decomposition and gas-phase nucleation. Thus, hydrides have to be used at lower temperatures and smaller partial pressures, as compared to chlorides. Thus, as a function of their properties, each precursor is attractive for particular film specifications and deposition conditions; today, chlorides are used for thick epi at high temperature, usually above 1000°C, and hydrides for thin epi at lower temperatures, below 900°C.

In pioneer RTCVD reports, DCS [15], silane [16], and to a less extent disilane [17] were studied for low-temperature applications. As an illustration, Figure 5.3 reports the Arrhenius plots of Si growth rate obtained with DCS and silane. First, we note that each system exhibits two deposition regimes. At high

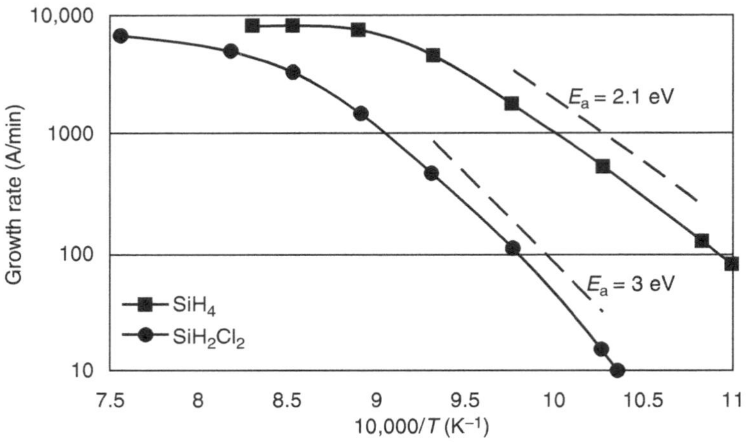

FIGURE 5.3 Arrhenius plot for silicon growth rate from silane and dichlorosilane in similar conditions. In the low-temperature domain, silane chemistry presents a smaller activation energy and a much higher kinetics compared to dichlorosilane.

temperatures, the growth rate is almost insensitive to temperature and, in most cases, is controlled by precursor gas-phase mass transport. At low temperatures, the growth rate is strongly dependent on temperature, and the variation is attributed to a thermally activated process. As this process is supposed to take place on the wafer surface, this domain is often referred to as "surface rate-limited." Second, as the curves correspond to deposition carried out in a given reactor and with similar conditions of gas flow, we conclude that the silane deposition kinetics is much faster than that of DCS, and with a smaller activation energy. The advantage of silane corresponds to more than one decade in terms of growth rate, or 100 K in temperature, and to a less temperature-sensitive process. These significant changes will enable the process to be more easily controlled and the thermal budget, which could be capital in some applications, to be reduced. As a consequence, DCS is often used for selective depositions because Cl atoms help at selectivity, as discussed in "Selective Epitaxy" but today silane is preferred and has been adopted for most of the nonselective epitaxies.

Considering the epi kinetics from silane at low temperature, the Arrhenius plot given in Figure 5.3 reveals an exponential dependence of GR with an activation energy of about 46 kcal/mol, which is in accordance with the majority of values reported in the literature for various experimental conditions [16,18–20]. This value also accords very well with the activation energy for hydrogen desorption from a Si⟨1 0 0⟩ surface (47 kcal/mol) [21]. Thus, growth kinetics is supposed to be correlated with equilibrium hydrogen surface coverage as depicted in Figure 5.4: the reactive adsorption of silane molecules produces adsorbed hydrogen atoms whose desorption is not immediate; this hydrogen surface coverage regulates further SiH_4 adsorption [18]. Consequently, hydrogen desorption is the mechanism that limits the growth rate (via silane adsorption), which is ultimately independent of silane pressure. Initially developed in a domain of very low pressure, this model was then applied in a number of experiments. It was also further refined; for example, taking the adsorption of molecular hydrogen quantitatively into account [16]. Finally, these calculations agree extremely well with experimental data covering a wide range of pressures and temperatures, suggesting a high level of understanding in the growth kinetics of silicon epitaxy using silane. A simplified form of reaction pathway can be written as

$$SiH_4 + 2\_ \rightarrow \underline{Si} + 2\underline{H} + H_2 \tag{5.3}$$

$$\underline{H}^* + \underline{H} \rightarrow H_2 + 2\_ \tag{5.4}$$

$$H_2 + 2\_ \rightarrow 2\underline{H} \tag{5.5}$$

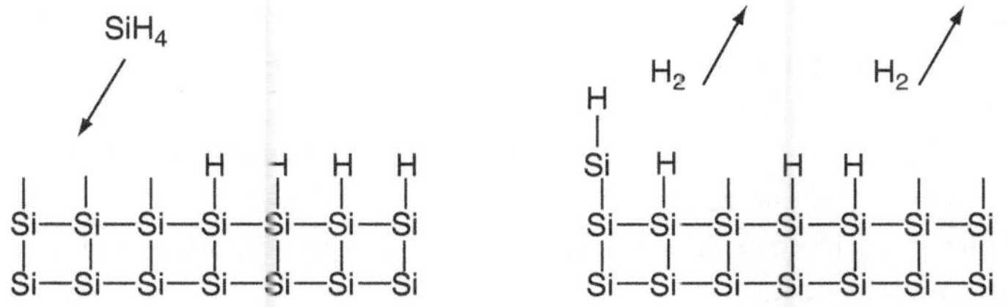

FIGURE 5.4 Schematic of low-temperature silane-based epitaxy. Silane molecules are supposed to adsorb on two Si sites leaving two adsorbed H atoms that passivate the surface. Growth kinetic corresponds then to the balance between silane adsorption and hydrogen desorption.

where _ denotes a site that is available for adsorption, X a specie that is adsorbed, H* corresponds to an excited state of hydrogen (following Ref. [21] hydrogen desorption is first-order relative to atomic adsorbed hydrogen) and no difference is made between the Si on surface and in bulk.

On the other hand, there are some precautions to be taken when using silane. First, the presence of impurities in the gas phase, even at very small concentrations, will not be purged in a compressed gas (silane) in the same way they are in a liquid (DCS). Second, the low stability of silane molecules makes it susceptible to gas-phase nucleation of particles, which can "rain down" and become incorporated in the growing film. However, this limitation is not so severe since silane is chosen for thin epi and low processing temperatures.

In conclusion, it should be noted that this hydrogen coverage of the silicon surface during low-temperature growth is of considerable importance. It will control or directly influence a number of epitaxy characteristics such as: dopant incorporation, film morphology, differential poly or mono growth, and the structural quality of epitaxy, etc. The author also considers this phenomenon, which has an important surfactant-like effect, to be the main differentiation between CVD and MBE techniques (to the advantage of CVD).

Germanium Incorporation

Solid Si and Ge have both the diamond crystal structure, and they form a solid solution that is almost ideal (negligible mixing enthalpy) and stable in the entire composition domain. These Si_xGe_{1-x} alloys have a lattice parameter that varies almost linearly with the Ge content from 5.431 Å (a_{Si}) to 5.667 Å (a_{Ge}) when described by the diamond cubic lattice:

$$a_{Si_{1-x}Ge_x} = a_{Si} + x(a_{Ge} - a_{Si}) \tag{5.6}$$

At the same time, the incorporation of Ge in the Si lattice leads to a significant bandgap narrowing. This bandgap narrowing is of interest for the creation of devices based on bandgap variations or energy band offsets. Thus, SiGe alloys opened the way of bandgap engineering and strained heterostructures to the silicon technology.

On a more general point of view, SiGe alloys present a variety of properties and characteristics that can be used in silicon technology. One can list: bandgap variations, band offsets, dopant diffusion reduction, chemical properties (selective etching), dopant activation improvement, strain management, optical properties, etc. SiGe epitaxies receive then an impressive attention for all their possible applications.

Effect on Kinetics

In RTCVD, SiGe alloys are deposited using either DCS, SiH_4, Si_2H_6, or even Si_3H_8 as the Si precursor and almost exclusively GeH_4 as the Ge precursor. Like in Si epitaxy, and for the same reasons, it is admitted that the deposition kinetics increases from DCS to Si_3H_8. And these precursors with different reactivities give rise to a variety of epi domains. Thus, SiH_4 is the standard gas for NSEG and DCS for SEG. The disilane and, more recently, the trisilane are considered for very low thermal budget applications [22]. The GeH_4 molecule that exists as a compressed gas, stable enough (more than 1 year in a cylinder), and reactive enough to allow significant (>10 nm/min) deposition kinetics at low temperature (below 500°C) [12] is really convenient for epi via CVD. GeH_4, even expensive, is then commercially available and widely used in industry. As a consequence, other molecules as $GeCl_4$ did not receive much attention up to now.

In terms of kinetics, the main characteristic of SiGe alloy deposition is the dramatic increase of growth rate, as compared to pure silicon. This effect is illustrated in Figure 5.2 where the kinetics of SiGe and Si depositions from SiH_4–GeH_4–H_2 chemistry at reduced pressure have been plotted as Arrhenius plots. In the low-temperature domain (550°C to 750°C) the SiGe deposition is thermally activated, with kinetics much higher than that of Si, about one decade around 600°C, and much lower apparent activation energy. At this point, one has to note that the SiGe kinetics reported in Figure 5.5 corresponds to a fixed germane/silane ratio but not to a constant solid composition. Indeed, increasing the temperature while fixing the germane and silane partial pressures causes a decrease in the Ge content of the alloy. In such a case, the kinetics limitation cannot be considered as a single and simple mechanism, and then the curve does not strictly correspond to an Arrhenius law.

As it will provide some important "keys" for the epi engineer, it is interesting to look little more closely at this kinetics enhancement. In gas phase, germane is supposed to behave like silane, and to undergo a similar reactive adsorption on two neighboring sites (same notations as Equations 5.3 to 5.5)

$$GeH_4 + 2\_ \rightarrow \underline{Ge} + 2\underline{H} + H_2 \tag{5.7}$$

In fact GeH_4 molecules are much more reactive than SiH_4molecules, and Equation 5.7 is more rapid and corresponds to a higher (×5) "sticking" coefficient as compared to Equation 5.3. This results in a deposition that is much more Ge-rich than the gas mixture. As soon as Ge atoms are incorporated in solid, they act as preferential desorption sites for H atoms. This comes from the fact that Ge—H bonds are less robust than those of Si—H (smaller binding energy). In case of SiGe deposition, Equation 5.4 is then more rapid and less limiting for the subsequent deposition as compared to silicon case. Note also that the easy H desorption from Ge is consistent with the very small activation energy of the Ge deposition kinetics reported in Ref. [12]. Other mechanisms like sticking coefficient variations, Ge

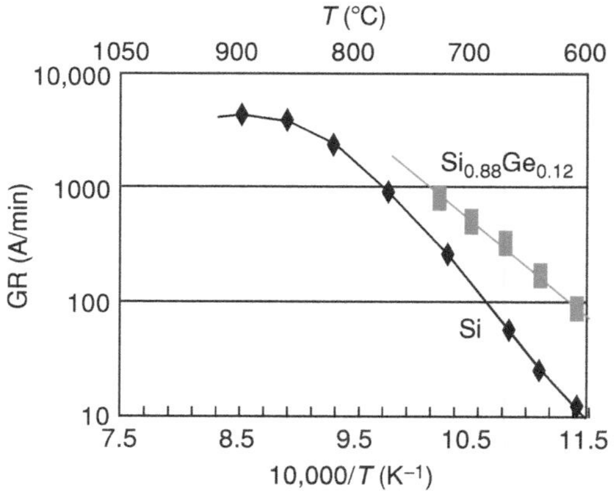

FIGURE 5.5 Pseudo-Arrhenius plot for SiGe growth rate from silane–germane–hydrogen compared to Si growth rate obtained in similar conditions. In the low-temperature domain, SiGe deposition presents a smaller apparent energy of activation and a much higher kinetics compared to silicon.

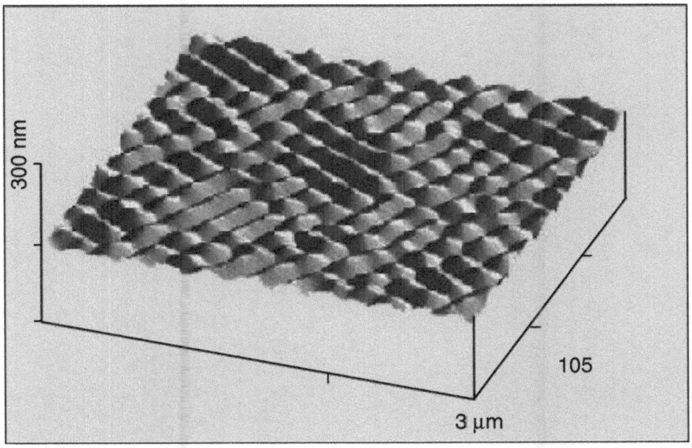

FIGURE 5.6 Stranski–Krastanov growth mode in SiGe–Si epitaxy observed by AFM. Note the ⟨1 0 0⟩-aligned undulations.

segregation, H-atom site exchanges (from Si to Ge), etc. may also take place but are not essential to get a good picture of SiGe deposition.

Strain Effects

Because of the lattice parameter variation given in Equation 5.6, SiGe epitaxies that are grown pseudomorphically on silicon substrates exhibit a large biaxial compressive stress (negative strain). The mechanical energy of these films increases (as the product thickness $*x_{Ge}^2$), and if critical values of x_{Ge} and thickness are exceeded, the film quality can be degraded by Stranski–Krastanov (SK) growth or misfit dislocations.

Figure 5.6 illustrates the surface morphology specific to the SK growth mode. This mode exhibits quasiperiodic undulations that grow exponentially with time–film thickness, and very rapidly with stress–x_{Ge}; very simple models demonstrate that their amplitude has to grow as x_{Ge}^6–x_{Ge}^8 function of the model simplifications. These surface undulations are generated without any extended crystalline defects like dislocations or stacking faults, and are aligned along the two ⟨1 0 0⟩ directions on a (0 0 1) Si surface. It is a kind of elastic relaxation: the top of the undulations has released a part of its stress and the mechanical energy of the system has been reduced.

The plastic relaxation corresponds to the nucleation and propagation of misfit dislocations. As illustrated in the Figure 5.7 (top view), misfit dislocations are easily identified as they usually glide in {1 1 1} crystalline planes. They usually lie near the SiGe–Si interface and allow the above-lying SiGe film to be relaxed. However, since the dislocation core corresponds to some excess energy, dislocations can extend only when the relaxation gain is larger than the dislocation loss. This energy balance corresponds to the well-known critical thickness for stability against dislocation formation due to the lattice mismatch. For a single $Si_{0.80}Ge_{0.20}$ layer (without any silicon capping layer), the equilibrium critical thickness is less than 20 nm. Films with thickness smaller than this value are stable; films with larger thickness are metastable, relaxation can occur if they are exposed to high temperatures, especially if efficient dislocation sources are present in the film.

The relative severity of the two relaxation mechanisms depends on the nature of the film (x_{Ge}) and experimental conditions of deposition (temperature and chemistry). However, for both mechanisms, the lower the temperature, the larger the x_{Ge}–thickness process window. At this point, it is important to note that the dramatic kinetics increase induced by the Ge incorporation allows keeping a certain process window against these mechanisms: the growth of Ge-rich films requires lower temperatures, and lower temperatures can be used owing to the Ge-catalyzed kinetics.

FIGURE 5.7 Top view of misfit dislocations observed by optical microscopy after chemical decoration.

Carbon Incorporation

Background

Solid carbon presents different crystalline structures: diamond, graphite, etc. And even if the diamond structure is similar to that of silicon, the lattice parameter presents a large size offset, $a_{Si}/a_{diamond} = 1.52$. On the other hand, according to the Si—C phase diagram, there is a compound that is very stable, stochiometric SiC, and the carbon solubility in solid silicon is very low, a few 10^{17} atoms/cm^3 at melting temperature. Consequently, the incorporation of carbon in Si or SiGe crystal is more complex than that of germanium in Si, and its incorporation in substitutional sites is by nature a nonequilibrium process. However, it was demonstrated that, by using low-temperature (<650°C) depositions, it is possible to incorporate significant amounts (percent range) of substitutional carbon in silicon lattice [23] to form metastable $Si_{1-y}C_y$ and $Si_{1-x-y}Ge_xC_y$ epitaxies.

The incorporation of carbon in silicon and SiGe layers was first investigated to extend the capabilities of the SiGe system. Indeed, in the previous section, pseudomorphic SiGe layers on Si were found to have fortuitous properties that made them very successful. They however suffer from significant limitations. First, the concept of critical thickness is not really compatible with applications that would require thick nondislocated SiGe layers on silicon, for light emission or detection as examples. Second, almost the entire bandgap offset was reported in the valence band, the SiGe–Si structures do not exhibit any significant conduction band offset and electron modulation is impossible in simple structures.

In the case of substitutional carbon incorporation, the lattice parameter of SiGeC material can be estimated using the following equation:

$$a_{Si_{1-x-y}Ge_xC_y} = a_{Si} + x(a_{Ge} - a_{Si}) + y(a_C - a_{Si}) \quad (5.8)$$

Equation 5.8 is a first estimation, and nonlinear variations may have to be taken into account. Nevertheless, it shows that the incorporation of C atoms, smaller than Si, can serve to reduce or even eliminate the strain in SiGe layers. Calculations as well as experiments show that carbon atom roughly compensates for about ten germanium atoms. It means that if x/y ratio is fixed at this value the ternary layer is perfectly matched to the silicon, and the critical thickness is no more a limitation. Carbon incorporation is also of particular interest as it leads to significant conduction-band offset and to strain effects that are reverse to the Ge-induced ones.

On the other hand, the presence of substitutional carbon atoms at relatively low concentrations, as low as 0.1%, has been found to dramatically reduce both diffusion and transient-enhanced diffusion of impurities like boron in Si and SiGe [24]. This fortuitous effect is widely used to fabricate highly boron-doped bipolar bases that are very thin. Note that this effect could or should be used in CMOS transistor for channel doping or shallow junction control as well.

Incorporation Kinetics

In RTCVD, the most common chemistry used for SiGeC epitaxy is SiH_4–GeH_4–$SiCH_6$ diluted in H_2 at reduced pressure. Other carbon precursors like C_2H_4 were also found to be effective, to a less extent, in producing substitutional carbon incorporation [25], but single alkanes like CH_4 that seem to not decompose the right way are not well suited for this application. Other silicon sources like disilane or trisilane seem attractive, but they are not currently used because of some difficulties such as cost, molecule stability, liquid source control, etc., as compared to silane.

Figure 5.8 presents an example of carbon incorporation in Si and SiGe layers [26]. The total carbon concentration $[C_{tot}]$ measured by SIMS is given as a function of the methylsilane (MS)/silane flow ratio for a variety of process conditions like different silane flows, Ge contents, and deposition temperatures. Note that the gas flow ratio corresponds to the flows through the mass flow controllers and do not correspond exactly to the partial pressures above the wafer surface. One observes that all the experimental points roughly follow a proportional variation with the gas flow ratio. This proportionality in a large experimental domain suggests that MS would undergo a surface-sorption similar to that of silane that requires two adjacent available sites (see Equation 5.3). The fact that the carbon concentration in solid is smaller than the gas ratio also suggests that MS would have a sticking coefficient slightly smaller than silane. Note that it has also been reported that the carbon incorporation increases when the total pressure is decreased and when the temperature is increased [12]. Such trends that are opposite to the germane behavior are also expected from a molecule with a smaller sticking coefficient (germane coefficient is higher).

On the other hand, contrarily to the effect of germane, the presence of MS partial pressure does not have significant effect on the deposition kinetics; accurate measurements allowed us to detect only a small growth rate reduction and a negligible x_{Ge} variations in the experimental conditions explored [26].

Nature of Incorporation

Contrarily to Ge, and for the basic reasons given above, carbon atoms are not necessarily incorporated in substitutional sites. They can easily form interstitial complexes or even react with silicon to form stochiometric SiC precipitates.

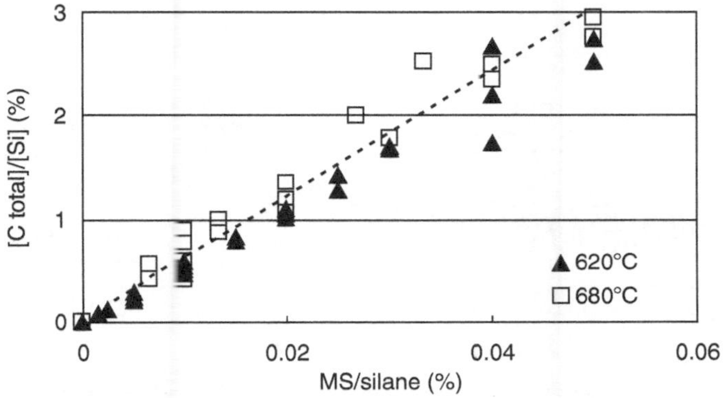

FIGURE 5.8 Total carbon incorporation as a function of the methylsilane–silane flow ratio for a variety of process conditions: different silane flows, germanium contents, and deposition temperatures.

For a long time, carbon contamination on a silicon wafer surface has been known to cause a large defectivity in epi. Indeed, at high temperature, during the hydrogen bake for example, carbon atoms form nanoprecipitates of SiC that induce dislocations or stacking faults. For the same reason, high-quality SiGeC epitaxies require relatively low deposition temperatures (well below 900°C): too high temperatures would allow the carbon precursor to react with the silicon surface to form SiC nuclei. In addition, even if such SiC precipitates are not formed, there is still an important challenge, that is to incorporate the carbon in substitutional sites; i.e., to avoid or minimize the incorporation of carbon in interstitial sites, as well as the formation of interstitial complexes or any other microdefects.

Substitutional carbon concentration can be easily quantified from its influence on the lattice parameter. Figure 5.9 shows the typical x-ray diffraction (XRD) rocking curves from pseudomorphic SiGeC grown on Si (1 0 0). The present diagrams correspond to layers grown with increasing MS/silane gas ratios, all other conditions being constant. The angular shift between the Si substrate peak and the SiGeC layer peak is a direct measure of the lattice mismatch of the epitaxy. And, as the presence of MS in the gas phase does not induce significant changes in Si and Ge deposition kinetics, the shift is directly related to the content of substitutional carbon. In the present case, we observe that the compressive stress induced by a Ge content of 14% is reduced by carbon incorporation in substitutional sites, or even eliminated when the C content is around 1.5%. In the same way, similar incorporation of carbon in pure silicon would have induced important tensile stresses.

The crystalline perfection of SiGeC epitaxies is an important concern in a number of applications. The experimental domain where high-quality epitaxies can be achieved is then of great interest, it has to be determined and optimized. As a first approach, XRD is a convenient, and routinely used, technique to determine the crystal quality of pseudomorphic layers on silicon. This way, the maximal carbon incorporation in terms of crystal quality has been explored as a function of temperature and germanium content, and the results obtained at 620°C and 680°C are given as examples in Figure 5.10a and b, respectively. In the figure, high-quality epitaxies (closed symbols) and poor-quality epitaxies (open symbols) are represented in the plane (x_C, x_{Ge}); the quality criterion being the discrepancy between measured XRD diagram and theoretical simulations (a significant broadening of the main SiGeC peak or the disappearance of the higher order oscillations). At each temperature, there are two domains, a "good quality" one at the bottom left and a "poor quality" one at the top right. It is observed that, for given experimental conditions, there is a maximal carbon concentration beyond which the epitaxial quality is not maintained, and this maximal limit is found to be reduced as the x_{Ge} is increased. The comparison of the two figures also indicates that the good-quality domain is enlarged by a temperature reduction. In other words, the lower the temperature, the larger the process window. Such a trend is also valid in other techniques like MBE and in various experimental conditions.

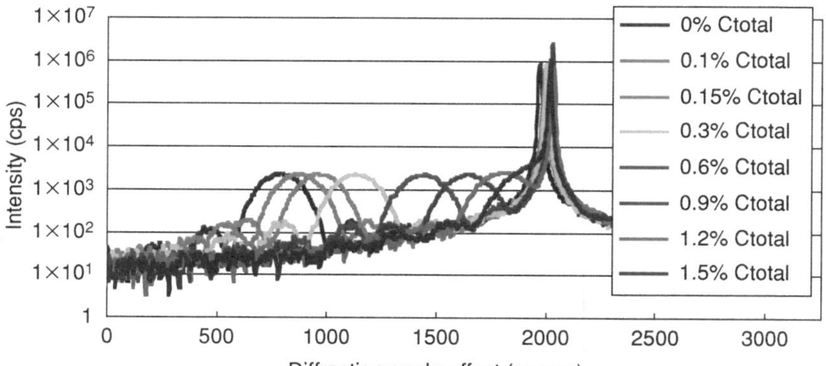

FIGURE 5.9 X-ray rocking curves from pseudomorphic SiGeC grown on Si (1 0 0). Diagrams correspond to layers with a fixed Ge content (about 14%) and C contents increasing from 0% to 1.5%. They show the reduction of the lattice parameter induced by substitutional carbon incorporation.

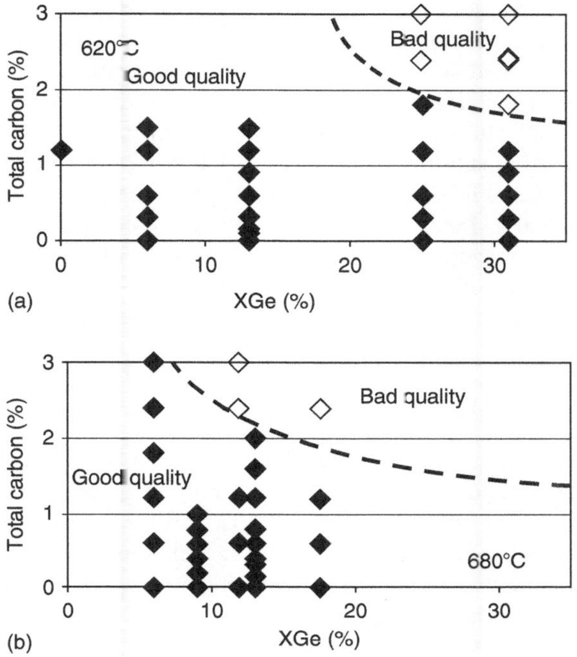

FIGURE 5.10 Quality of SiGeC epitaxies deposited by RTCVD as a function of germanium and carbon contents: (a) 620°C and (b) 680°C. Low temperatures lead to larger process windows in term of carbon incorporation.

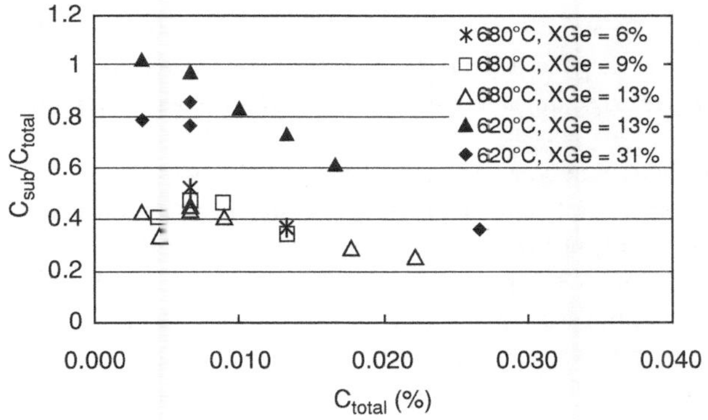

FIGURE 5.11 Proportion of substitutional carbon as a function of the carbon content in SiGeC epi grown at 680°C and 620°C.

Another important challenge of SiGeC epitaxy is how to minimize the number of interstitial atoms and to incorporate all carbon atoms in substitutional sites. The presence of interstitial atoms is indeed not desirable as they are admitted to induce detrimental effects in a number of material characteristics like minority carrier lifetime, mobility, and maybe impurity diffusion, etc. It is then crucial to know the variation of the $C_{subtitutional}/C_{total}$ ratio as a function of the main growth parameters such as growth temperature, partial pressures, etc. The most important trends are illustrated in Figure 5.11 that gives the evolution of the ratio in RTCVD epitaxies as a function of the total carbon incorporation, and for different temperatures and different germanium contents. The substitutional carbon was deduced from XRD measurement and total concentration from SIMS. It is observed that whatever the temperature, the ratio of substitutional carbon decreases with increasing total carbon incorporation, and that the substitutional carbon increases with decreasing deposition temperature and with decreasing Ge content. All these trends are well admitted and have been qualitatively reported for different techniques. They have even been modeled in the case of MBE with simple first-order kinetics of thermally activated interstitial incorporation [27]. However, if we compare the present results with MBE results, we note an important difference. In MBE, substitutional incorporation requires deposition temperatures below 550°C, 450°C preferred, whereas in RTCVD, a significant substitutional ratio is obtained with temperatures as high as 680°C to 700°C. In the opinion of the author, this dramatic change has to be attributed to the presence of an important hydrogen surface coverage that blocks the movement of carbon atoms or eventually improves the stability of the substitutional position. One has also to note that such a crucial role of surface coverage, sometimes referred as a surfactant-like effect, is consistent with all the trends reported above.

Selective Epitaxy

Selective epitaxial growth (SEG) is very desirable because, as a fully self-aligned process, it leads to fully self-aligned device architectures that are simpler, or even to novel device structures (see Section 5.4). By essence, SEG has to be run on patterned wafers, and the epi characteristics depend on the process and on the precise nature of the patterned wafers as well. Namely, we can list: epitaxy perfection, selectivity limit, thickness and doping in the different windows, and exact morphology of deposition (mainly faceting at the edge of deposition). As a consequence, in addition to complex process developments, SEG process may also present important integration problems, and some of them are discussed in Section 5.3.

The concept of SEG is based on the fact that silicon has to undergo a nucleation step before deposition over a dielectric and not over silicon. This nucleation step corresponds to a certain amount of energy, and the barrier is large enough to produce selective deposition if the process is well chosen. Various SEG processes have been published via RTCVD, but the most common ones are based on the DCS–HCl–H_2 chemistry. Apart from the DCS choice, the addition of a certain flow of HCl is generally required to totally inhibit any deposition on the dielectric, especially when using Si_3N_4 and utilizing an industrial reactor working at pressures exceeding a few Torr. As such, DCS-based epitaxy and DCS–HCl-based SEG are presented below.

DCS chemistry is widely used in conventional epitaxy and has been studied for SiGe deposition at low temperature. As such, experimental data on kinetics are available in literature, however the modeling and understanding of this process is not as clear as for silane. Following the Arrhenius plot of the DCS–H_2 epitaxy kinetics given in Figure 5.3, for temperature of interest here (<850°C) deposition is kinetically limited. Although they are supposedly predominant, surface reactions are still uncertain. It is supposed that DCS molecules adsorb directly on the silicon surface and undergo a reactive dissociation into Si, H, and Cl, and that these species can desorb, depending on the temperature, as H_2, HCl, or $SiCl_2$. A simplified form of the reaction pathway proposed first by Coon et al. [27] may be written as follows:

$$SiH_2Cl_2 + 4\_ \rightarrow Si + 2\underline{H} + 2\underline{Cl} \tag{5.9}$$

$$\underline{H}^* + \underline{H} \rightarrow H_2 + 2\_ \tag{5.4}$$

$$\underline{H} + \underline{Cl} \rightarrow HCl + 2\_ \tag{5.10}$$

$$2\underline{Cl} \rightarrow Cl_2 + 2\_ \tag{5.11}$$

$$Si + 2\underline{Cl} \rightarrow SiCl_2 + 2\_ \tag{5.12}$$

(same notations as Equation 5.3 to Equation 5.5). However, as epi is often performed in hydrogen ambient, and as hydrogen is known to increase the GR of silicon compared to Ar, N_2, or He, we prefer to include its influence with the following reaction suggested in [28]:

$$2\underline{Cl} + H_2 \rightarrow 2HCl + 2\_ \quad (5.13)$$

The apparent activation energy of DCS–H_2 kinetics given in Figure 5.3 is about 70 kcal/mol; a value very similar to the activation energies of the desorption reactions (5.10) and (5.13) that are supposed to be the limiting steps. Kinetics limitations in DCS–H_2 system are reported in Figure 4 of Ref. [29]. Silicon growth rate has been plotted as a function of DCS flow (or partial pressure) for different temperatures, all other parameters being constant. For low DCS flows, GR is proportional to the flow, but above a given flow, GR exhibits a very clear saturation indicating that reactions (5.10) and (5.13) would be very effective. At 750°C, for example, GR is roughly constant (about 13 nm/min) for DCS pressures exceeding 0.05 Torr. Thus, the surface coverage would regulate DCS adsorption and growth rate. And based on Hierlemann calculations [27], chlorine coverage would dominate in most of the experimental conditions, bringing about an important variation as compared to silane chemistry.

Silicon SEG based on DCS–HCl–H_2 chemistry has given rise to a large amount of work, but SEG is a complex process with a number of parameters (any comparison is very difficult as experimental conditions are different); no (or almost no) conclusive attempts have been made to relate the SEG process to the underlying surface reaction physics. Reactions (5.9) to (5.13) are still valid but a significant HCl adsorption, with reactions reverse to reactions (5.10) and (5.13), has to be taken into account. And consequently, the growth rate is depressed because of an increased Cl surface coverage, and reaction (5.12) is much more effective in removing Si adatoms from the dielectric surface, compared to the DCS–H_2 chemistry. A perfect selectivity against dielectrics like SiO_2 or Si_3N_4 can be then obtained in the DCS–HCl–H_2 system by HCl flow adjustment. This is illustrated in Figure 5.12, where the nucleus density on full-sheet Si_3N_4 CVD films on 200 mm wafers has been plotted as a function of the HCl flow. Deposition process was carried out in a commercial epitaxy reactor, and nuclei were measured with an optical inspection tool that enables the full wafer to be scanned. We observe a relatively abrupt selectivity threshold (a 10% variation in HCl flow changes the level of defects by 2 to 3 decades) and a very high degree of selectivity (not published to the author's knowledge) as nucleus densities or defectivity lower than 0.1/cm^2 are measured. To conclude the silicon SEG part, and as discussed in Ref. [29], we can draw the following general rule: the HCl flow requested to obtain full selectivity has to be increased with any of the following parameters: DCS flow (but the ratio DCS–HCl can be decreased), temperature, pressure, doping level (boron), and Si coverage.

Germanium and carbon can be incorporated in selective epi via the addition of GeH_4 and MS in the previous chemistry.

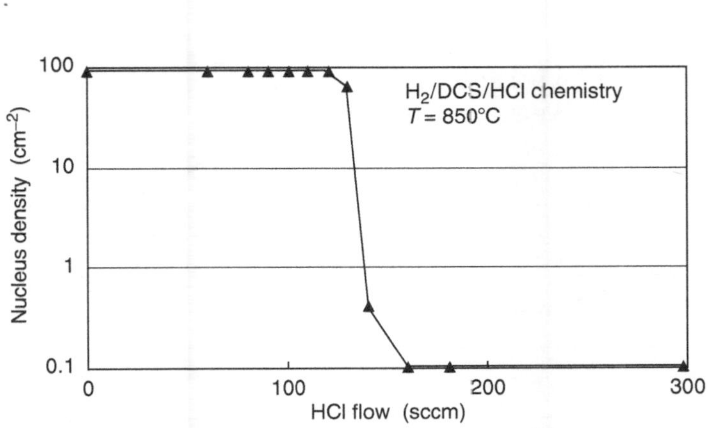

FIGURE 5.12 Selectivity threshold of silicon deposition on CVD Si_3N_4. The nucleus density was measured as light point defects using a Surfscan that allows the entire 200 mm wafer to be inspected. In the present conditions, a perfect selectivity is obtained when using 140 sccm of HCl and more. Note that the points located on the bottom and top axes correspond to densities below 0.1 and above 100 LPD/cm^2.

In terms of kinetics, the main characteristic of SiGe alloy deposition is the dramatic increase of growth rate, as compared to pure silicon. As for hydrogen, it is supposed that the much lower Cl—Ge binding energy, as compared to Cl—Si, facilitates the Cl desorption and allows much more rapid kinetics. Note that a part of this kinetic enhancement may also come from a chemical reduction of DCS radicals by atomic hydrogen that is produced by GeH_4 decomposition. However, up to now, these mechanisms were not properly modeled. Indeed, the kinetics, already not so simple for pure Si, becomes much more complex, especially in the low-temperature domain where a number of surface reactions should be taken into account. On the other hand, as x_{Ge} (new parameter) has a high influence on kinetics, the deposition domain becomes very extended and the number of experiments to explore it very huge.

As an example of SiGe SEG kinetic, Figure 5.13 gives the variations of GR and germanium content as functions of the DCS partial pressure, all other conditions being constant. We observe that when the GeH_4/DCS gas ratio is reduced by a factor of 5 via a DCS increase, the germanium content roughly decreases from 25% to 20% and the GR decreases from 17.5 to 11 nm/sec. In other words, the germanium content follows a really nonlinear variation, and the GR variation is opposite to the main active gas (DCS) partial pressure change: dependencies that are not straightforward for x_{Ge} or even counterintuitive for the GR.

In terms of selectivity, the germanium incorporation plays a very positive role. It is observed that SiGe deposition is much more selective as compared to pure silicon. Whatever the dielectric, SiO_2 or Si_3N_4, the Ge-rich epi, the more selective. Different possible explanations were proposed in literature, but none of them is really convincing. In the opinion of the author, this increased selectivity would come from the interface–surface energies larger for SiGe and then less favorable to nucleation, as compared to Si (this is still under experimentation). Indeed, a larger SiO_2–SiGe interfacial energy (and to a less extent a larger SiGe surface energy) would induce a much higher nucleation barrier and then would lead to an easier intrinsic selectivity.

On the other hand, substitutional carbon incorporation in chloride chemistry is more delicate as this incorporation requires low temperatures and chloride kinetics higher deposition temperatures, as compared to silane. However, some interesting compromises do exist as selective SiGeC epi were already reported in the literature [30,31]. In both cases, the depositions allowed to fabricate high-performance HBTs, and this point will be illustrated in Section 5.5.

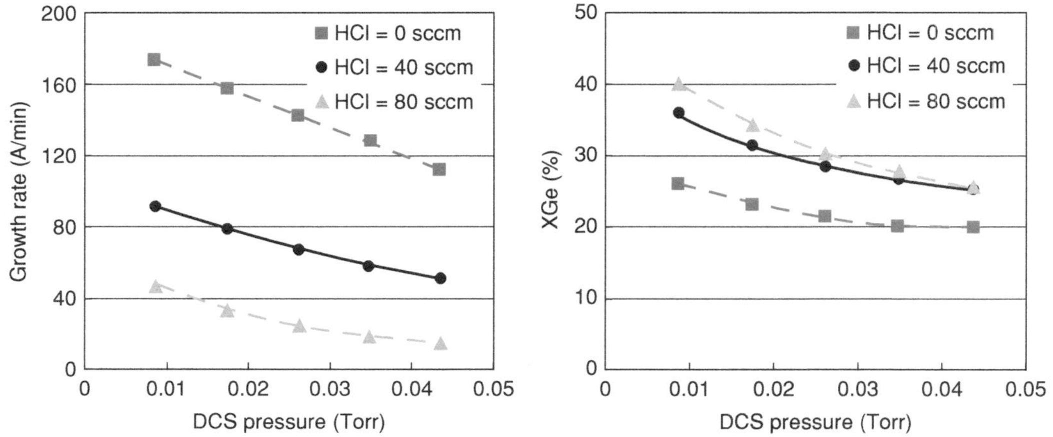

FIGURE 5.13 SiGe SEG kinetics as a function of the DCS partial pressure (flow). Note that the nonlinear variation of x_{Ge} and the counterintuitive variation of GR that decreases with increasing active gas (DCS) flows.

5.4 Epitaxy Integration

Compared to standard high-temperature blanket epi, new epi processes are facing new challenges. Some difficulties come from the properties of the epitaxial layer that has to be deposited: metastable-strained layers (SiGe), nonstable materials (substitutional carbon), etc. and have been discussed previously. Others difficulties come from the device application requirements, and from integration in modern technologies. On one hand, new devices like the base of HBTs require very complex stacks that correspond up to 8 to 18 parameters that have to be controlled. In addition, as the base is an active part of the device that is very critical, it has to be well controlled. On the other hand, as a part of the devices is already present, the global thermal budget of epi is strictly limited and the epi has to be run on patterned wafers.

Today, new (established or potential) applications are very numerous, and all of them are not precisely specified. Thus, it is difficult to detail all possible issues in an exhaustive way. However, the major ones can be tentatively listed as:

- Very limited thermal budget (H_2 bake)
- Low deposition temperature (complex kinetics and structural quality problem)
- Abrupt dopant profiles (at the nanometer scale)
- Strained epitaxies: SiGe, SiGeC
- Metastable materials: SiGeC
- Different doping types and levels
- Cleaning of structured substrates
- Deposition on structured substrates (very different as compared to prime Si wafers)
- Epitaxy in presence of various materials: TEOS, Si_3N_4, SiO_2, poly or amorphous silicon, etc.
- Epitaxy in very complex structures
- Epitaxy after etching and implant steps
- Selective epitaxy: Si, SiGe, SiGeC, etc.
- Control of facets or poly–epitaxy interfaces
- Local or global loading effects (LE)
- Complex film stacks
- Differential epi or poly growth rate
- Control of poly roughness in nonselective epitaxy

In the following, we discuss some general difficulties that are present in LTE by RTCVD for new applications, namely, thermal budget limitation, LE, deposition morphology, and pattern-induced defectivity.

Thermal Budget Limitation

As a very general approach, one could consider that the thermal budget has to be reduced at each technology node according to the design size reduction and to the gradients required. Based on the boron diffusion that is a limitative mechanism in a number of applications, the thermal budget reduction from one technology node to the next one can be estimated to 15°C to 20°C ($D_{\text{boron}}/2$ or diffusion length $\times$ 0.7). This rule holds for established epi applications that would follow a constant integration scheme.

For a new application, the thermal limitation can be even more demanding and severe. As there is no reference, any request is possible; the general trend being the more advanced the technology and the later (in the process flow) the epi integration, the lower the allowed thermal budget. This way, in most recent applications we had to move from the "RCA + high-temperature H_2 bake" clean strategy to the "HF-last + moderate-temperature H_2 bake". Taking the bipolar transistor that is demanding in terms of epi quality as an example, collectors (buried layers) are grown with the first strategy for decades, whereas the more recent epitaxial bases of HBTs are generally deposited using the second one. Note also that in some cases, the process flow, and then the device architecture, has to be designed taking into account for epi limitations. In the case of HBTs in BiCMOS, the epi base is usually run before CMOS S/D implantation. This way a significant thermal budget can be used to optimize the surface cleaning and epitaxial deposition perfection.

In some other applications like the deposition of elevated S/D in the most advanced CMOS or SOI-CMOS technologies (65 to 45 nm node and below), as a function of the process flow we may have to go to the "HF-last + low-temperature H_2 bake" strategy. For very demanding applications, even the deposition temperature has to be reduced and SiGe depositions may be preferred to silicon epi because of their much higher kinetics. As an example, hydride chemistry at 600°C with fixed conditions gives typical GR of 1, 10, and 50 nm/min for germanium contents of 0%, 13%, and 26%, respectively. Thus, SiGe depositions can be considered for industrial applications, with significant throughputs, at temperatures much lower than 600°C.

In such a case of thermal budget reduction, we have to keep in mind that, as stated previously, the surface cleaning may be not efficient enough, and any surface imperfection can induce important epi defectivity. Indeed, any particle or precipitate will not be eliminated and can be the source of crystalline defects, especially in strained epi. Figure 5.14 gives an example of such a high defectivity observed in Si–SiGe epi deposited with a low thermal budget. The image of the epitaxial zone (800 × 800 μm^2) has been obtained by room temperature photoluminescence [32]: dark areas outside the box are due to polycrystal deposition, and dislocations randomly appear in the box as dark lines and points. In this case, a poor surface preparation associated with a low-temperature H_2 bake has left a high density of defects such as particles or precipitates that have induced crystalline defects and misfit dislocations.

On the other hand, the deposition temperature reduction may also affect the quality of the epitaxy. Indeed, as soon as the partial pressure of active gases is increased and the temperature decreased, minority carrier transport and photoluminescence properties are usually affected, supposedly because of punctual defects. At the limit, if the deposition conditions are pushed further, we may even cross the amorphous or crystalline deposition transition.

As a conclusion, the epi process that determines the film quality has to be chosen in proper accordance with the application requirements. And, as an example, epitaxial bases of HBT and elevated S/D of CMOS with different requirements probably should have to be deposited with different conditions.

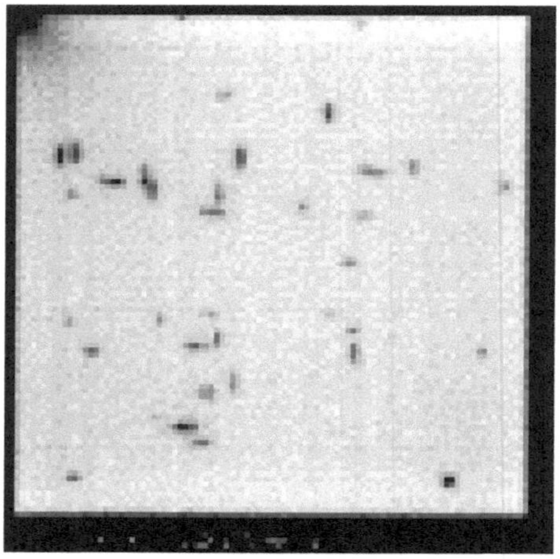

FIGURE 5.14 Room temperature photoluminescence image of a Si–SiGe epitaxial structure. Dark areas outside the box are due to polycrystal deposition, and dislocations appear in box as dark lines or points. The important defect density is due to an improper surface preparation associated with a low-temperature H_2 bake.

Loading Effects

In RTCVD, the growth rate of Si (or SiGeC) in the active areas of patterned wafer usually depends on the pattern. Differences that are observed by changing the absolute size of exposed silicon areas on a given substrate are referred to as "local loading." Differences from one wafer to the other, due to the total exposed silicon surface or to any other wafer change, is referred to as "global loading." It means that a fixed process recipe (chemistry, temperature target, pressure, etc.) does give different depositions on different substrates or in different locations. These LE are an important difficulty in industrial applications where epi depositions have to be precisely (specifications smaller than a few percentages) controlled in any devices and on any wafer. On the other hand, as a function of the root cause of these differences, temperature variations or chemical changes (most probably in the gas-phase composition), one usually classifies the effects as "thermal" LE or "chemical" LE.

At the origin, LE referred almost exclusively to thickness variations found in Si SEG as these ones were strong and easily detected. Such a view is, however, too restrictive and has to be enlarged: the LE concept is valid, and has to be considered, for the incorporation of any element that is present in deposit. This way, LE may affect thickness but also other parameters like germanium content, carbon incorporation, and dopant level. In addition, as it is not yet reported in literature, it is important to note here that LE may also induce growth rates and dopant incorporations that are not constant within the film thickness.

Thermal Loading

In existing RTCVD epi tool, the wafer rests on a SiC-coated graphite susceptor and both pieces are lamp-heated inside a cold-wall chamber possibly under low hydrogen pressure. As temperature measurements on wafers with patterns and complex film stacks are very delicate, only the susceptor temperature is used to control the process. However, in such equipment, which do not follow the blackbody emission, the actual temperature of an object is a function of its absorptance or emissivity. As a consequence, the wafer temperature T does not match exactly the susceptor temperature T_{susc}, and a temperature offset ΔT can be written as [29]:

$$\Delta T = [\varepsilon^{\text{lamp}} P - \varepsilon\sigma T^4_{\text{susc}} - K_2(T_{\text{susc}} - 298)]/(8\varepsilon\sigma T^3_{\text{susc}} + K_1 + K_2) \tag{5.14}$$

where $\Delta T = T - T_{\text{susc}}$, $\varepsilon^{\text{lamp}}$ is the emissivity of wafer corresponding to the lamp wavelengths, P, the radiative power density from lamps, ε, the silicon emissivity corresponding to its own emission, K_1, the constant of heat conduction between the wafer and the susceptor backside, K_2, the conduction coupling between the wafer and the chamber, and the other letters have their usual meaning. Note that susceptor emissivity was taken to unity. In practical cases, we have relations : $K_1 \gg K_2$ and $K_1 \gg 8\varepsilon\sigma T^3_{\text{susc}}$, and Equation 5.14 can then be simplified as

$$\Delta T = [\varepsilon^{\text{lamp}} P - \varepsilon\sigma T^4_{\text{susc}} - K_2(T_{\text{susc}} - 298)]/K_1 \tag{5.15}$$

During epi process, T_{susc} is well controlled and can be considered as constant and repeatable. However, as ΔT is not necessarily equal to zero, Equation 5.14 and Equation 5.15 establish that, in a general way, the precise temperature of process T varies with wafer emissivity, top and bottom lamp power repartition (via P), susceptor design and gas conductivity (via K_1). Note that, in equipments, which use thermocouple placed at the susceptor bottom for temperature control, the situation is even worse as an additional temperature offset, between susceptor and thermocouple, is introduced. As a consequence, a fixed process (chemistry, temperature target, pressure, etc.) does not give the same deposition on different substrates.

Figure 5.15 gives the SiGe growth kinetics as a function of the germane gas flow, measured on blanket Si wafers and on typical product wafers. The comparison shows clearly that growth rate is lower when using structured substrates compared to full-sheet wafers. Basically, this change can be attributed to two different causes: a global "thermal" LE or a global "chemical" LE. The actual causes of the GR variations

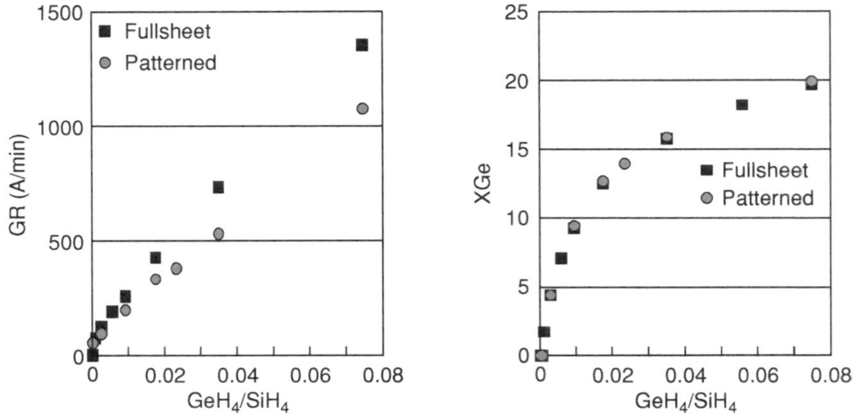

FIGURE 5.15 SiGe growth kinetics as a function of the germane–silane flow ratio for silicon full-sheet and device wafers. In the present conditions, growth rate variations are mainly due to thermal loading effects.

presented in Figure 5.15 have been studied in detail [33]. In these specific conditions, it was clearly established that:

- Silicon deposition did not present any significant chemical LE and the observed GR variation is due to the thermal effect (about 6°C).
- SiGe growth did present both a thermal LE (to the same extent in Celsius than the silicon one), and a chemical LE that is smaller.

The present results are specific to the conditions used for deposition. Nevertheless, they demonstrate that, in a general way, the deposition on device wafers is different from an eventual calibration using full-sheet wafers. Because of the presence of polycrystalline silicon on field oxide, the emissivity of product wafers is significantly smaller than that of blanket wafers, and the first term of Equation 5.14 and Equation 5.15 is decreased more significantly than the second one. Thus, the wafer temperature is decreased as compared to prime wafers, in agreement with the result of Figure 5.15. Note that the thermal LE are supposed to vary during deposition as wafer emissivity varies with the poly thickness. Finally, despite severe simplifications made, Equation 5.14 and Equation 5.15 explain perfectly how the actual process temperature varies with the wafer optical properties, giving rise to the thermal LE.

On the other hand, thermal LE reported in Figure 5.15 correspond to global effects. However, local thermal LE play a certain role each time that important surface emissivity nonuniformities are present on the wafer at large lateral scales (greater than a few mm); any lateral thermal gradient at smaller scales is effectively smoothed by the high thermal conductivity of the silicon substrate.

It is also interesting to note that thermal LE, illustrated here for NSEG, are operative for any process, and that LE reported for SEG as chemical LE usually are in fact a mix between thermal and chemical effects.

Chemical Loading

Chemical LE, often referred as "loading effects," are reported for a long time in silicon SEG. They are, however, almost not modeled even in the case of pure silicon deposition, as the kinetics with DCS–HCl–H_2 chemistry is not perfectly understood. Because of the additional parameters, namely the silicon surface coverage and the size of the windows, complex variations may be found as a function of silicon coverage or HCl partial pressure (see as an example Figure 10 of Ref. [29]). Indeed, in different experimental conditions, LE may present opposite variations, a behavior attributed to the fact that the kinetics is dominated by DCS or HCl molecules.

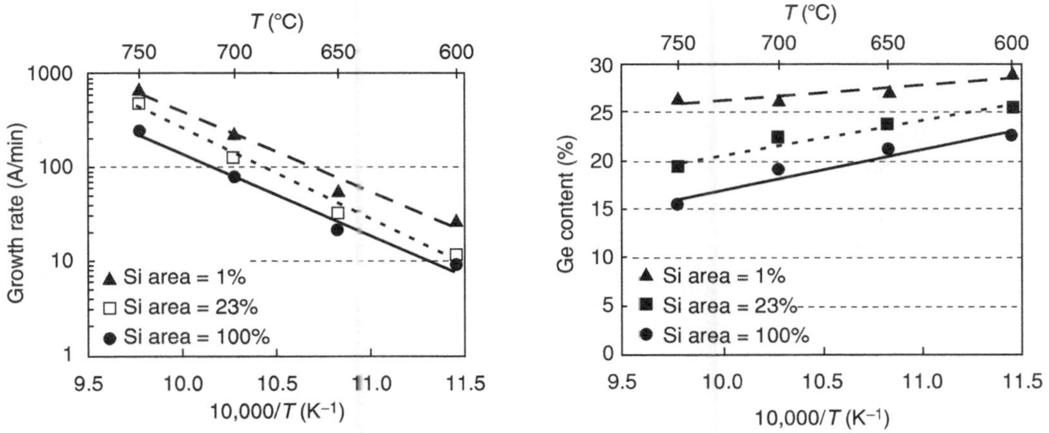

FIGURE 5.16 Global loading effect in DCS–GeH_4–HCl–H_2-based SiGe SEG. Growth rates are plotted as a function of reciprocal temperature for wafers with silicon coverages of 1%, 23%, and 100%.

In the case of SiGe or SiGeC deposition, the situation is even more complex as the deposition domain is enlarged very much by the addition of parameters (germane or MS partial pressures). Figure 5.16 gives the growth rate of SiGe SEG as a function of reciprocal temperature on wafers with different silicon coverages. The DCS–GeH_4–HCl–H_2 chemistry was used and the deposition conditions were chosen close to the selectivity threshold. In such a case, the kinetics and LE are dominated by DCS and germane (and not HCl), and we observe that whatever the deposition temperature, the smaller the silicon coverage, the higher the growth rate. In the same way, we also observed that the smaller the silicon coverage, the higher the Ge content. In these experiments, special care has been taken in order to eliminate any thermal contribution of the different patterned or blanket wafers resulting from the optical properties of wafers and the results only report the actual chemical LE [34]. The present interpretation is that the germane depletion (that increases with the silicon coverage) with the subsequent Ge content reduction is the main cause for the significant GR decrease (a factor of 3 to 4). Note at this point that a more important global LE would be expected with smaller Ge contents as the relative germane depletion would increase and as the kinetics would be more sensitive to x_{Ge}.

Local LE are also observed in these deposition conditions. Figure 5.17 gives the growth rate and Ge content of SiGe SEG as a function of the silicon window area [34]. We observe that the smaller the window, the higher the growth rate and the Ge content. In the same way as for global LE, a local germane depletion would induce locally both the GR and x_{Ge} decreases. We also note that for smaller and smaller silicon windows, the GR and x_{Ge} seem to saturate. This trend has been confirmed by SEM cross sections where no significant offset was found between submicron and larger windows. It means that optical measurements carried out in windows as large as several tens of microns are representative of epi in submicron devices. At this point, we have to note that the results are reported as a function of the window area. However, this parameter is fundamental only for isolated windows, and in a more general design all the surrounding patterns will play a role and have to be taken into account.

In this section, chemical LE were presented in the case of SiGe SEG. For the sake of simplicity, we have presented simple variations that are obtained in an experimental domain properly chosen. However, we have to keep in mind that the volume of the experimental domain, with a number of parameters, is huge, and that more complex variations of global and local LE are usually found in larger or not well-chosen domains. On the other hand, even if often neglected, NSEG also present chemical LE that can be important. As an example, boron doping presents both local and global LE in Si or SiGe epi as a consequence of an important "differential" incorporation (a ratio of 2) between polycrystalline and epitaxial deposition.

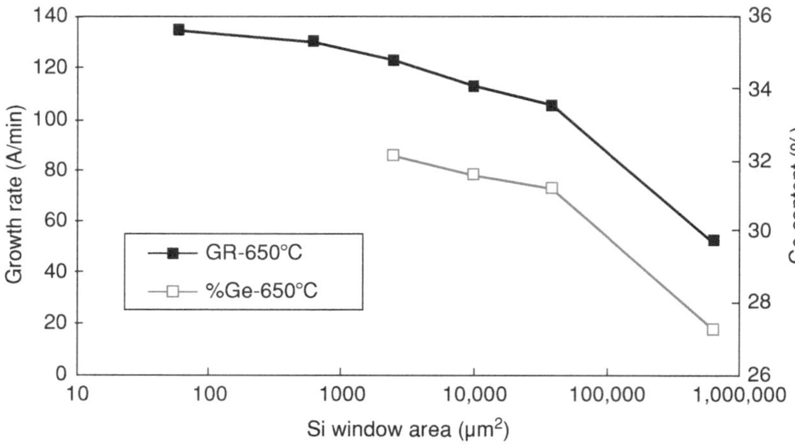

FIGURE 5.17 Local loading effect in DCS–GeH_4–HCl–H_2-based SiGe SEG. Growth rate and Ge content are plotted as a function of the silicon window area (silicon coverage of wafer was about 1% of the total surface).

Deposition Morphology

The epi morphology of blanket depositions can be usually described in a simple way: thickness being usually the unique parameter. Only in a few cases, when exhibiting a Stranski–Krastanov growth mode for example, a more complete description is required. On patterned wafer, the situation is different and epitaxial films always present a morphology that is more or less complex. This point is illustrated in the following.

In SEG, LE can induce depositions that vary from one epi window to the other, or from the window edge to the center; these effects were discussed previously. On the other hand, facets that are well-known problems are usually observed. They come in the form of flat surfaces that correspond to certain atomic plans. They are caused by the important variations of the growth rate as a function of crystal orientation. Indeed, dense atomic planes like {1 1 1}, {3 1 1}, {1 1 0}, etc., which may present significant nucleation barriers, exhibit slower GR as compared to {1 0 0}, and become apparent in convex-growing zones just like in a Wulf construction.

In fact, facets do not correspond to a minimization of the crystal energy but are rather the crystal response, dominated by kinetics effects, to a certain environment. As a consequence, their development depends on several dominant factors that are:

- Structure conformation: orientations of dielectric edge and surface, etc.
- Nature of dielectric: SiO_2, Si_3N_4, etc.
- Crystal orientation: {1 0 0} Si wafers are usual
- Nature of deposition (Si, SiGe, etc.)
- Process conditions (T, P, gas flows, etc.)

These factors can be combined to give rise to a huge number of different possibilities. All of them cannot be detailed or discussed here, and only the effects of the process conditions and of deposit nature will be illustrated in the following.

Figure 5.18 illustrates the influence of temperature on the morphology of silicon SEG grown between SiO_2 walls. In both cases, epi was carried out by RTCVD using the SiH_2Cl_2–HCl–H_2 chemistry. SiGe markers were introduced in epi and chemically decorated for X-SEM observation in order to analyze the growth surface evolution. These markers were chosen thin and with a low Ge content in order to minimize their influence on the growth. On the left picture (Si epi at 850°C) we observe very clear {3 1 1} facets, defined by an angle of 25.2° with the (0 0 1) plane, that appear at the beginning of the growth. Such experiments also allow to estimate a growth rate ratio between {3 1 1} and {1 0 0} planes of about

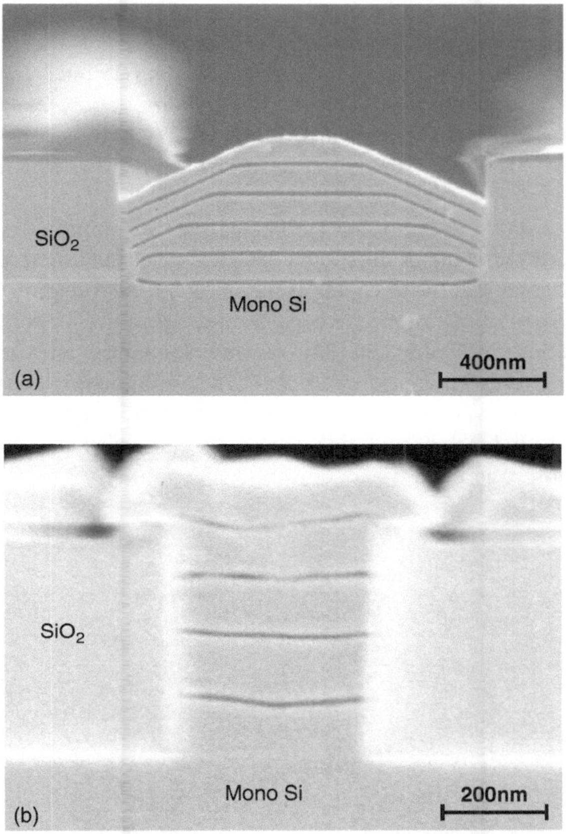

FIGURE 5.18 SEM cross sections of Si SEG grown at 850°C (a) and 750°C (b) with the SiH_2Cl_2–HCl–H_2 chemistry and between SiO_2 walls. Thin SiGe markers were introduced and chemically etched (dark) in order to visualize the growth surface evolution.

0.53, close to the values reported in the literature. On the contrary, no facet can be detected inside the oxide walls on the right picture that corresponds to a Si epi at 750°C. In our opinion, faceting is eliminated because the growth rate ratio (between {3 1 1} and {1 0 0} planes) is significantly increased (as a consequence of the different activation energies), and may be also because of the dramatic surface diffusion decrease due to the lower temperature and the subsequent Cl surface coverage.

Figure 5.19 shows the morphology of SiGe SEG grown at 750°C in conditions similar to those of Figure 5.18 except that germane was added. In this case, Si markers (bright on the picture) were used to visualize to growth morphology. Compared to the silicon growth, SiGe behaves differently. Two systems of facets are clearly visible: the ⟨1 0 0⟩ growth is bordered by {3 1 1} facets, and {1 1 1} facets are present alongside the SiO_2 walls. By an important temperature reduction, the authors were able to avoid first (*T* around 650°C) the {3 1 1} planes and further (*T* around 600°C) the {1 1 1} ones [35].

In another way, from these conditions it is also possible to delay the facet formation by modifying the surrounding environment and the condition process [35,36]. As an example, with the integration of nitride instead of oxide and with HCl partial pressure reduction, facets do not appear at the beginning of the growth but after 1000 Å or more. In such a case, for applications that are based on films thinner than that, faceting would be no more an issue.

When NSEG is carried out on patterned wafers, a technique also referred as "differential deposition," polycrystalline material is grown on top of dielectric (or polysilicon if present) and epi is grown on the monocrystalline Si regions. In this case, in addition to the epi thickness, a number of parameters like the thickness of poly, its nature (grain size and texture), roughness, and the shape or orientation of the poly or mono interface are of interest. Indeed, the poly deposition very often plays a role in the structure and its characteristics have to be taken into account for the application. As an example, in the most common HBT structure that is based on NSEG, the Si–SiGeC poly is a part of the extrinsic base and some of its properties (resistivity, thickness, proclivity to form salicide, etc.) are crucial for the device performances.

The poly–epi thickness ratio is an important parameter of a differential deposition. Indeed, once the epi characteristics (thickness, dopant profiles, etc.) are chosen, usually in accordance with the device performances, this ratio determines the thickness of the simultaneously deposited poly. The variation of this poly–epi ratio is given as a function of the deposition temperature for the SiH_4–H_2 chemistry and given experimental conditions in Figure 5.20 [37]. We observe that the ratio as high as 2.4 in the low-temperature domain decreases down to values close to 1. This variation is supposed to be correlated with the deposition structure that transits from an amorphous nature to a crystalline one. It means that, for a fixed epitaxial structure, the poly thickness (and structure) can be adjusted within a certain domain by a proper choice of deposition conditions.

FIGURE 5.19 SEM cross section of SiGe SEG grown at 750°C with SiH_2Cl_2–GeH_4–HCl–H_2 chemistry and between SiO_2 walls. Thin Si markers were introduced and chemically decorated (bright) in order to visualize the growth surface evolution.

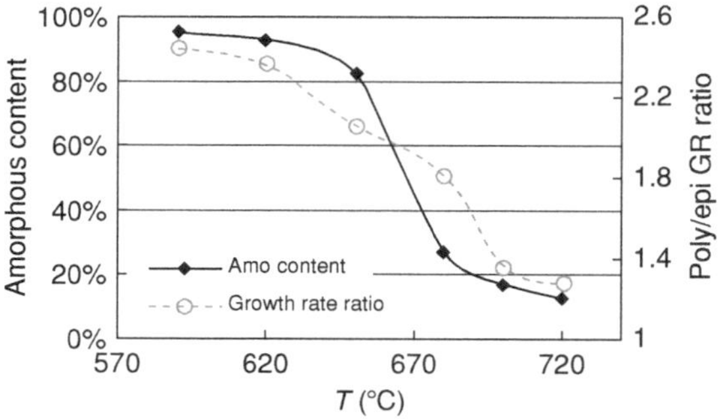

FIGURE 5.20 "Poly"–epi growth rate ratio and amorphous content in "poly" as a function of growth temperature. "Differential deposition" was carried out on Si–SiO_2 patterns by RTCVD and with the SiH_4–H_2 chemistry.

On the other hand, we also believe that this ratio plays an important role in the morphology or orientation of the poly–epi interface. When the poly or amo growth rate is much higher than that of epi, the poly growth will push the poly–epi interface toward the epi region to produce a very inclined interface. This interface usually looks like a facet, and in some cases it may correspond or be very close to well-defined facets like the {3 1 1} ones. When the ratio is around 1, the development of the two parts will be more balanced and the interface will be more vertical (perpendicular to the surface). As an illustration, Figure 5.21 gives two very different developments of poly–epi interface. At low temperature, with a large growth rate ratio, the poly–epi interface is indeed very inclined. At a higher temperature, the interface is less inclined. And with optimized conditions at 820°C, the interface is almost vertical (not reported here) Between the two extreme conditions (590°C and 820°C), the rule would be: the higher the ratio, the more inclined the poly–epi interface.

At this point, we have to note that among the different characteristics of the differential deposition, we preferred to discuss this poly–epi interface orientation because it is a very important characteristic although almost never reported in literature, and also because we are convinced it is somewhat related to the previously discussed faceting effect observed in SEG.

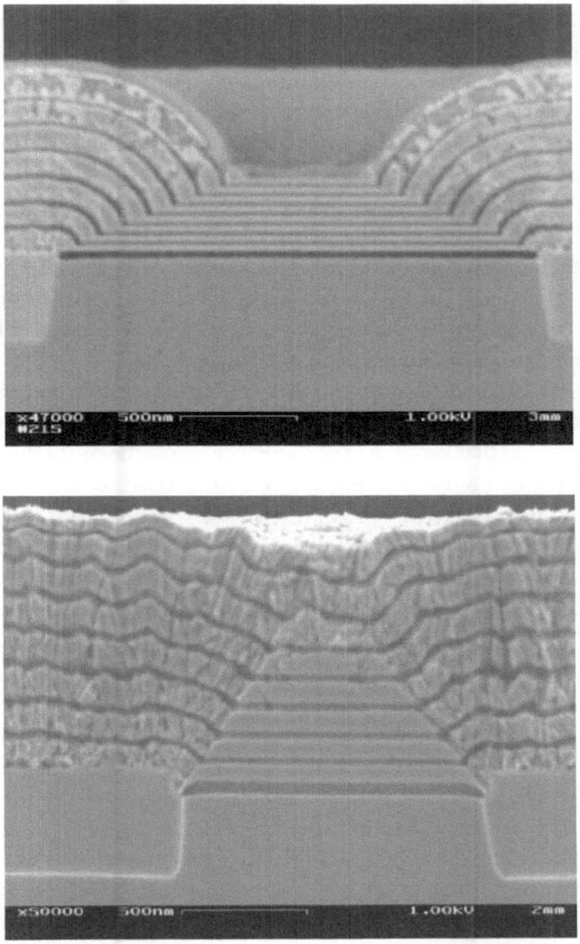

FIGURE 5.21 SEM cross sections of silicon NSEG (differential deposition) grown at 590°C (left) and 720°C (right) with SiH_4–H_2 chemistry (P_{SiH_4} = 0.35 Torr). Dark lines correspond to thin $Si_{0.9}Ge_{0.1}$ markers submitted to a chemical decoration.

Pattern-Induced Defectivity

As compared to full-sheet, patterns can generate additional epi defectivity via several mechanisms:

- Possible outgassing from the different materials present on the surface
- Growth surface can be forced in nonfavorable planes, {3 1 1}, {1 1 1}, etc.
- Stress field induced by dielectric patterns (STI, etc.)
- Presence of edges (facets or poly–epi interfaces) causing local shear stress
- Presence of edges as effective sources of defects

Let us review briefly these points, and note that they may be effective at the same time.

As materials and dielectrics present on the surface are deposited increasingly at low temperatures, one may get a significant outgassing of species, like H_2O from TEOS films for example, that are capable of oxidizing the silicon surface (during the moderate temperature bake or temperature stabilization) and to induce crystalline defects. This mechanism is even supposed to be more severe in batch systems where the outgassing surface is larger and time longer, and especially in UHVCVD system because the silicon surface may be more sensitive, as compared to single-wafer systems. On the other hand, in some configurations, the epitaxial growth may have to progress via low-kinetics crystalline orientations. As an example, if high aspect-ratio trenches have to be filled on a ⟨1 0 0 ⟩ wafer, the epi may have to grow via facets, {3 1 1} or {1 1 1} as a function of the material or conditions (see Figure 5.18 and Figure 5.19). In such a case, crystalline defects like stacking fault, etc., or even polycrystalline deposition are more easily generated, at least in nonoptimized deposition conditions. These two first points are met in any epi, for pure silicon, for SiGeC alloys as for any other semiconductor materials as well.

On the contrary, the two last mechanisms of the list are more specific to strained epi like SiGe or SiGeC on silicon. Indeed, as explained previously, in strained epi there is a strong tendency to relax the mechanical energy of the film by generation and development of misfit dislocations. This behavior has been extensively studied in SiGe–Si blanket wafers and has led to the concept of critical thickness. This concept is twofold. There is the critical thickness that borders the stability domain, a domain that corresponds to the balance between the energy required to extend a misfit segment and the mechanical energy saved by this extension. In practice, one can grow strained epi beyond this limit, and another critical thickness is often defined as the limit of the metastable domain. Note that this limit is not unique and depends on the epi process, especially on the thermal budget. In this domain, it is well established that once nucleated, one misfit is capable to extend on long distances, i.e., the wafer size in some cases. However, there is still one point that is not so clear: where and how the misfit dislocations are generated? In high-quality epi, defects that are not present cannot be supposed to nucleate dislocations, and then wafer edges that are more defective and that present shear stress are supposed to play a major role. In Ge-rich epi, dislocation loops generations, possibly favored by a SK growth if any, may occur.

In the case of patterned wafers, the situation is dramatically changed. Patterns are necessarily associated with edges that correspond to poly–epi interfaces in case of NSEG or to facets in case of SEG, and in both cases the edges will play an important role in dislocation generation. As an example, Figure 5.22 gives a photoluminescence image of a Si–SiGe NSEG measured in a $800 \times 800\ \mu m^2$ window. The Ge content is 22% and the thickness is well above the critical thickness. We observe that all the dislocations (propagating along {1 1 0} directions) are connected to the pattern boundary. Our interpretation is that all dislocations were generated at the epi–poly interface and the poly has then to be considered as an efficient misfit source. As a consequence, in case of NSEG the critical thickness corresponding to the metastable domain is reduced on patterned wafers as compared to full-sheets. It also means that, for a given thickness, patterns will increase the probability to a higher extent to find a misfit in a given epi surface.

In SEG, the epi edges certainly also play an important role. However, as their nature is fundamentally different as compared to NSEG, a different behavior can be expected. The NSEG–SEG comparison in terms of misfit dislocation apparition and stress relaxation has been done in certain RTCVD conditions [38]. Figure 5.23 reports the density of misfit dislocations for both types of deposition as a function of the film thickness and that summarizes the results. It appears that a significant dislocation density

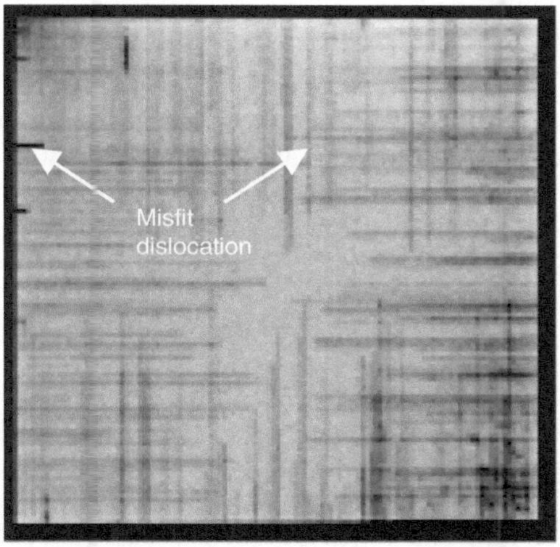

FIGURE 5.22 Polycrystal-induced misfit dislocations in SiGe epi observed by room temperature photoluminescence (box area = 800 × 800 μm^2).

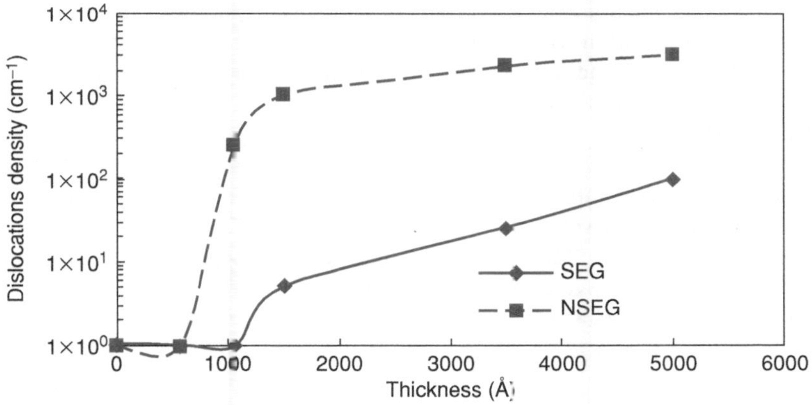

FIGURE 5.23 Dislocation density observed in SiGe layers as a function of epi thickness: comparison between NSEG and SEG. Despite a higher deposition temperature, SEG is more robust than NSEG against misfit dislocations.

(50 dislocations/cm) is obtained for <100 nm in case of NSEG and for about 400 nm in case of SEG. Thus, despite a higher deposition temperature (by about 100°C), SEG is much more robust that NSEG in terms of misfit dislocation generation and plastic relaxation. This dramatic change can be attributed to the different morphology that would give rise to a smaller shear stress (as a consequence of faceting?) and to the absence of poly that would be a dislocation source. This way, SEG is generally observed to provide much larger critical thickness as compared to NSEG. At this point, we have to note that the third point of the list (stress field in substrate) can also play a role in the dislocation appearance, and that it is often mixed with the two last ones in practical applications. However, in the experiments reported above, both epi types were carried out on similar structures and the variations observed have to be attributed then only to the nature of the border, facet or poly.

On the other hand, misfit dislocations can also nucleate and propagate during the subsequent operations of the technology as soon as crystalline defects, additional stress or thermal budget are provided. And SEG is expected to be more resistant also against this relaxation.

5.5 Recent Applications

At this point, it is interesting to illustrate the possibilities of SiGeC epi by RTCVD with the description of some actual applications of RTCVD. It should help the reader in keeping in mind where and how LTE can be used in practice in electronics. However, as this section contributes to a material chapter, it will be very short. A more complete discussion can be found in Ref. [29]. On the other hand, a number of applications will be detailed in the following chapters. For these reasons, we chose to briefly present only three recent applications we worked on.

SEG-Based Bipolar

SiGe LTE has become an accepted technique for the fabrication of the bipolar bases. Initially, HBT technologies were based on SiGe NSEG whereby the epi grown on single-crystal regions forms the intrinsic base and a part of extrinsic base, and the polycrystal deposited on top of dielectric forms a part of the base electrode.

More recently, SiGeC was substituted to SiGe for the base formation in order to block boron diffusion during the subsequent thermal budgets. As an illustration of this boron-blocking effect, Figure 5.24 reports the SIMS boron profiles of different samples before and after annealing (1000°C for 30 sec). Samples were grown with a 10-nm thick boron-doped layer and with different carbon contents, keeping all the other growth parameters constant. The sample measured as-deposited (e) is for control and for SIMS transfer function estimation, and the sample of pure SiGe, without C, measured after annealing (a) gives a reference for diffusion. As compared to pure SiGe, there is dramatically less diffusion in SiGeC samples: in the present conditions the higher the C concentration, the lower the B diffusion. This way, npn HBTs with thinner bases and higher performances have been developed and integrated in complex technologies. And today, all major semiconductor companies already offer a collection of SiGeC-based BiCMOS technologies (see Chapter 3, *Fabrication of SiGe HBT BiCMOS Technology*).

In addition to carbon incorporation, fully self-aligned (FSA) emitter–base structures are a further step toward high performances. SEG, well suited for the fabrication of a FSA-emitter–base structure [39], has been studied for a long time at STMicroelectronics, first with SiGe epi and more recently with SiGeC [30]. Figure 5.25 shows a TEM cross section of such a self-aligned SEG-based SiGeC HBT. The SiGeC epi was grown by RTCVD in a commercially available reactor (Applied Material Centura™), and, in the

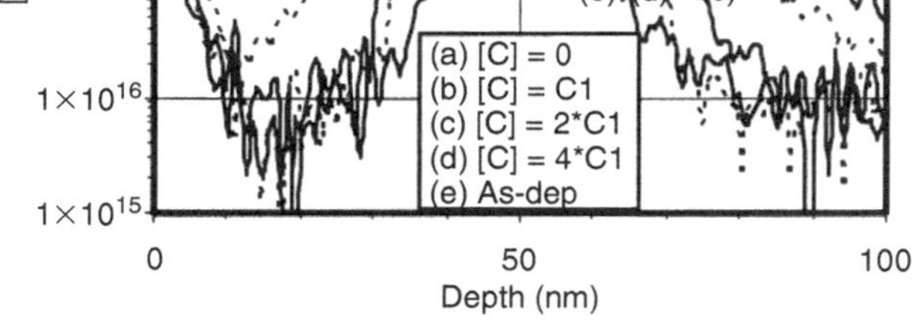

FIGURE 5.24 Boron SIMS profiles after a 30 sec annealing at 1000°C in different SiGeC films.

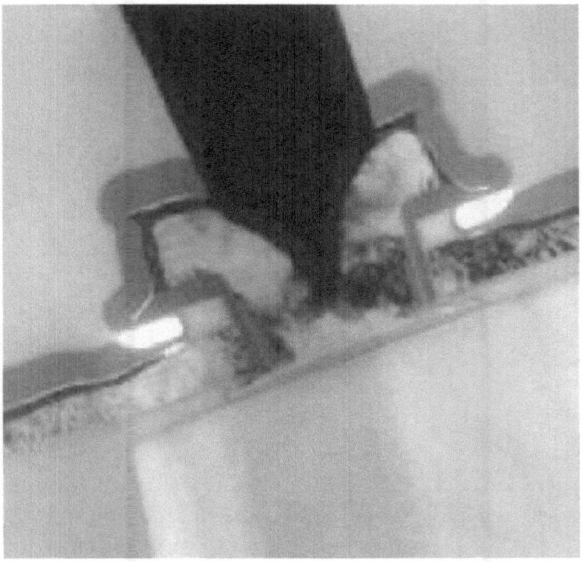

FIGURE 5.25 TEM cross section of a SiGeC SEG-based HBT.

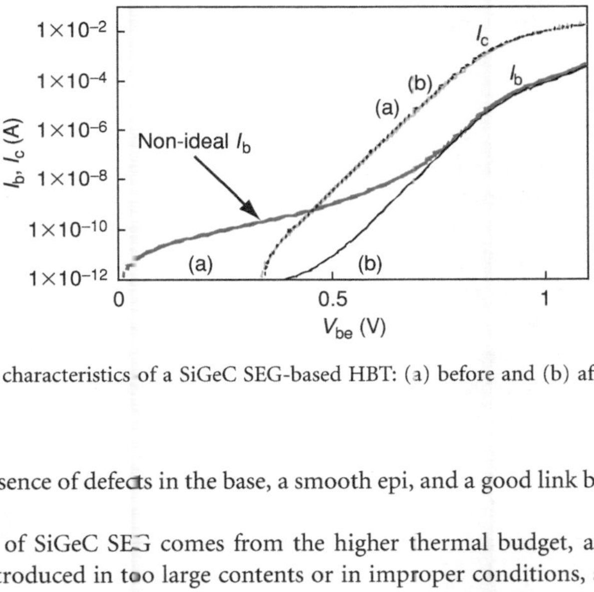

FIGURE 5.26 Gummel characteristics of a SiGeC SEG-based HBT: (a) before and (b) after epi optimization.

picture, we note the absence of defects in the base, a smooth epi, and a good link between the epi and the polycrystal.

The main difficulty of SiGeC SEG comes from the higher thermal budget, as compared to NSEG. Indeed, if carbon is introduced in too large contents or in improper conditions, significant amounts of carbon atoms will be incorporated in nonsubstitutional sites (interstitial, clusters, etc.) and the epitaxial base will then have poor electronic properties. The first detrimental effect usually detected is an excess base current due to increased recombination. This effect is illustrated by Figure 5.26, which reports HBT Gummel plots before and after selective SiGeC epi optimization. When using nonoptimized epi process, the base current at low levels deviates very much from its ideal variation: interstitial carbons or SiC clusters induce more recombination, especially if they are present in the depleted emitter–base region [30]. On the other hand, base current below 10^{-11} A/μm^2 at $V_{be} = 0.5$ V can be achieved with optimized base epitaxy.

High-frequency measurements of HBT are another way to characterize our SiGeC SEG process in terms of boron diffusion, as well as to test the validity of the FSA structure in terms of parasitics.

		207.6	212.3	214.5		
	219.0	215.4	215.5	217.8	218.7	
217.2	213.5	213.2	212.6	208.4	215.5	220.0
217.2	208.8	207.9	209.0	206.6	206.7	215.8
215.8	209.5	209.3	207.5	207.7	206.8	212.3
219.3	210.1	209.3	209.3	208.3	208.7	213.1
	210.6	209.8	208.5	210.7	211.8	
		213.3	213.5	212.5		

Min	Max	Mean	Stdev
206.6	220.0	212.1	3.9

FIGURE 5.27 f_T cartography (GHz) of a $0.17 \times 3.6\,\mu m^2$ bipolar at $V_{CB} = 0.5$ V and $V_{BE} = 0.92$ V.

The results obtained for f_T (220 GHz) and f_{max} (240 GHz) [40], and other DC characteristics not reported here, demonstrated that the neutral electrical base was very thin and kept inside SiGe. Thus, despite the significant activation annealing used for CMOS compatibility, the boron diffusion was effectively blocked in the selectively deposited SiGeC base. In addition, the simultaneous high values of f_T and f_{max} demonstrate the low parasitic components of this architecture.

On the other hand, SEG processes are often feared for their micro- or macroloading effects even though they can be efficiently solved. Indeed, the f_T cartography presented in Figure 5.27 shows a standard deviation of less than 2% (4.8% for f_{max}). These values demonstrate that any process uniformity or macro-LE issue has been solved in the present process. In the same manner other results showed that it does not suffer from micro-LE as well. Thus, no blocking point was detected toward industrialization.

Double Gate MOS

Double gate MOS (DG-MOS) transistors present important potential advantages: excellent I_{on}/I_{off} trade-off, good control of short-channel effects (SCE) and drain-induced barrier lowering (DIBL), and higher channel conductivity. They may be the alternative device structure for the ultimate CMOS nodes.

The DG devices presented here have been processed in a planar configuration owing to the utilization of the silicon-on-nothing (SON) process and the main operations are detailed in Chapter 12. After classical STI formation, a two-step epitaxy is done: a selective epi of SiGe (20 to 40 nm) followed by a thin NSEG of silicon. This Si NSEG leads to a mono-Si layer on active zones with a well-controlled thickness of the conduction channel, and to a poly-Si deposition on STI areas. Then, after source–drain–channel definition by photolithography and etching, the underlying SiGe is selectively etched, thanks to a specific isotropic plasma processing, leaving a silicon bridge (future DG-MOS channel) over the active zone. The SEM view given in Figure 5.28 illustrates the situation. One major advantage of this structure on some others reported in literature is related to the fact that the thickness of the silicon channel is defined by epitaxy that is capable of nanometric thickness with excellent uniformity. The remaining silicon bridge is oxidized to form a 20-Å thick gate oxide. *In situ* phosphorus-doped polysilicon is deposited all around the silicon bridge, and top gate and bottom gate are patterned by standard RIE gate etching. After the formation of spacers, extensions are implanted, and because thin Si films are used, elevated S/D were formed by Si SEG before standard 8 nm Co salicidation. A TEM cross section of a finished device with $L_{gate} = 40$ nm and $T_{Si} = 15$ nm is presented in Figure 5.29. This picture shows that the morphology of the device is well controlled, whereas additional high-resolution TEM observations proved that the Si channel was a perfect monocrystal. We can also note that we optimized the cleaning step before elevated S/D in order to minimize the TEOS consumption under the offset spacer (as located by the white circle), and that the elevated S/D SEG exhibited some faceting but without any detrimental effect on electrical device characteristics.

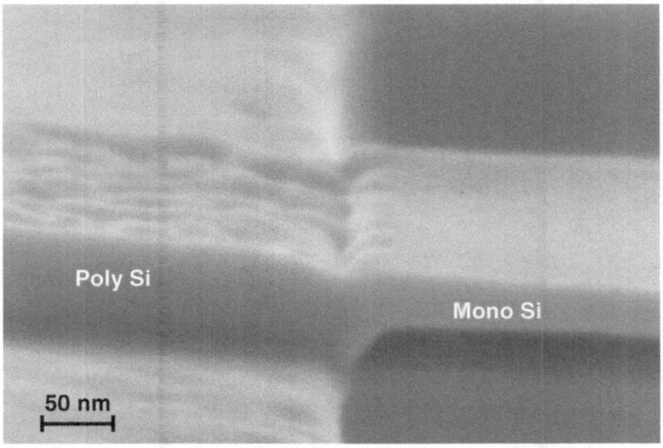

FIGURE 5.28 Silicon bridge over active zone after SiGe selective etching.

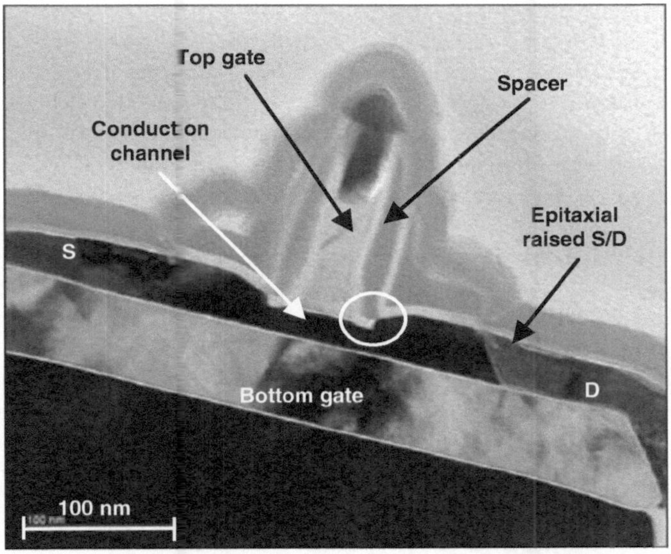

FIGURE 5.29 Along-channel TEM cross section of final DG device, $L_{gate} = 40\,nm$, $T_{Si} = 15\,nm$.

Electrical results confirm that the DG architecture improves considerably the device performances with very high drain currents (more than a factor of 2 as compared to conventional pMOS), small SCE, well-controlled DIBL, and subthreshold slope almost ideal [41]. The best results in terms of drive currents were obtained for a 70 nm gate length coupled with a 30 nm T_{Si} with $I_{on} = 1954\,\mu A/\mu m$ ($I_{off} = 283\,nA/\mu m$) for nMOSFETs (more detailed results are reported in Ref. [41]). On the other hand, the geometry of DG devices provides an excellent electrostatic integrity. This effect is illustrated in Figure 5.30 which reports the variation of DIBL as function of the gate length and for different silicon channel thicknesses. We observe that the thinner the channel, the lower the DIBL, and that DIBL is excellent as soon as the channel thickness does not exceed a third of the gate length.

Ultrathin Si films (10 nm), which can be controlled by epi deposition in the present technology, are thus mandatory for future CMOS technologies with short (32 to 45 nm) gates.

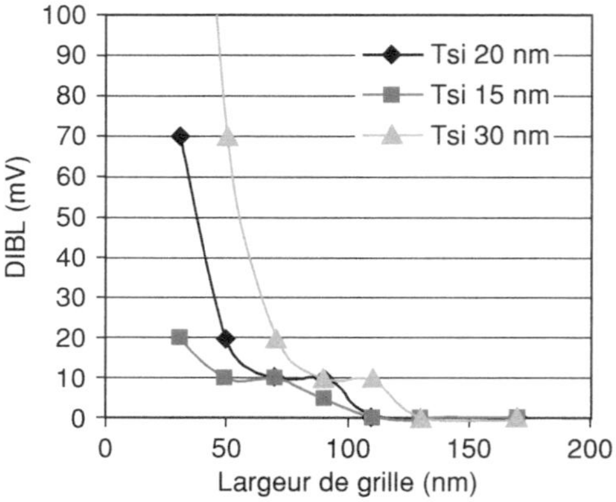

FIGURE 5.30 Drain-induced barrier lowering (DIBL) measured for different SOI thicknesses.

FD-SOI

Silicon on insulator (SOI) substrates have some traditional advantages like reduced junction capacitance, lower power consumption, and higher RF performances, as compared to bulk silicon. In addition to that, ultrathin SOI with an improved electrostatic integrity should dramatically improve SCE that are important issues in advanced digital CMOS with ultrashort channels. Thus, ultrathin film (UTF) approach is very suited to achieve future CMOS and these substrates may become a standard for these applications.

When fabricating CMOS on thin SOI films, the formation of S/D requires that these regions have been thickened. During salicidation, a certain excess of silicon is indeed necessary to insure good channel–silicide continuity. Elevated S/D deposited by SEG are then mandatory, and the situation is illustrated by Figure 5.31 that gives the cross section of such a FD-MOS fabricated on thin SOI.

However, the manipulation of UTF requires special attention, as they are not stable with temperature. The difficulty is illustrated in Figure 5.32a, where a 6-nm-thick SOI film has been submitted to a

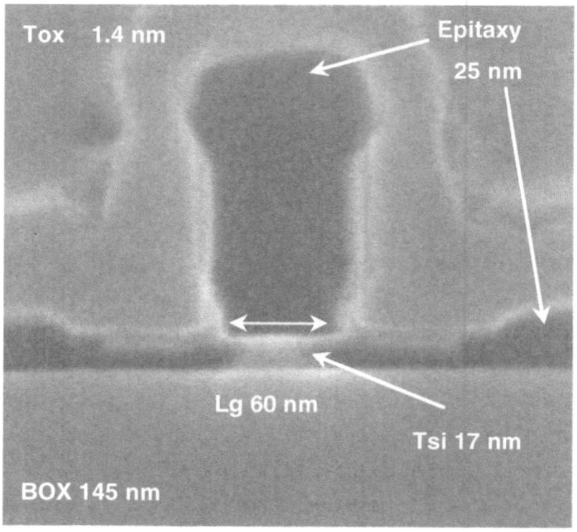

FIGURE 5.31 Along-channel TEM cross section of final FD-SOI device with elevated S/D: $L_{gate} = 60$ nm, $T_{Si} = 15$ nm, S/D elevation = 25 nm.

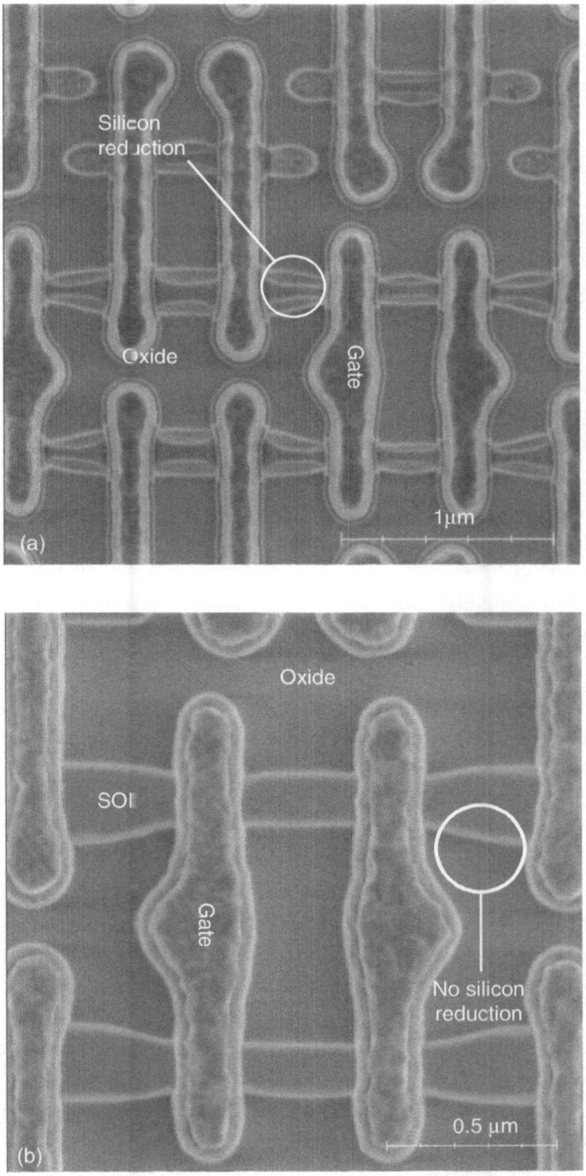

FIGURE 5.32 SEM top view of a 40-nm thick SEG on ultrathin (6 nm) FD-SOI films: (a) standard process and (b) improved process.

standard Si SEG process. In this case, the H_2 bake was carried out at a moderate temperature (<850°C) and selective deposition at 750°C with DCS–HCl chemistry, and we observe an important lateral constriction (about 50 nm) of SOI in active areas, as indicated in white circle. Some experiments were undertaken to examine or quantify if the missing SOI part was removed or displaced by evaporation (Equation 5.1 may be effective) or by agglomeration. On the other hand, via a dramatic thermal budget reduction we managed to carry out a 40-nm-thick Si SEG without any film restriction, as shown in Figure 5.32b. By further process optimization, we even demonstrated Si SEG on SOI films as thin as 3 nm, results not reported here.

This application is well illustrative of future material developments that will concern excessively thin films and that will present new issues.

5.6 Summary

SiGe(C) materials and low-temperature epitaxy are today expanding the capabilities and applications of Si technologies. They have already pushed the Si bipolar technologies to tremendous performances, the f_T/f_{max} values (220 GHz/240 GHz) above-mentioned and obtained in a fully industrial environment, and even more aggressive values reported in literature. On the other hand, LTE is entering the CMOS fabrication domain to give important improvements of MOS performances, using strained channels as an example [42], as well as to develop new innovative architectures like the DG-SON architecture illustrated in the previous section.

SiGe(C) materials make the concept of bandgap engineering available in Si technologies. In association with other techniques like the selective etching (see Chapter 13), these materials also give a number of new device architectures possible. As a consequence, their (potential) applications are numerous and different.

Thus, rather than try an exhaustive list of these applications, an almost impossible task, we prefer to detect what are the main trends in epi science. In our opinion, the most pressing requests are: new metastable materials (epitaxial SiGeC bases of HBT, highly doped epi, strained CMOS channels, etc.), complex stacks of very thin elementary epi films (base–emitter structure of HBTs), and extremely low thermal budgets (elevated source–drains in future CMOS, etc.). These materials requests can be translated into equipment or process requirements.

- Low deposition temperatures: this may be achieved by a very smart process or technology optimization, but new chemistries or new precursors would be welcome.
- Advanced gas panel functionality: new processes require a number of gases in the same time or in sequence; the gas panel has to be able to flow every gas needed in a step while it prepares or purges those required for the next step.
- Efficient gas control: as the elementary films are thinner and thinner, with deposition times possibly short, any lag time in gas flow has to be minimized. The gas pressure of the different gas panel parts has to be matched, and the gas panel design has to be compact with minimal dead volumes.
- High-temperature agility: with metastable materials and with very sensitive structures, temperature stabilization time may be detrimental and has to be minimized.
- Low-temperature cleaning: hydrogen bake cannot be carried out on very temperature-sensitive structures; integrated pre-clean (wet, plasma, new chemistries, etc.) will be mandatory if high-quality epi is required.
- Efficient tool: LTE will have to be less expensive for mass production, reactors have then to present robust cleanliness, no memory effect and high uptime.

Note that we do not include the quality of process gases and of reactor atmosphere because the author believes that, although the epi temperature will have to be decreased, these criteria already fulfilled today will not be much more severe in future.

As a final note, we believe that LTE will be a key technique to pass the very important roadblocks that are present on the silicon technology roadmap. And, for these reasons and as predicted [29], the present decade is very exciting, especially in SiGeC epi by RTCVD.

Acknowledgments

We wish to thank the many people of ST Crolles who made this contribution possible, especially the persons involved in epitaxy: P. Ribot, L. Georjon, T. Ouptier, N. Loubet, and F. Brossard; and those in advanced device applications: H. Baudry, A. Chantre, P. Chevalier, C. Fenouillet, M. Laurens, M. Marty, S. Monfray, A. Monroy, J. Mourier, T. Skotnicki, B. Szelag, G. Troillard, and A. Vandooren.

References

1. J.F. Gibbons, C.M. Gronet, and K.E. Williams. Limited reaction processing: silicon epitaxy. *Appl. Phys. Lett.* 47, 721, 1985.
2. J.C. Sturm, P.V. Schwartz, E.J. Prinz, and H. Manohara. Growth of $Si_{1-x}Ge_x$ by rapid thermal chemical vapor deposition and application to heterojunction bipolar transistors. *J. Vac. Sci. Technol. B* 9, 2011, 1991.
3. FUV4 from Jipelec—France. See for example: *Thin Solid Films* 222, 52, 1992.
4. J.C. Sturm, H. Manoharan, L.C. Lenchyshyn, M.L.W. Thewalt, N.L. Rowel, and D.C. Houghton. Well-resolved band-edge photoluminescence of excitons confined in strained $Si_{1-x}Ge_x$ quantum wells. *Phys. Rev. Lett.* 66, 1362, 1991.
5. D. Dutartre, G. Bremond, A. Souifi, and T. Benyattou. Excitonic photoluminescence from Si-capped strained $Si_{1-x}Ge_x$ layers. *Phys. Rev. B* 44, 11525, 1991.
6. C.A. King, J.L. Hoyt, C.M. Gronet, J.F. Gibbons, M.P. Scott, and J. Turner. $Si/Si_{1-x}Ge_x$ heterunction bipolar transistor produced by limited reaction processing. *IEEE Electron Dev. Lett.* 10, 52, 1989.
7. D. Dutartre, P. Warren, I. Sagnes, P.A. Badoz, J.C. Dupuis, G. Prudon, and A. Perio. Epitaxy and doping of Si and $Si_{1-x}Ge_x$ at low temperature by rapid thermal chemical vapour deposition. *J. Vac. Sci. Technol. B* 11, 1134, 1993.
8. McD. Robinson and L.H. Lawrence. Characterization of high growth rate epitaxial silicon from a new single wafer reactor. In *Semiconductor Fabrication: Technology and Metrology*, ASTM STP 990, Dinesh C. Gupta, Ed. American Society for Testing and Materials, Philadelphia, 1989.
9. J.C. Bean. Growth techniques and procedures. In *Germanium Silicon: Physics and Properties, Semiconductors and Semimetals*, Vol. 56, R. Hull and J.C. Bean, Eds. Academic Press, San Diego, CA, 1998, p. 8.
10. W. Kern and D.A. Puotinen. Cleaning solution based on hydrogen peroxide for use in semiconductor technology. *RCA Rev.* 31, 187, 1970.
11. T. Hsu, B. Anthony, R. Qian, J. Irby, D. Kinosky, A. Mahajan, S. Banerjee, C. Magee, and A. Tash. Defect microstructure in low temperature epitaxial silicon grown by RPCVD. *J. Electron. Mater.* 19, 4, 1990.
12. D.J. Meyer. Si-based alloys: SiGe and SiGe:C. In *Silicon Epitaxy, Semiconductors and Semimetals*, Vol. 72. D. Crippa, D.L. Rode, and M. Masi, Eds. Academic Press, New York, 2001, pp. 345–395.
13. B.S. Meyerson. Low-temperature silicon epitaxy by ultrahigh vacuum/chemical vapour deposition. *Appl. Phys. Lett.* 48, 797, 1990.
14. See for example the integrated wet clean referred as "OASIS CleanTM," AMAT Inc, 2003.
15. J.C. Sturm, P.V. Schwartz, E.J. Prinz, and H. Manoharan. Growth of $Si_{1-x}Ge_x$ by rapid thermal chemical vapour deposition and application to heterojunction bipolar transistors. *J. Vac. Sci. Technol. B* 9, 2011–2016, 1991.
16. D. Dutartre, P. Warren, I. Berbezier, and P. Perret. Low temperature silicon and $Si_{1-x}Ge_x$ epitaxy by rapid thermal chemical vapour deposition using hydrides. *Thin Solid Films* 222, 52, 1992.
17. G. Pares, J.L. Regolini, J. Mercier, D. Dutartre, and D. Bensahel. Kinetics aspects of epitaxial silicon growth using disilane in a rapid thermal processing system. *J. Appl. Phys.* 68, 4885, 1990.
18. M. Lier, C.M. Greelief, S.R. Kasi and M. Offenberg. Kinetics of silicon epitaxy using SiH_4 in a rapid thermal chemical vapor deposition reactor. *Appl. Phys. Lett.* 56, 629, 1990.
19. M. Racanelli and D.W. Greve. Temperature dependance of growth of $Si_{1-x}Ge_x$ by ultra high vacuum chemical vapor deposition. *Appl. Phys. Lett.* 56, 2524, 1990.
20. D.J. Robbins, J.L. Glasper, A.G. Cullis, and W.Y. Leong. A model of heterogeneous growth of $Si_{1-x}Ge_x$ films from hydrides. *J. Appl. Phys.* 69, 3729, 1991.
21. K. Sinniah, M. Sherman, L. Lewis, W. Weinberg, J. Yates Jr, and K.C. Janda. New mechanism for hydrogen desorption from covalent surfaces: the monohydride phase on Si (1 0 0). *Phys. Rev. Lett.* 62, 567, 1989.

22. M. Todd and D. Weeks. Low temperature, high growth rate epitaxial silicon and SiGe alloy films. First International SiGe Technology and Device Meeting, ISTDM2003, Nagoya, Japan, 2003, p. 153.
23. J.P. Posthill, R.A. Rudder, S.V. Hattangady, C.G. Fountain, and R.J. Markunas. On the feasibility of growing dilute C_xSi_{1-x} epitaxial alloys. *Appl. Phys. Lett.* 56, 734, 1990.
24. L.D. Lanzerotti, J.C. Sturm, E. Stach, R. Hull, T. Buyuklimanli, and C. Magee. Suppression of boron outdiffusion in SiGe HBTs by carbon incorporation. 1996 IEDM Tech. Digest, 1996, p. 249.
25. P.S. Chen, S.W. Li, Y.H. Liu, M.H. Lee, M.J. Tsai, and C.W. Liu. UHVCVD of hetero-epitaxial SiGe:C and SiC thin films on Si(1 0 0) with ethylene (C_2H_4) precursor as a carbon source. Second International SiGe Technology and Device Meeting, ISTDM2004, Frankfurt/Oder, Germany, 2004, p. 196.
26. F. Deleglise, C. Fellous, and D. Dutartre. Carbon incorporation study for SiGeC epitaxy optimization in RTCVD. First International SiGe Technology and Device Meeting, ISTDM2003, Nagoya, Japan, 2003, p. 151.
27. P.A. Coon, P. Gupta, M.L. Wise, and S.M. George. Adsorption and desorption kinetics for SiH_2Cl_2 on Si (1 1 1) 7 × 7. *J. Vac. Sci. Technol. A* 10, 324, 1993.
28. M. Hierlemann, A. Kersch, C. Werner, and H. Schaefer. A gas-phase and surface kinetics model for silicon epitaxial growth with SiH_2Cl_2 in a RTCVD reactor. *J. Electrochem. Soc.* 142(1), 259. 1995.
29. D. Dutartre. Silicon epitaxy: new applications. In *Silicon Epitaxy, Semiconductors and Semimetals*, Vol. 72, D. Crippa, D.L. Rode, and M. Masi, Eds. Academic Press, New York, 2001, pp. 397–457.
30. C. Fellous, F. Deleglise, A. Talbot, and D. Dutartre. Selective SiGeC epitaxy by RTCVD for high performance self-aligned HBT. 33rd European Solid-State Device Research Conference, Estoril, Portugal, 2003, p. 481.
31. H. Schaefer, J. Boeck, R. Stengl, H. Knapp, K. Aufinger, M. Wurzer, S. Boguth, M. Rest, R. Schreiter, and T.F. Meister. Selective epitaxial growth of SiGe:C for high speed HBTs. *Appl. Surf. Sci.* 224, 18, 2004.
32. N. Laurent, C. Fellous, and D. Dutartre. Non-destructive defect inspection in SiGe HBT using room temperature photoluminescence. Second International SiGe Technology and Device Meeting Proceedings, 2004, Frankfurt/Oder, Germany, 2004, p. 49.
33. C. Fellous, F. Romagna, and D. Dutartre. Thermal and chemical loading effects in non selective Si/SiGe epitaxy. *Mater. Sci. Eng. B* 89(1), 323, 2002.
34. P. Ribot and D. Dutartre. Low-temperature selective epitaxy of silicon with chlorinated chemistry by RTCVD. *Mater. Sci. Eng. B* 89(1), 306, 2002.
35. A. Talbot, J. Arcamone, C. Fellous, F. Deleglise, and D. Dutartre. Investigation of facet formation in RTCVD Si/SiGe selective epitaxie. In SiGe Materials, Processing and Devices, ECS 2004, Honolulu, Hawaii, p. 601.
36. R. Loo and M. Caymax. Avoiding loading effects and facet growth: key parameters for successful implementation of selective epitaxial SiGe deposition for HBT-BiCMOS and high-mobility heterochannel pMOS devices. *Appl. Surf. Sci.* 224, 24, 2004.
37. A. Talbot, J. Arcamone, C. Fellous, F. Deleglise, and D. Dutartre. Study of the poly/epi kinetics ratio in SiH4-RTCVD for new CMOS architectures. Second International SiGe Technology and Device Meeting, ISTDM2004, Frankfurt/Oder, Germany, 2004, p. 200.
38. C. Fellous, F. Romagna, and D. Dutartre. Comparison of dislocation generation in selective and non-selective SiGe epitaxy. Fifteenth International Symposium on Chemical Vapor Deposition (CVD-XV), May 14–19, 2000, Toronto, Canada, 2000, p. 442.
39. F. Sato, T. Hashimoto, T. Tasumi, H. Kitahata, and T. Tashiro. Sub 20 ps ECL circuits with 50 GHz f_{max} self-aligned SiGe HBTs. IEDM 1992. Tech. Digest, 1992, p. 397.
40. P. Chevalier, C. Fellous, L. Rubaldo, D. Dutartre, M. Laurens, T. Jagueneau, F. Leverd, S. Bord, C.T. Richard, D. Lenoble, J. Bonnouvrier, M. Marty, A. Perrotin, D. Gloria, F. Saguin, B. Bartalat, R. Beerkens, N. Zerounian, F. Aniel, and A. Chantre. Accepted to the 2004 Bipolar/BiCMOS Circuits and Technology Meeting, Montreal, Canada, September 13–14, 2004.

41. S. Harrison, P. Core [illegible]. Leverd, R. Cerutti, R. Palla, D. Delille, S. Borel, S. Julian, R. Pantel, S. Descombes, D [illegible], Y. Morand, MP. Samson, D. Lenoble, A. Talbot, A. Villaret, S. Monfray, P. Mazoyer, J [illegible] H. Brut, A. Cros, D. Munteannu, J-L. Autran, and T. Skotnicki. Highly performant [illegible] MOSFET realized with SON process. IEDM 2003. Tech. Digest, 2003, p. 449.
42. K. Mistr [illegible]n, C. Auth, S. Cea, T. Coan, T. Ghani, T. Hoffmann, A. Murthy, J. Sandford, R. Sha [illegible]ski, K. Zhang, S. Thompson, and M. Bohr. Delaying Forever: Uniaxial Strained Sili [illegible] a 90 nm CMOS Technology. 2004 Symposium on VLSI Technology, Hawaii, [illegible] 50.

6

MBE Growth Techniques

Michael Oehme
and Erich Kasper
University of Stuttgart

6.1 Introduction

In the past few years, the development of modern semiconductor devices utilizing structures with atomic dimensions has been advanced by vertical design. Layer structures with nanometer dimensions featuring quantum effects have become dominant. This would be applicable in some SiGe semiconductor devices such as in the resonant tunneling diode [1] or high electron mobility transistor (HEMT) [2].

Modern epitaxy techniques like the molecular beam epitaxy (MBE) method [3,4] allow the growth of very thin crystalline silicon germanium layers with well-defined dopants by very low growth temperatures. In this temperature regime volume diffusion is negligible. Furthermore, the critical thickness of SiGe heterostructures is increasing. However, in this low-temperature growth regime, surface segregation of dopant or alloy atoms has turned out to be a dominant mechanism for profile smearing. Knowledge of segregation behavior is necessary for device design and realization based on low-temperature epitaxy. Especially in the case of SiGe MBE surface segregation is observed for alloy atoms and dopants like boron, phosphorus, or antimony [5].

The purpose of this chapter is to describe the SiGe MBE growth techniques with the individual technical components such as the ultrahigh vacuum (UHV) system, the evaporation sources, the substrate heating, and the *in situ* monitoring of the process.

6.2 Requirements on an SiGe MBE System

Epitaxy describes the oriented growth of a single crystalline film on a substrate. The growing layer follows the orientation of the substrate. The driving force for the crystalline growth is minimizing of energy because the perfect crystal is the energetically favorable solid. A requirement for epitaxy is a clean substrate surface, otherwise the information about orientation of the substrate is disordered or even destroyed and the consequences are crystalline defects up to polycrystalline or amorphous growth.

MBE growth is carried out under conditions far from thermodynamic equilibrium and is dominated mainly by the kinetics of the surface processes occurring when the impinging beams react with the top atomic layer of the substrate.

An MBE machine involves the generation of molecular beams of matrix material such as silicon or germanium and doping species and their interaction with the substrate surface to form a single-crystal deposit under UHV conditions. The atomic or molecular fluxes of elemental constituents are evaporated or sublimated in special electron beam evaporators (EBEs) or in radiatively heated effusion cells. For a precise control of the beam fluxes over the substrate all sources exhibit rapidly acting mechanical shutters, and the flux of the deposition materials can be measured directly with a flux monitor, for example, a quadrupole mass spectrometer (QMS) or indirectly over the effusion cell temperature (after calibration). This enabled the growth of SiGe layers with controlled Ge content between 0% and 100% with an arbitrary doping structure. Another advantage is the growth of very sharp doping transitions over several decades with special doping strategies.

The quality of the vacuum conditions and the purity of the constituent fluxes are fundamental specification of the MBE because the semiconductor properties are sensitive to impurities. Crystal quality and defect density are drastically limited by insufficient partial pressure [6]. For the growth of epitaxial layers with high-purity and crystalline perfection UHV equipment with a basis pressure less than 1×10^{-10} mbar is needed. Really important are the vacuum conditions during the layer growth, where very hot components in the reactor need special materials with a low vapor pressure such as tantalum or molybdenum. From the above discussion, the essential features of an SiGe MBE system are

1. UHV conditions during growth
2. Clean Sources of atomic and molecular beams
3. Substrate-cleaning scheme with low thermal budget
4. *In situ* analysis of process conditions and real-time monitoring for close-loop control

In the following sections, we discuss these topics with emphasis on technological aspects. Figure 6.1 shows a picture of a commercial 200 mm SiGe MBE equipment (with permission of DCA instruments). The machine can be upgraded to wafers with 300 mm diameter.

6.3 UHV Conditions

In order to overcome the exchange between air and UHV, an MBE machine consists of at least two chambers, a growth chamber and a load lock. According to different application needs, designs of MBE systems with two [7], three [8], or more [9] chambers are known. A schematic view of the basic unit is shown in

FIGURE 6.1 Picture of a commercial 200 mm SiGe MBE equipment (with permission of DCA instruments). The machine can be upgraded to wafers with 300 mm diameter.

6

MBE Growth Techniques

Michael Oehme
and Erich Kasper
University of Stuttgart

6.1 Introduction

In the past few years, the development of modern semiconductor devices utilizing structures with atomic dimensions has been advanced by vertical design. Layer structures with nanometer dimensions featuring quantum effects have become dominant. This would be applicable in some SiGe semiconductor devices such as in the resonant tunneling diode [1] or high electron mobility transistor (HEMT) [2].

Modern epitaxy techniques like the molecular beam epitaxy (MBE) method [3,4] allow the growth of very thin crystalline silicon germanium layers with well-defined dopants by very low growth temperatures. In this temperature regime volume diffusion is negligible. Furthermore, the critical thickness of SiGe heterostructures is increasing. However, in this low-temperature growth regime, surface segregation of dopant or alloy atoms has turned out to be a dominant mechanism for profile smearing. Knowledge of segregation behavior is necessary for device design and realization based on low-temperature epitaxy. Especially in the case of SiGe MBE surface segregation is observed for alloy atoms and dopants like boron, phosphorus, or antimony [5].

The purpose of this chapter is to describe the SiGe MBE growth techniques with the individual technical components such as the ultrahigh vacuum (UHV) system, the evaporation sources, the substrate heating, and the *in situ* monitoring of the process.

6.2 Requirements on an SiGe MBE System

Epitaxy describes the oriented growth of a single crystalline film on a substrate. The growing layer follows the orientation of the substrate. The driving force for the crystalline growth is minimizing of energy because the perfect crystal is the energetically favorable solid. A requirement for epitaxy is a clean substrate surface, otherwise the information about orientation of the substrate is disordered or even destroyed and the consequences are crystalline defects up to polycrystalline or amorphous growth.

MBE growth is carried out under conditions far from thermodynamic equilibrium and is dominated mainly by the kinetics of the surface processes occurring when the impinging beams react with the top atomic layer of the substrate.

An MBE machine involves the generation of molecular beams of matrix material such as silicon or germanium and doping species and their interaction with the substrate surface to form a single-crystal deposit under UHV conditions. The atomic or molecular fluxes of elemental constituents are evaporated or sublimated in special electron beam evaporators (EBEs) or in radiatively heated effusion cells. For a precise control of the beam fluxes over the substrate all sources exhibit rapidly acting mechanical shutters, and the flux of the deposition materials can be measured directly with a flux monitor, for example, a quadrupole mass spectrometer (QMS) or indirectly over the effusion cell temperature (after calibration). This enabled the growth of SiGe layers with controlled Ge content between 0% and 100% with an arbitrary doping structure. Another advantage is the growth of very sharp doping transitions over several decades with special doping strategies.

The quality of the vacuum conditions and the purity of the constituent fluxes are fundamental specification of the MBE because the semiconductor properties are sensitive to impurities. Crystal quality and defect density are drastically limited by insufficient partial pressure [6]. For the growth of epitaxial layers with high-purity and crystalline perfection UHV equipment with a basis pressure less than 1×10^{-10} mbar is needed. Really important are the vacuum conditions during the layer growth, where very hot components in the reactor need special materials with a low vapor pressure such as tantalum or molybdenum. From the above discussion, the essential features of an SiGe MBE system are

1. UHV conditions during growth
2. Clean Sources of atomic and molecular beams
3. Substrate-cleaning scheme with low thermal budget
4. *In situ* analysis of process conditions and real-time monitoring for close-loop control

In the following sections, we discuss these topics with emphasis on technological aspects. Figure 6.1 shows a picture of a commercial 200 mm SiGe MBE equipment (with permission of DCA instruments). The machine can be upgraded to wafers with 300 mm diameter.

6.3 UHV Conditions

In order to overcome the exchange between air and UHV, an MBE machine consists of at least two chambers, a growth chamber and a load lock. According to different application needs, designs of MBE systems with two [7], three [8], or more [9] chambers are known. A schematic view of the basic unit is shown in

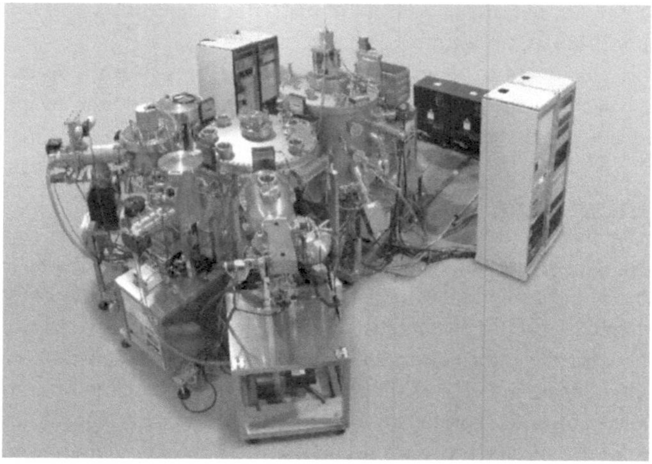

FIGURE 6.1 Picture of a commercial 200 mm SiGe MBE equipment (with permission of DCA instruments). The machine can be upgraded to wafers with 300 mm diameter.

Figure 6.2. It consists of two independently pumped UHV chambers connected with a UHV gate valve. At the left side there is the load lock with the wafer-transfer mechanism. This chamber accommodates a cassette-type magazine for silicon wafers. In modern MBE systems this magazine meets an industrial standard for 25 wafers with diameters of 150 or 200 mm. The load lock is pumped rapidly with a turbomolecular pump or with a cryopump. A typical system achieves a pressure in the range of 10^{-8} mbar within 90 min. The wafer transport between the storage chamber and growth chamber is computer controlled, and performed completely under UHV conditions. To minimize the contamination during transfer of a substrate from the load lock into the deposition chamber, multiple wafer loading is often used. These wafers are transferred into a parking stage within a central chamber for storage under UHV with a pressure in the range of 10^{-10} mbar.

Figure 6.2 (right side) shows a schematic view of a growth chamber. This chamber accommodates all the installations necessary for MBE growth. These are substrate holder, substrate heater, and substrate rotation; evaporation sources for matrix materials and dopants; and monitoring and analyzing facilities. Typically, the growth chamber is evacuated with a pumping cascade existing of an oil-free prevacuum pump, a turbomolecular pump, and a high-speed sorption pump, e.g., a sublimation pump cooled with liquid nitrogen. The growth chamber is always under UHV except for maintenance work or source refilling. After venting the chamber, a conditioning of the MBE system requires bakeout up to 300°C over more than 24 h. For this heater elements in a jacket are welded to the chamber walls from outside. For a minimum of preparation time to the start of bakeout, a layer of thermal isolation fiber is permanently sandwiched between the chamber wall and an aluminum cover. Under growth conditions, cooling water pipes welded to the chamber walls effectively reduce outgassing. Also, all effusion cells exhibit cooled walls to reduce the thermal radiation.

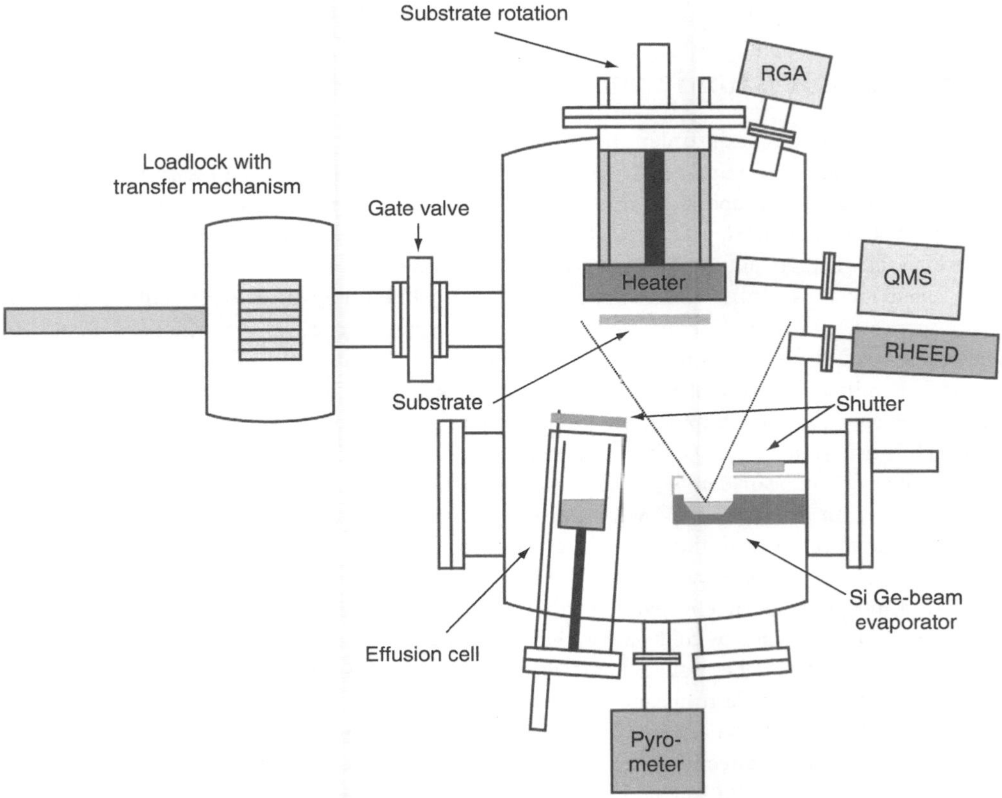

FIGURE 6.2 Schematic illustration of an SiGe MBE system; the pumping system is not shown.

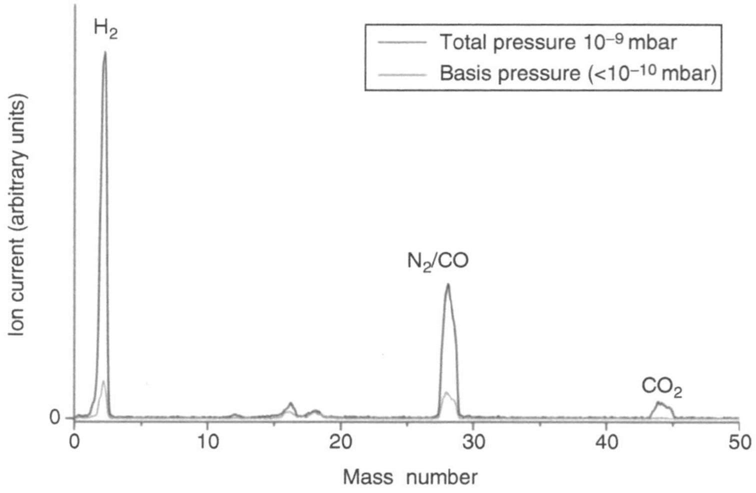

FIGURE 6.3 Residual gas spectrum of well-conditioned MBE chamber before (basis pressure $<10^{-10}$ mbar) and during (total pressure 10^{-9} mbar) the growth of an SiGe layer.

The outgassing of equipment materials and surfaces limits the vacuum pressure. Figure 6.3 shows a typical residual gas spectrum of a well-conditioned MBE system before and during the growth of an SiGe layer. Hydrogen (H_2) is the main detectable residual species in the UHV. Small increases in the heights of residual H_2O, N_2—CO, and CO_2 occur during growth, but the pressure is also dominated by H_2.

6.4 Sources of Atomic and Molecular Beams

For typical Si growth rates of the order of nanometers per second flux densities in the range of 5×10^{14} to $5 \times 10^{15}\,\mathrm{cm^{-2}\,sec^{-1}}$ are required. Si temperatures of about 1700°C are necessary when these flux densities are generated by evaporation from a Si-source distance of about 30 cm from the substrate. At such temperatures, molten Si is very reactive and it would cause excessive contamination from the crucible material [10]. A solution for this problem is heating of Si with an electron beam. Figure 6.4 shows a schematic view of the major components of a Si EBE. Tungsten filament emits electrons and these are accelerated in an electric field (electron gun) with typical energies between 4 and 12 keV. In a magnetostatic field, the electrons are deflected to an angle of 270° to avoid contamination by tungsten atoms, which emitted from the filament, as metallic impurities are undesirable in the molten Si.

The high-energy electrons impinge on a single-crystal, high-purity (resistivity $> 1000\,\Omega\,\mathrm{cm^{-1}}$) Si ingot, which is placed in a water-cooled Cu-crucible. The focused electron beam is partially melting the Si surface. The electron beam can be scanned over the whole crucible. Due to the thermal properties of Si, the molten region is restricted to the area directly heated by the electron beam.

Several problems result from the EBE application; electron- and ion-beam-stimulated desorption of gas molecules adsorbed on the system walls and cold shrouds, damage of the grown film by impinging electrons and ions and interference with *in situ* monitoring instruments. In order to minimize these problems, the EBE is screened by cold walls with defined aperture, which limit the beam cone and reduce the thermal load of the chamber walls. All metal surfaces are covered with high-purity Si plates to reduce the risk of interaction with electrons and ions.

The flux monitoring of Si is performed by a QMS. The QMS measures the intensity of the Si isotope, with the mass number 30, directly in the Si molecular beam because this mass is free of fragments of other residual species. The QMS signal is proportional to the flux, respectively the growth rate on the substrate surface.

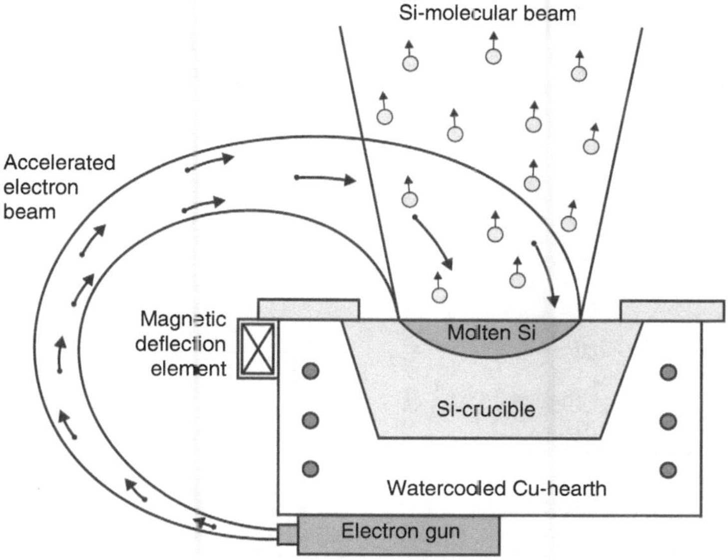

FIGURE 6.4 Schematic view of the major components of an Si electron beam evaporator.

The Si EBE is also a source for silicon ions. In the area above the silicon melt, the flux of evaporated atoms is ionized partially by the interaction with the electron beam. A potential up to several kV can be applied to the isolated substrate holder to focus and accelerate the Si ions. Thus, it is possible to use special growth strategies such as doping by secondary ions (DPI) [11], generation of point defects [12], or dislocation annihilation [13].

For the evaporation of Ge there are two possibilities, with an EBE or a single effusion cell. The construction of Ge-EBE is similar to the Si-EBE. However, the turbulence in Ge is much more severe as in Si, because Ge has much smaller thermal conductivity than Si, resulting in complete melting of the Ge reservoir. Contact of the molten Ge with the water-cooled Cu crucible triggers heavy turbulence. To overcome this problem, a liner made from high-purity, single-crystal Si is used. Possible enrichment of Si in the Ge reservoir is acceptable, since usually SiGe layers are grown and both fluxes are monitored simultaneously by a mass spectrometer.

The other possibility is the Ge evaporation from a single effusion cell [14]. The particular property of this cell is the co-evaporation of Ge and carbon with a flux ratio of roughly 1000, because the crucible consists of pyrolithical carbon. Carbon is partly dissolved in molten germanium up to the saturation limit. Saturation is obtained by heating the crucible with the germanium–carbon ingot above the melting point. Dissolved carbon evaporates much easier and the germanium/carbon ratio is reduced to three orders of magnitude, which is exactly the right ratio for SiGe-HBT applications with suppressed boron out-diffusion [15]. A technical solution is shown in Figure 6.5. Ge is melting in a graphite crucible with a large area opening (4 cm^2). The crucible and the graphite support are specially designed to withstand the strong tensile forces exerted by the increase of Ge volume during solidification. The diamond lattice is an open structure with a considerable volume increase both for Si and Ge during solidification. The electrical current heater surrounds the crucible, the outward heat flux is shielded by a graphite cage, tantalum- and Si-shields, and a cold wall. A thermocouple is mounted near the bottom of the crucible. A clear problem of this source after long-time operation seems to be the formation of a thin carbon film, which covers part of the molten germanium like a skin. This carbon skin is probably created by carbon precipitation during cooling down the melt from the operating temperature (typically 1300°C) to the standby temperature chosen slightly above the Ge melting point of 938°C. This carbon skin reduces with time the effective area for evaporation for both germanium and carbon. (Remember:

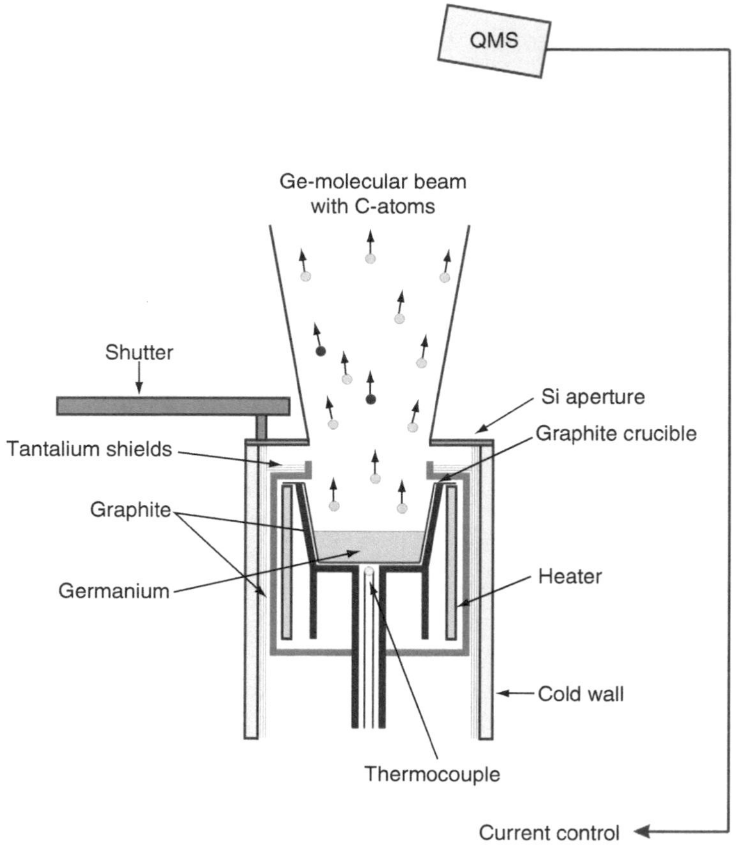

FIGURE 6.5 Effusion cell for Ge—C co-evaporation with quadrupole mass spectrometer (QMS) flux control.

The evaporation from carbon itself is negligible at the source temperatures.) The source, therefore, cannot be operated at the usual mode of an effusion cell with thermocouple control. Instead a flux control with a QMS is necessary for power regulation. The single effusion cell for germanium evaporation will considerably reduce the effort of carbon doping of MBE-grown HBT structures.

In nanoelectronic devices based on Si or SiGe, carbon doping plays an important role for two reasons. First, carbon doping delivers a heterostructure barrier for electrons [16] and may be used for quantum wells with electron confinement. Second, the retarded boron diffusion [17] is utilized in SiGe-HBTs with improved process stability. The sublimation of carbon requires very high temperatures up to 2300°C. Special high-temperature sublimation cells achieve C concentrations up to several percent in $Si_{1-x-y}Ge_xC_y$ layers [18].

p-Type and n-type doping over a wide range of concentration and with abrupt doping transitions is necessary for novel device structures. High and moderate levels of dopants of both types are needed to grow devices. In an SiGe MBE several doping techniques are used.

Boron is commonly used for p-type doping as it has a larger equilibrium solid solubility and a very low surface segregation. In modern SiGe MBE systems, boron is evaporated using a high-temperature source with special graphite crucibles. Temperatures between 1500°C and 2000°C are required to achieve doping levels of 10^{15} cm^{-3} up to several 10^{20} cm^{-3}.

n-Type doping during SiGe MBE growth is problematic because of low sticking coefficients and surface segregation. Thus, a limited range of dopant concentrations is available. In an SiGe MBE antimony and phosphorus can be used.

Antimony is the more attractive n-type doping material because it can be directly co-evaporated from a controlled source during the growth. However, the extreme temperature sensitivity of surface segregation requires special doping strategies such as prebuildup, flash-off techniques [19], or the DSI [20]. DSI allows doped SiGe layers with very sharp doping transitions over few decades.

Elemental phosphorus cannot be used due to its high vapor pressure, which gives rise to high background contamination in an MBE system. For P-doping special sources exist. Possible realizations of such a source are an arch made of phosphorus-doped Si [21] or the doping from a GaP source [22]. Comparison with Sb-doping shows that the segregation behavior in the lower temperature regime is surprisingly similar. In the higher temperature regime, however, surface segregation of Sb is much stronger [5].

6.5 Substrate Heating and Cleaning

The substrate temperature is one of the most important growth parameters because it influences all adatom processes of the surface, the crystalline growth, the surface morphology, the abruptness of doping transitions, and the relaxation processes in heterostructures.

In an SiGe MBE, the substrate is heated radiatively from the backside by a current-powered meander. A typical construction is shown in Figure 6.6. The heater is made of meander-shaped pyrolytical graphite. A reflector ring under the substrate and the thickness profiling of the heater element guarantee a homogenous temperature distribution between room temperature and 1000°C. A radiation shield of pyrolytical graphite minimizes the thermal losses. This construction is mounted in a water-cooled steel case with a tantalum heat shield in the inner part. The temperature of the graphite meander T_{th} is monitored by a thermocouple behind the heater. The substrate temperature is measured with a near-infrared pyrometer. The pyrometer measurement is used to calibrate the heater temperature.

However, at the typical growth temperatures, only a small overlap of the spectral emission of the radiative heater and the spectral absorption of silicon is given. At 700°C, the maximum of the thermal radiation for a black emitter has a wavelength of 3 μm. In this wavelength range intrinsic silicon is transparent. The heating of substrate is caused mainly by radiative absorption of the free charge carrier. Figure 6.7 shows the temperature dependency of a high- and a low-doped substrate as a function of heater temperature measured with a pyrometer. For temperatures above 650°C no differences for this two types of substrates can be seen because the intrinsic carrier concentration in Si is 10^{18} cm^{-3} at 700°C [23]. However, at lower heater temperatures, the temperature difference between high- and low-doped

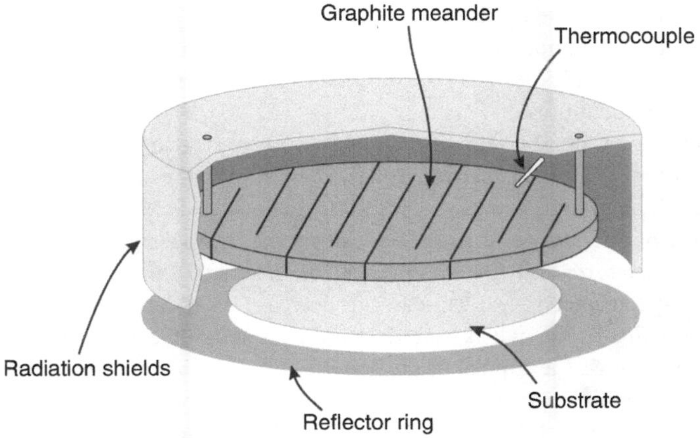

FIGURE 6.6 Schematic view of a typical radiative heater system.

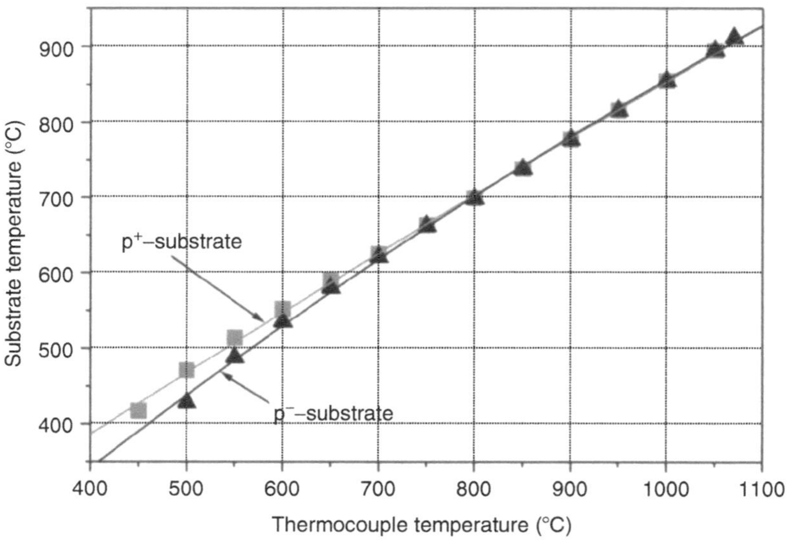

FIGURE 6.7 Temperature dependency of a high- and low-doped substrate as a function of heater temperature measured with a pyrometer.

substrates increases. At $T_{th} = 500°C$ the difference is 40 K. The consequence is that different calibration curves for high- and low-doped substrates have to be used.

High-quality growth needs interfaces with low contamination levels [24]; otherwise strain fields, dislocations, stacking faults, unintentional doping, and reduced carrier lifetimes may result. A very common cleaning scheme employs an *ex situ* etch with H_2O_2—NH_4OH and H_2O_2—HCl (RCA etch) followed by an *in situ* heating at 860°C to 900°C. The thin protective oxide (SiO_2) is desorbed as SiO after the reaction $Si + SiO_2 \rightarrow 2SiO$ [25] has taken place during heating. Wafer manufacturers deliver their substrates already covered with the thin protective oxide. In an SiGe MBE system, the SiO_2 layer is completely desorbed at substrate temperatures above 900°C [8]. If it is possible, the growth starts with an intrinsic Si buffer, which overgrows residual contaminations at the substrate interface such as carbon or SiC cluster.

6.6 *In Situ* Analysis

Current process parameters are determined in an MBE system with different *in situ* analyzing methods such us vacuum monitoring, temperature monitoring and controlling, flux calibration and controlling, or the measurement of layer thickness. These parameters are used for process control or for regulating sources and substrate temperatures.

High-temperature compatible thermocouples are used for measuring and controlling the temperatures of the effusion sources and the substrate. For the partial pressure analyzing a residual gas analyzer (RGA) is integrated in the growth chamber. This system consists of a QMS with multiplier. The RGA allows the monitoring of different residual species during the complete growth process.

In modern SiGe systems, *in situ* flux monitoring of the matrix materials like Si and Ge can be performed by a mass spectrometer with Faraday cup detector. After calibration, the mass spectrometer signal gives directly the flux of the matrix materials and is used as a feedback parameter for controlling the EBEs. A multiplexer allows the control of the fluxes of two or more different masses simultaneously, for example, the mass number 30 for Si and 74 for Ge, which is used for the growth of Si—SiGe structures. The flux measurement can also be performed with other methods such as quartz crystal resonators or electron-induced emission spectroscopy [26,27].

Silicon ions from the EBE can be measured with a special ion detector [28]. This monitoring system measures the distribution of ion and electron densities over the substrate depending on the substrate voltage.

Standard pyrometric measurements cannot be applied during film growth, since interference effects create artifacts and obscure the evaluation via Planck's law. A combination of a pyrometer and a reflectometer will result in a real-time correction of emissivity ε, thus allowing the determination of temperatures even for arbitrary multilayer stacks. This measurement method is called reflection-supported pyrometric interferometry (RSPI) [29]. RSPI has been proven to be capable of providing *in situ* real-time information concerning temperature and film thickness for numerous applications in semiconductor manufacturing. Temperature measurements range from around 450°C to over 900°C with a resolution of about 0.1°C. In the very low temperature regime (150°C to 450°C) the heater is calibrated with a Pt—PtRh thermocouple incorporated in the substrate, and then operated with fixed heater power [30].

Reflection high-energy electron diffraction (RHEED) is an electron diffraction technique that can yield information on surface structure, cleanliness, smoothness, and growth rate [4]. It is particularly useful in MBE because of the determination of surface reconstruction, a re-ordering of the top atomic layer of the crystal by reducing the energy of the free surface.

6.7 Summary

This chapter has presented an overview of topical SiGe MBE growth techniques. A wide substrate temperature regime (typically 150°C to 600°C) with low thermal budget, a simple *in situ* cleaning procedure, precise control of doping fluxes, doped heterostructures with abrupt profiles, and sophisticated *in situ* process monitoring are the main advantages of MBE. Several technical problems also have been discussed, for example, the influence of the epitaxial growth by the quality of the vacuum conditions or the dependence of substrate temperature from the substrate doping. A single Ge—C source for MBE growth of C-doped SiGe base layers was presented. This source will considerably reduce the effort for carbon doping of MBE-grown HBT structures.

References

1. K. Eberl, R. Duschl, O.G. Schmidt, U. Denker, and R. Haug. *J. Cryst. Growth* **227** 770 (2001).
2. F. Schaeffler. *Semicond. Sci. Technol.* **12** 1515 (1997).
3. R.A.A. Kubiak, E.H.C. Parker, and S.S. Iyer. In *Silicon Molecular Beam Epitaxy*, E. Kasper and J.C. Bean, Eds. CRC Press, Boca Raton, FL (1988), p. 5.
4. M.A. Herman and H. Sitter. *Molecular Beam Epitaxy.* Springer-Verlag, Berlin (1996).
5. H. Jorke. In *Properties of Silicon Germanium and SiGe:Carbon*, E. Kasper and K. Lyutovich, Eds. INSPEC, Institution of Electrical Engineers, London, 2000, p. 287.
6. E. Kasper and H. Kibbel. *Vac. Technol.* **33** 13 (1983).
7. E. Kasper, H. Kibbel, and F. Schaeffler. *J. Electrochem. Soc.* **136** 1154 (1989).
8. E. Kasper, M. Bauer, and M. Oehme. *Thin Solid Films* **321** 148 (1998).
9. T. Rupp, J. Messarosch, and I. Eisele. *J. Cryst. Growth* **183** 99 (1998).
10. C.W. Pearce. In *VLSI Technology*, chapter 1, S.M. Sze, Ed. McGraw-Hill, Singapore (1983).
11. H. Jorke, H.-J. Herzog, and H. Kibbel. *Appl. Phys. Lett.* **47** 511 (1985).
12. W.-X. Ni, G.V. Hannson, I.A. Buyanova, A. Henry, W.M. Chen and B. Momemar. *Appl. Phys. Lett.* **68** 238 (1996).
13. M. Bauer, M. Oehme, K. Lyutovich, and E. Kasper. *Thin Solid Films* **336** 104 (1998).
14. M. Oehme, M. Bauer, and E. Kasper. *Mater. Sci. Eng.* **B89** 332 (2002).
15. H. Rücker, B. Heinemann, W. Röpke, G. Fischer, G. Lippert, H.J. Osten, and R. Krups. Technical Digest, International Conference Simulates on Semiconductor Processes and Devices, IEEE, Piscataway, NJ, 1997, p. 281.

16. R. Duschl, O.G. Schmidt, and K. Eberl. In *Properties of Silicon Germanium and SiGe:Carbon*, E. Kasper and K. Lyutovich, Eds. INSPEC, Institution of Electrical Engineers, London (2000), p. 158.
17. H.J. Osten, G. Lippert, P. Gaworzewski, and R. Sorge. *Appl. Phys. Lett.* **71** 1522 (1997).
18. K. Eberl, S.S. Iyer, S. Zollner, J.C. Tsang, and F.K. Legoues. *Appl. Phys. Lett.* **60** 3033 (1992).
19. S.S. Iyer, R.A. Metzger, and F.G. Allen. *J. Appl. Phys.* **52** 5608 (1981).
20. H. Jorke and H. Kibbel. *J. Electrochem. Soc.* **133** 774 (1986).
21. E. Friess, J. Nützel, and G. Abstreiter. *Appl. Phys. Lett.* **60** 2237 (1992).
22. R. Duschl, O.G. Schmidt, G. Reitemann, E. Kasper, and K. Eberl. *Electron. Lett.* **35** 1111 (1999).
23. S.M. Sze. *Physics of Semiconductor Devices.* Wiley Interscience, New York (1969), p. 28.
24. P.J. Grunthaner. *Thin Solid Films* **183** 197 (1989).
25. S.T. Ahn, H.W. Kennel, W.A. Tiller, and J.D. Plummer. *J. Appl. Phys.* **65** 2957 (1989).
26. C.A. Gogol and C. Cipro. *Electrochem. Soc. Proc.* **85** 415 (1985).
27. R.A. Kubiak, S.M. Newstead, A.R. Powell, E.H.C. Parker, T.E. Whall, T. Naylor, and K. Bowen. *J. Vac. Sci. Technol.* **9** 2423 (1991).
28. G. Eifler, E. Kasper, Kh. Ashurov, and S. Morozov. *J. Vac. Sci. Technol.* **A 20** 945 (2002).
29. H. Möller, F.G. Böbel, B. Hertel, T. Lindenberg, and G. Ritter. *J. Cryst. Growth* **157** 327 (1995).
30. M. Bauer, M. Oehme, M. Sauter, G. Eifler, and E. Kasper. *Thin Solid Films* **369** 228 (2000).

7

UHV/CVD Growth Techniques

Thomas N. Adam
IBM Semiconductor Research and Development Center

7.1 Introduction

The aggressive scaling of silicon-based high-performance integrated circuits [1] has led to device dimensions and junction depths that require ultralow thermal processing budgets. For example, state-of-the-art CMOS gate lengths have shrunk below 100 nm over the past few decades. The current International Technology Roadmap for Semiconductors (ITRS) projects physical gate lengths in microprocessors and ASICs of 40 nm in the near future and 10 nm beyond 2015. Without considering suppression and enhancement effects, such as substitutional carbon or transients, the classical dopant diffusion alone (e.g., boron: $D_0 \approx 0.67\,cm^2/sec$, $E_A \approx 3.46\,eV$ [2]) exhibits comparable $2\sqrt{Dt}$ values at 900°C for minute-scale durations. Growth techniques that employ low temperatures (≤700°C) for homo- and hetero-epitaxy are necessary to maintain abrupt doping profiles and prevent, nearly eliminate, dopant and alloy diffusion, especially in high-performance BiCMOS integration schemes. In particular, pseudomorphically grown silicon–germanium has emerged as an important alloy due to its bandgap properties and strain-induced effects on carrier mobility [3]. Among others, substrate-compliant epitaxial SiGe is utilized in the fabrication of heterojunction diodes [4], resonant tunneling diodes [5], heterojunction bipolar transistors [6], modulation-doped field-effect transistors [7], MOS field-effect transistors [8], and photodetectors [9]. Today, the most commonly employed *low-temperature* epitaxial deposition methods are molecular beam epitaxy (MBE) [10], gas-source MBE (GSMBE) [11], and chemical vapor deposition (CVD) [12]. The deposition using chemical vapors is practical in a commercial manufacturing environment because it offers high wafer-throughputs by batch processing, outstanding film uniformities, excellent control over alloy composition and dopant concentration, and selective epitaxial growth. However, chemical vapor deposition at atmospheric pressures and in low vacuum was traditionally restricted to high substrate temperatures. Although improvements in gas purity, chamber cleanliness and design, precursor chemistry, and interface cleanliness soon enabled further temperature reductions, only a concurrent and significant reduction in deposition pressure provided the key to defect-free low-temperature epitaxy and manufacturing-friendly batch processing of silicon and silicon–germanium [13,14].

In this chapter, ultrahigh vacuum chemical vapor deposition (UHV/CVD) is compared to other silicon and silicon–germanium epitaxial deposition methods with emphasis on the ultralow temperature regime below 700°C and on 200 to 300 mm substrate size. Substrate surface cleaning and advantages and drawbacks of UHV/CVD are discussed.

7.2 Chemistry

The deposition of Si and SiGe from chemical vapors is based on the thermal decomposition of appropriate precursors (pyrolysis), most commonly silanes (SiH_4, Si_2H_6, Si_3H_8, etc.) or chlorosilanes, and germanes (GeH_4, Ge_2H_6, etc.) or chlorogermanes. Diluted (He, H_2, or N_2 [15]) or undiluted precursors are introduced into a closed system via gas injectors to form a well-controlled gas stream with flow dynamics that are engineered to prevent turbulences and uniformly deliver precursors. The wafer is placed into the gas stream and heated to temperatures at which chemical reactions occur with the transported gas molecules. In general, a reactor may be of batch-type (boat, barrel, platen, etc.) or single-wafer design, while the process of CVD is traditionally divided into pressure regimes, such as atmospheric-pressure CVD (APCVD: 760 Torr), reduced-pressure CVD (RPCVD: $\approx$1 to 100 Torr), low-pressure CVD (LPCVD: $\approx$100 mTorr to 1 Torr), and UHV/CVD (10^{-2} to 10^{-4} Torr). Wafer heating is achieved by lamp arrangements or resistive coils, and samples are placed either on flat susceptors, pins, or in boats. Doping is realized by adding gaseous dopant compounds, such as diborane, phosphine, arsine, etc. to the precursor–carrier-gas mix. The overall reaction is commonly described in steps: (a) decomposition of precursors in the gas phase, (b) mass transport of precursors and by-products from the gas stream to the substrate surface, (c) adsorption at the surface, (d) surface diffusion, (e) dissociation, reaction, and bonding with surface atoms, and (f) desorption of by-products and excess precursors into the gas stream. Depending on the process pressure, substrate temperature, reactor design, and choice of precursor mixture, different steps may emerge as the rate-limiting mechanism. In the remainder of the chapter, the UHV *low-temperature* reaction-limited silane (disilane) deposition characteristics are discussed in detail. In this low-pressure, low-temperature extreme, gas-phase reactions can be neglected. An excellent introduction to UHV/CVD and its applications can be found in Refs. [12,13].

During the thermal decomposition of silanes on single-crystalline silicon surfaces, various reaction pathways are feasible, involving different simple and multistep hydrogen and group dissociations and trapping. A description of the exact reaction kinetics is beyond the scope of this chapter and the reader is referred to published theories such as the transition state theory (TST) [16], Rice–Ramsperger–Kassel–Marcus (RRKM) [17], or quantum-Ramsperger–Kassel–Marcus (QRKM) theory [18]. Competing temperature and pressure-dependent adsorption, decomposition, and desorption rates of precursor and intermediate molecules complicate an accurate prediction of growth rate and morphology. However, various research groups have found that the primary adsorption species upon exposure of Si (1 0 0) to SiH_4 at low temperatures were SiH_3 and Si—H, with small amounts of SiH_2 [19], while Si_2H_6 adsorbs and reacts mainly by breaking the Si—Si bond to form SiH_3 fragments, which readily bond to the surface sites [20,21].

$$2SiH_4(gas) \xrightarrow{\text{low } T} 2SiH_3(ads) + 2H(ads) \text{ for silane} \tag{7.1}$$

$$Si_2H_6(gas) \xrightarrow{\text{low } T} 2SiH_3(ads) \text{ for disilane} \tag{7.2}$$

$$2SiH_3(ads) \xrightarrow{\text{low } T} [\text{H}-\underset{/\,\backslash}{\text{Si}}-\underset{/\,\backslash}{\text{Si}}-\text{H}] + 2H_2(gas) \tag{7.3}$$

In Equations 7.1 to 7.3, “ads” denotes adsorption, while / and \ indicate bonds to the bulk silicon. As lumped into Equation 7.3, the pyrolysis of silane and disilane can be conceptionalized as a series of detachments of atomic hydrogen from the precursor molecule adsorbing on dangling bonds, and H coadsorption, with eventual release of molecular hydrogen. The individual reactions can be traced in vacuum by temperature-programmed desorption (TPD), spectroscopy, RHEED observations, STM, or ellipsometry. It was shown on an atomic scale by real-time STM that, upon heat treatment, the two nearest neighboring atoms of a clean hydrogen-terminated (1 0 0) silicon surface pair up under hydrogen desorption to form monohydride dimers (bracketed term in Equation (7.3)). Dimerization was found to reduce the number of dangling bonds and surface energy [22], giving rise to three-dimensional growth. Additionally, this pairing materialized in a well-defined row configuration [23] with rows terminated where atomic steps occurred. As a consequence of the tetragonal bonding nature of silicon, the subsequent atomic layer has dimer rows oriented perpendicular to the previous set of rows. Epitaxy from silanes on monohydride dimer surfaces at low temperatures requires breaking the dimer bond to supply the necessary number of dangling bonds. STM studies revealed different growth modes in the low-temperature regime depending on the exact substrate temperature and amount of precursor exposure, although in most of these studies, disilane rather than silane was used due to a higher sticking coefficient [21]. The reduction of adatom surface mobility at low temperatures suppressed step-flow growth and caused island nucleation and partial passivation, although multilayer growth was still observed [24]. The findings suggest that very low temperatures (<500°C) not only cause drastically reduced growth rates but also result in increased film disorder and roughness, a concern for *ultralow* temperature epitaxy.

The desorption of hydrogen—the primary reaction by-product—from the surface silicon bonds is thermally activated with energies reported between 2.04 and 2.47 eV [16,25,26], while the precursor bond-dissociation energies of Si—Si, Si—C, Si—H, and Si—Cl bonds were measured ranging between 2.95 eV (Si_4H_{10}) and 4.81 eV ($SiCl_4$) [27]. The adsorption ("sticking") coefficients of silane, disilane, and trisilane molecules under UHV conditions were found to be dissimilar but only weakly temperature-dependent [28]. It is therefore generally accepted that the desorption of hydrogen is the rate-limiting step at sufficiently low temperatures. This yields the well-observed linear relationship between logarithmic growth rate and inverse temperature in a classical Arrhenius plot, such as depicted in Figure 7.1. Epitaxial silicon deposition rates are plotted for RPCVD, LPCVD, and UHV/CVD compiled from published silane data with the remarkable result, that independent of process pressure, precursor dilution, and reactor design, the hydrogen desorption is the apparent *absolute* limitation in growth rate. Care was taken to compare only true crystalline epitaxial growth. As published by Beers and Bloem

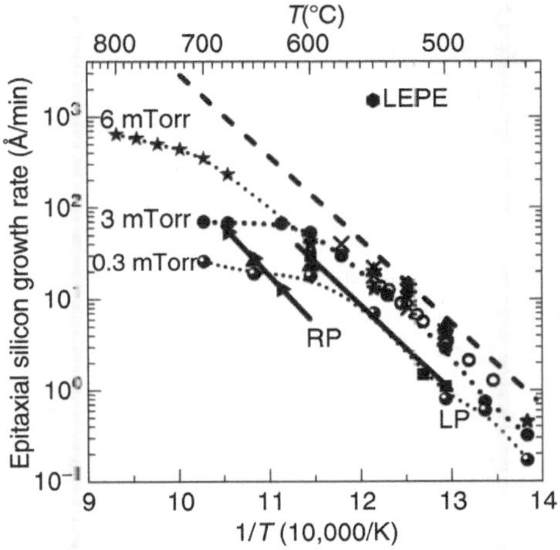

FIGURE 7.1 Arrhenius plot of the epitaxial silicon deposition rate as a function of inverse temperature. Only single-crystalline film growth based on diluted and undiluted silane is compared. Representative values published for RPCVD (✕ [15], ▶ [52]), LPCVD (■—batch-type hot-wall reactor, data provided by A. Little (Kokusai USA, www.ksec.com) and M. Miyamoto (Japan, www.h-kokusai.com)—personal communication), UHV/CVD (●, ● [26], ★ [35], ★ [37], ◆ [40], and ▲ [64]), and low-energy plasma-enhanced (LEPE) CVD (● [31]) are included as closed symbols and dotted lines. Open circles denote values obtained employing a UHV/CVD batch reactor.

[29], silane pyrolysis can be driven into higher deposition rate regimes, but the deposited films were polycrystalline or amorphous. It is understood that this may be due to insufficient surface diffusion of the adsorbed SiH_x species to find energetically favorable bonding positions at high deposition rates. UV assistance [30] and more recently, plasma enhancement [31] were demonstrated to boost growth rates while retaining crystallinity due to enhanced precursor gas-phase dissociation, stimulated surface migration, and promoted hydrogen desorption.

7.3 Low Temperature and Low Pressure

As invented by Meyerson [14], UHV/CVD is performed at pressures of 10^{-2} to 10^{-4} Torr, while maintaining background pressures of less than 10^{-8} Torr. To achieve ultimate background pressures in the UHV range, prevent hydrocarbon backstreaming, and allow unthrottled pressure equilibration in the low-milliTorr range for typical process gas flows (below 200 sccm); the use of turbopumps backed by a stack of rootsblower and mechanical pumps is typically required. If necessary, higher deposition pressures can be accomplished using a throttle valve coupled to a closed-loop pressure control circuit. The use of low process pressures has various favorable consequences. The rate of precursor gas-phase decomposition in the heated reactor volume is significantly reduced, as it is linearly dependent on precursor pressure below 10 Torr [13]. Therefore, thickness nonuniformities originating from spatial variations in silane concentration (and its fragments) are prevented. Undiluted and carrier-gas-free precursors are typically used, which reduces gaseous contamination, such as water vapor, unintentionally introduced by large amounts of carrier gases in other pressure regimes. Of course, contamination is also present in precursors but their partial pressure is well below the values detrimental for epitaxy [32,33]. It is interesting to note that for a given temperature, precursor-to-layer conversion efficiencies in UHV/CVD, i.e., the number of Si atoms deposited (including tube and boat) per number of silane molecules admitted per minute, remain comparable to those of optimized carrier-gas-based regimes, such as RP or LPCVD. Furthermore, the mean-free path of the precursor molecules at 10^{-2} to 10^{-4} Torr is tens of centimeters and compared to higher pressure regimes, gas-flow dynamics are simple and without convectional effects. Combined with a small sticking coefficient of silane and long reactor residence times, this results in a homogeneous precursor delivery to all exposed vacuum surfaces and holds great potential for uniform deposition rates in batch-type reactors. Contamination originating from the reactor's inner periphery is minimized by evacuating the chamber to UHV pressures. Some contamination of the wafer surface from a finite vacuum is unavoidable but can be mitigated by flushing with inert gases during wafer loading and unloading, provided that the reactor leak rate is larger than the rate at which contamination is introduced by the flush gas. The value of necessary background pressure before growth is governed by required minimum partial pressures of O_2 and H_2O at the target growth rate and temperature [32,33]. Typically, chamber pressures in the mid-10^{-9} Torr range were sufficient for a wide range of substrate temperatures as partial pressures of contaminants were then in the 10^{-11} Torr range, measured using a residual gas analyzer. Pressures in the mid-10^{-8} Torr range proved satisfactory for higher temperatures and faster growing SiGe.

The clean reactor environment, in turn, enables an extended range of growth temperatures. Very low growth rates of silicon and silicon–germanium (below 1 Å/min) at ultralow temperatures are thus feasible. Temperatures as low as 450°C were demonstrated to yield clean epitaxial silicon films employing purified silane [26]. However, without substrate rotation and modified gas injection systems, temperatures in excess of 600°C to 650°C can cause substantially degraded thickness uniformities due to gas depletion in the reactor (transport-limited regime). As depicted in Figure 7.2 for different substrate temperatures, a saturation of silane-based growth rate occurs in UHV/CVD reactors at low temperatures and pressures, and at lowest temperatures only slightly increasing growth rates were observed with increasing silane pressures. In contrast, no saturation was observed for comparable partial pressures of silane at much higher total pressures. A strong dependence on flow rate (or pressure) is indicative of gas depletion across the reactor, boat, and wafer and results in degraded within-wafer and

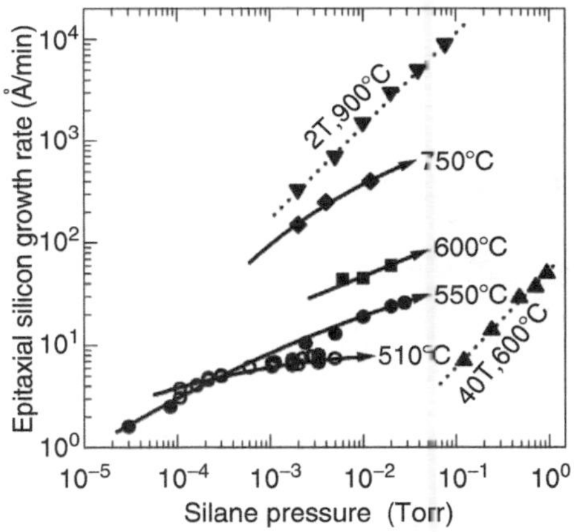

FIGURE 7.2 Epitaxial silicon growth rate as function of (partial) silane pressure in UHV/CVD (solid lines: ■ [36,37], ● [37], and ◆ [65]) and RPCVD reactors (dotted lines: ▲ [15], ▼ [66]). Steep lines indicate a strong dependence on partial pressure and are undesirable for batch-type configurations. Gas depletion over the wafer and boat can spatially modify the precursor partial pressure, and thus, create deposition inhomogeneities. Best uniformities are obtained using lowest temperatures and undiluted precursors.

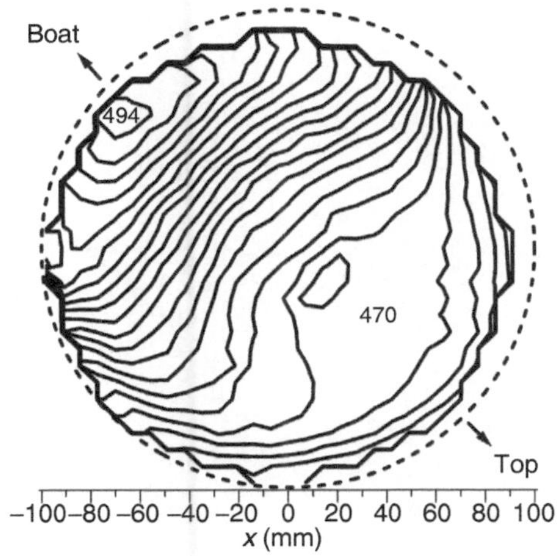

FIGURE 7.3 Silicon thickness contour plot of the center wafer deposited in a low-temperature UHV/CVD batch reactor. The target thickness was 500 Å and local maximum and minimum values are included (in Å). A total temperature variation of 2.88°C across the wafer (2 mm edge exclusion) was inferred from thickness variations obtained using spectroscopic ellipsometry. The orientation of the wafer inside the reactor, i.e., top and bottom (boat), is indicated. The small offset between the wafer and measurement grid was caused by automated stepping with disabled pattern recognition.

within-boat thickness uniformities (transport-limited operation). Operating the reactor in saturation (reaction-limited deposition) ensures excellent uniformities, and one-sigma wafer-to-wafer, within-wafer, run-to-run, and tool-to-tool variations of germanium fractions in SiGe within the range of 3% to 4% and of combined Si and SiGe thicknesses within the range of 1% to 3% are typically achieved in 200 mm UHV/CVD batch reactors. However, intrinsic silicon thickness maps of the center wafer, as displayed in Figure 7.3 revealed radially increasing values that were eccentric relative to the wafer and wafer-boat geometry but concentric relative to the reactor–heater axis, in addition to locally modified thicknesses in the vicinity of the structural support from the boat, totaling ≈6.1% variation. Combined with the measured activation energy and preexponential factor of silane-based silicon epitaxy in the same reactor (≈1.74 eV and 7.5×10^{11} Å/min), a total temperature variation across the center wafer of ≈1.88°C was inferred from these measurements. It is apparent that good thickness uniformities and repeatability in UHV/CVD reactors therefore require superior temperature uniformities and stabilities as they are exponentially connected, while precursor delivery is intrinsically simple and noncritical. In contrast, uniformity in mass-transport-limited silicon epitaxy at higher substrate temperatures (most APCVD, RPCVD, and LPCVD reactors) is strongly dependent on even precursor delivery.

7.4 Alloys, Doping, and Selective Epitaxial Growth

Pseudomorphic alloys of silicon with other group IV elements, such as germanium and carbon, have gained much attention due to bandgap engineering, strain effects, and suppression of dopant diffusion. Epitaxial UHV/CVD of SiGe is typically achieved by adding diluted germane or digermane to the precursor gas flow. However, as depicted in the inset of Figure 7.4, the germanium fraction and SiGe deposition rate are nontrivial functions of flow rate (pressure) and substrate temperature, although the Ge fraction appears less sensitive to temperature changes. For batch reactors, the values of listed wafer

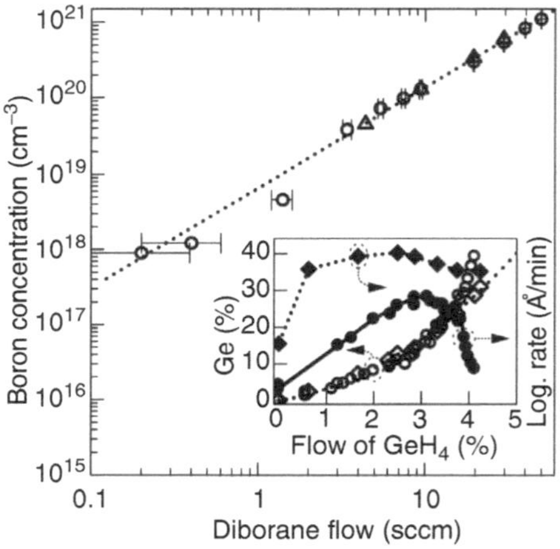

FIGURE 7.4 Boron concentration in films of silicon (o) and SiGe (▲) as a function of B_2H_6 flow (1000 ppm in He) in a UHV/CVD batch reactor, determined by SIMS measurements. The inset depicts the germanium fraction and logarithmic SiGe growth rate in the same batch-reactor as a function of germane flow percentage in $SiH_4 + GeH_4 +$ He for 525°C (dotted lines and diamonds) [40] and 510°C (solid lines and circles). Germanium fractions and layer thicknesses were extracted using XRD, SIMS, Auger, and spectroscopic ellipsometry. Ge fractions between 0% and 25 % appear much less sensitive to temperature variations.

temperature are somewhat arbitrary as large errors may be caused by sensing the vessel temperature using spring-loaded thermocouples. Even if the reactor wall temperature was known with sufficient accuracy, actual wafer temperatures are expected to be several degrees lower. Absolute temperature calibration to growth rates using standard silane flows and pressures should be employed to establish grounds for comparison and repeatability. In single-wafer tools, thermocouples can be placed in direct contact to or within the immediate vicinity of the wafer. Furthermore, eutectic reactions, such as with Au, may be observed through view ports in order to calibrate to the true wafer surface temperature. In general, the addition of GeH_4 increases the growth rate [13,34]. This is believed to occur due to a reduced hydrogen desorption activation energy for Ge—H (≈1.6 eV) as compared to Si—H (≈2.0 eV), and a lower binding energy of Ge—H (≈3.1 eV) compared to Si—H (≈3.4 eV) [16]. Depending on the substrate temperature, the growth rate of SiGe may reach values that are several times larger than that of silicon. As a consequence, care must be taken to avoid gas-depletion effects at low process pressures, i.e., silane flows.

In principle, $Si_{1-x}Ge_x$ with compositions between $x = 0\%$ and 100% are attainable by adjusting the flows of silane and germane. However, for increasing germanium fractions, both the apparent and metastable critical thicknesses reduce accordingly [40]. The use of low temperatures during growth, as enabled by UHV/CVD, is expected to result in better alloy metastability because relaxation (dislocation nucleation and glide) exhibits stronger thermally activated behavior as compared to growth ($E_A \approx 4.2$ to 4.75 eV [35]). Low deposition rates also ensure exceptional profile control, repeatability, and atomic abruptness while at the same time, yielding excellent uniformities in batch-type reactors. Sophisticated germanium, carbon, and dopant profiles, such as polynomial or exponential ramps, can be obtained by dynamic (time-dependent) programming of massflow controllers.

Typical n-type (p-type) dopants for silicon- and silicon–germanium-based devices are phosphorous, arsenic, and antimony (boron and gallium). B_2H_6, PH_3, AsH_3, and also SbH_3 are dopant compounds that are employed in chemical vapor deposition. It is generally accepted that the use of low temperatures (<600°C) results in (nearly) complete dopant activation in as-grown films over a wide range of concentrations, eliminating the need for high-temperature annealing steps. Furthermore, dopant diffusivities in both silicon and silicon dioxide are strongly dependent on temperature, with activation energies (3.4 to 4.2 eV in c-Si, 2.4 to 3.9 eV in poly-Si, and 1.3 to 4.7 eV in SiO_2 [2]) that are generally much larger than that of epitaxial growth (1.5 to 2.1 eV for silane-based growth [13,26,36] and ≈1.5 eV for disilane-based growth [30,37]). Therefore, dopant diffusion is nearly eliminated for growth temperatures below 600°C and typical deposition durations. However, the incorporation of dopants is dependent on various factors, such as temperature, precursor, chamber environment, and of course gas flow. Not all gaseous dopant sources allow abrupt profiles and high concentrations, even at low substrate temperatures. For example, PH_3 and AsH_3 have been shown to be unstable at high concentrations during silicon epitaxy [13,38] and concentration transients after switching the gas flow were detected in RPCVD, LPCVD, and UHV/CVD during silane-based growth. Segregation and surface poisoning were observed for these dopant sources. Phosphine stabilized in SiGe, and also, by using a Cl-containing chemistry that was argued to be caused by an enhanced desorption of P from the surface and formation of volatile phosphorous chlorides at higher temperatures. In addition, when disilane rather than silane was used, the surface saturation and doping transition effects were mitigated [37] possibly due to a higher reactivity and sticking coefficient of Si_2H_6. A small growth rate reduction was observed for phosphorous-doping in silane and disilane.

In contrast, boron incorporation is abrupt and stable at low deposition temperatures [39]. No change of concentration over time was observed during growth, except when boron was incorporated into SiGe. Although no "memory effect" was detected, a steady increase of concentration in SiGe and subsequent decrease in Si were observed by Lafontaine et al. [40]. The authors argued that this may be due to increased boron adsorption in the presence of germane. No noticeable change in growth rate was observed when adding B_2H_6 to silane [39] while disilane-based boron-doping exhibited a considerable enhancement of growth rate [37]. As depicted in Figure 7.4, high boron concentrations and large dynamic doping ranges are achievable in UHV/CVD.

Selective epitaxial growth (SEG) during CVD is achieved by adding an appropriate amount of hydrochloric acid to the precursor mixture, which prevents nucleation on oxide and nitride by etching. This method enables the selective intrinsic and *in situ* doped Si and SiGe epitaxy for arbitrary thicknesses. However, temperatures in excess of 700°C are typically required because the addition of HCl suppresses Si and SiGe growth rates [41]. More recently, low-temperature thickness-limited incubation-time-based SEG has regained attention due to less required total thicknesses and stricter thermal processing budgets in high-performance integration schemes. Kiyota et al. demonstrated SEG at temperatures as low as 660°C using dichlorosilane in LPCVD [42]. Up to 3000 Å of Si–SiGe were selectively deposited employing the incubation period of poly-Si–SiGe on SiO_2. The authors postulated that the chlorine content in SiH_2Cl_2 was still somewhat responsible for nuclei etching. Later, the concept of nucleation delay was extended to UHV/CVD in order to further reduce deposition temperatures. Typically, several hundred angstroms of SEG can be achieved in UHV/CVD using Cl-free silane or disilane precursor mixtures, and temperatures as low as 550°C have been demonstrated thus far for an 850 Å epi-base in self-aligned HBTs [43]. However, this process is not expected to be 100% selective, and small and isolated nuclei may be found on oxide and nitride areas.

7.5 Substrate Cleaning

An oxygen- and carbon-free substrate surface is essential for smooth, single-crystalline, and defect-free epitaxy. As much as one monolayer of residual surface oxide can cause polycrystalline or amorphous rather than single-crystalline film growth. Interface contamination gives rise to increased film surface roughness and various crystal defects. Although the exact mechanism is unclear, it is argued that surface contamination reduces the diffusivity of adsorbed SiH_x species [44], thus resulting in an increased initial three-dimensional growth. Boron surface contamination commonly occurs as a result of wafer handling in air, while oxide forms by reacting with oxygen and H_2O vapor during transport in air and water rinses. Carbon is found as contamination in native oxide and processing liquids, such as hydrofluoric acid or DI-water. "Backstreaming" hydrocarbons originating from hot and turbulent pump oils can cause significant accumulation of carbon on silicon substrates processed in vacuum equipment. Other organic and metal impurities are typically found on the wafer surface after polishing steps, lithography, storage in plastic containers, and handling in air [45].

Nayak et al. [44] have observed with *in situ* RHEED and AFM that islands evolve which grow in size during the initial stages of low-temperature growth on contaminated silicon surfaces. Upon further growth, these three-dimensional structures coalesce, and proceed as a disordered 2×1 reconstruction, but free of contamination. They concluded that cleaner interfaces yield smoother and more ordered film growth. Particularly, the pseudomorphic growth of strained layers (SiGe on Si, Si on relaxed SiGe) is affected by interfacial contamination due to enhanced dislocation nucleation [46] and three-dimensional growth, visible under the microscope as haze, pitting, or islands. In the past, traditional substrate cleaning involved a combination of *ex situ* wet chemical etching and *in situ* high-temperature bake. In a production environment, the *ex situ* wet chemical clean typically employs an RCA-type procedure [47,48]. Briefly, a solvating and oxidizing base mixture removes organic, particle, and some metals contamination. It is followed by an oxidizing acid-based solution in which metal contaminants are converted to soluble complexes. After rinsing, the wafers are then immersed in an HF solution to remove the surface oxide and terminate dangling silicon bonds with atomic hydrogen. This procedure has proven to result in a surface that is stable in air against re-oxidation for several minutes [13], with less than 1% coverage with residual oxygen, carbon, and boron (typical interface doses in the mid-$10^{12}\,cm^{-2}$). The surface can be purified further employing *in situ* cleans. Most commonly, high-temperature bakes in a reducing atmosphere or vacuum are used. Hydrogen bakes have been reported to be effective only at temperatures above 750°C [41,49–52] for durations of 30 sec to several minutes. HCl-based etching produces clean and smooth surfaces employing comparable temperatures [45,53]. Vacuum bakes are typically performed in two steps. A first step at temperatures of 250°C to 275°C desorbs carbon [54], while a subsequent high-temperature bake at 630°C to 1350°C for 30 sec to several minutes removes surface oxide [11,21,24,28,32]. High-temperature *in situ* precleans of this type are based on the reduction of SiO_2 to volatile SiO_x compounds. In vacuum bakes, this reduction is achieved by thermal diffusion of oxygen into the underlying substrate, while hydrogen atmosphere also promotes the reduction of the surface oxide to suboxide and water vapor. Lower substrate cleaning temperatures have been achieved employing controlled exposures to Ga molecular beams, where volatile mixed-oxide compounds were formed, which desorbed at temperatures as low as 800°C. Although clean surfaces were produced [55], applicability is limited to MBE or other ultrahigh vacuum single-wafer deposition methods. Morar et al. have shown that a submonolayer coverage of germanium yielded nonstoichiometric germanium-oxides, which desorbed at temperatures as low as 625°C [56], and proposed a low-temperature CVD batch-compatible oxide removal based on brief germane exposures. Similarly, the addition of small amounts of disilane during thermal annealing accelerates the surface oxide decomposition, and annealing temperatures as low as 780°C were successfully demonstrated [57]. Recently, ultralow temperature *in situ* cleans were developed, such as HF-vapor [58], downstream [59], or plasma-based cleaning. Hydrogen ECR cleaning was demonstrated to be effective in reducing interfacial contaminants at temperatures between room temperature and 600°C [60], while low-energy argon [61] and argon–hydrogen [31] ion bombardment at temperatures below 300°C and 100°C, respectively, yielded defect-free crystalline growth and proved successful in removing molecular contamination. However, higher ion energies, though more effective in removing contamination, resulted in increased surface roughness and argon incorporation [62]. Ma et al. suggested the use of broadband UV radiation in combination with O_2 and Cl_2 to react adsorbed hydrocarbons and metal contamination to carbon dioxide, water vapor, and volatile metal chlorides, at temperatures below 250°C [63].

The choice of a particular substrate cleaning method is dependent on various process parameters, such as allowable thermal budget, type of available reactor, or tolerable interface dose and surface roughness. For example, a production-environment hot-wall batch-type low-temperature UHV/CVD system cannot be thermally cycled due to time constraints and long-term tube integrity concerns. Therefore, high-temperature *in situ* precleans are not feasible, unless arranged in a cluster configuration. However, single-wafer UHV/CVD reactors can be equipped with a variety of cleaning accessories.

7.6 Reactors

As discussed in the previous section, the use of low processing pressures and substrate temperatures during UHV/CVD permits excellent uniformities in Si and SiGe layer thickness and composition, reduces thermal dopant diffusion, and maximizes as-grown dopant electrical activation. Autodoping is minimized so that ultra-shallow and hyperabrupt junctions and alloy profiles can be formed without growth interruption. Among others, applications include high-performance SiGe heterojunction bipolar transistors, tunneling diodes, and quantum wells. To date, a variety of single-wafer and batch-type reactors are available for substrates up to 200 mm in size. They are generally constructed of stainless steel (cold-wall) or quartz-glass (hot-wall) chambers, which are evacuated to UHV pressures ($<10^{-8}$ Torr) by turbopumps and a combination of rootsblower and mechanical pumps. In order to preserve a clean growth environment, the reactors are equipped with UHV-capable load locks and ultrapure precursor, doping, and flush gases are introduced using massflow controllers. Currently, 300 mm tools are predominantly of single-wafer type. Besides cost-related issues, incorrigible temperature gradients across the wafer and excessive chamber sizes may complicate the development of 300 mm reactors to batch-type systems. Single-wafer systems are appropriate for research and development purposes and various aspects are attractive for a manufacturing environment. For example, *in situ* wafer and chamber precleaning, wafer temperature sensing and uniformity by heater zoning, dynamic temperature control, substrate rotation, *in situ* film measurements (ellipsometry, interferometry, or spectroscopy), and advanced gas injection systems promise better repeatability, control, and tool failure rates. However, large-scale manufacturing generally requires wafer throughput values that cannot be matched using single-wafer platforms, unless profit margins allow the purchase and operation of multiple tools.

Acknowledgments

I would like to acknowledge the team of scientists, engineers, analysts, and technicians at IBM for their open and interdepartmental support. In particular, I would like to thank Paul Ronsheim, Andy Turansky, Harry Hovel, and Anita Madan for outstanding analytical work and leading edge film characterization, and Joseph Kocis and David Rockwell for processing support. Furthermore, I am grateful to Toshiaki Yanagisawa, Huajie Chen, Basanth Jagannathan, Jack Chu, Kevin Chan, Ryan Wuthrich, and Wade Hodge for technical assistance and many fruitful discussions on the various details of UHV/CVD. I wish to thank my colleagues in the BiCMOS and SiGe Technology Development department for help and encouragement, and Dave Ahlgren and Ken Stein for managerial support.

References

1. J.S. Dunn, D.C. Ahlgren, D.D. Coolbaugh, N.B. Feilchenfeld, G. Freeman, D.R. Greenberg, R.A. Groves, F.J. Guarín, Y. Hammad, A.J. Joseph, L.D. Lanzerotti, S.A. St.Onge, B.A. Orner, J.-S. Rieh, K.J. Stein, S.H. Voldman, P.-C. Wang, M.J. Zierak, S. Subbanna, D.L. Harame, D.A. Herman Jr., and B.S. Meyerson. *IBM J Res Dev* 47: 101–139, 2003.
2. S.M. Sze. *VLSI Technology*, 2nd ed. New York: McGraw-Hill, 1988.
3. P.M. Garone, V. Venkataraman, and J.C. Sturm. *IEEE Electron Dev Lett* 13: 56–58, 1993.
4. T.G. Jung, C.Y. Chang, C.S. Liu, T.C. Chang, H.C. Lin, W.C. Tsai, G.W. Huang, and L.P. Chen. *J Appl Phys* 76: 4921–4923, 1994.
5. L.-E. Wernersson, S. Kabeer, V. Zela, E. Lind, J. Zhang, W. Seifert, T. Kosel, and A. Seabaugh. *Electron Lett* 40: 83–85, 2004.
6. P.S. Chen, Y.T. Tseng, M.-J. Tsai, and C.W. Liu. Semiconductor Manufacturing Technology Workshop, December 10–11, 2002, pp. 145–148.
7. W. Lu, A. Kuliev, S.J. Koester, X.-W. Wang, J.O. Chu, T.-P. Ma, and I. Adesida. *IEEE Trans Electron Devices* 47: 1645–1652, 2000.

8. Y.M. Lin, S.L. Wu, S.J. Chang, S. Koh, and Y. Shiraki. *IEEE Electron Dev Lett* 24: 69–71, 2003.
9. A. Gupta, S.P. Levitan, L. Selavo, and D.M. Chiarulli. Proceedings of 17th International Conference on VLSI Design, 2004, pp. 957–960.
10. J. Kolodzey, P.R. Berger, B.A. Orner, D. Hits, F. Chen, A. Khan, X. Shao, M.M. Waite, S.I. Shah, C.P. Swann, and K.M. Unruh. *J Cryst Growth* 157: 386–391, 1995.
11. T.R. Bramblett, Q. Lu, T. Karasawa, M.-A. Hasan, S.K. Jo, and J.E. Greene. *J Appl Phys* 76: 1884–1888, 1994.
12. B.S. Meyerson. *Appl Phys Lett* 48: 797–799, 1986.
13. B.S. Meyerson. *Proc IEEE* 80: 1593–1608, 1992.
14. B.S. Meyerson. U.S. Patent 5,906,680, 1999.
15. P. Meunier-Beillard, M. Caymax, K. Van Nieuwenhuysen, G. Doumen, B. Brijs, M. Hopstaken, L. Geenen, and W. Vandervorst. *Appl Surf Sci* 224: 31–35, 2004.
16. M. Hierlemann and C. Werner. *Mater Sci Sem Proc* 3: 31–39, 2000.
17. P.J. Robinson and K.A. Helbrook. *Unimolecular Reactions.* New York: Wiley Interscience, 1972.
18. A.M. Dean. *J Phys Chem* 89: 4600–4608, 1985.
19. S.M. Gates, C.M. Greenlief, S.M. Kulkarni, and H.H. Sawin. *J Vac Sci Technol A* 8: 2965–2969, 1990.
20. S. Hong and M.Y. Chou. *Phys Rev B* 58: R13363–R13366, 1998.
21. P.-H. Wu and D.-S. Lin. *Phys Rev B* 57: 12421–12727, 1998.
22. W.K. Liu, S.M. Mokler, N. Ohtani, C. Roberts, and B.A. Joyce. *Surf Sci* 264: 301–311, 1992.
23. A.R. Laracuente and L.J. Whitman. *Surf Sci* 545: 70–84, 2003.
24. J.J. Boland. *Phys Rev B* 44: 1383–1386, 1991.
25. K. Sinniah, M.G. Sherman, L.B. Lewis, W.H. Weinberg, J.T. Yates, and K.C. Janda. *Phys Rev Lett* 62: 567–570, 1989.
26. C.M. Greenlief and M. Armstrong. *J Vac Sci Technol B* 13: 1810–1815, 1995.
27. R. Walsh. *Acc Chem Res* 14: 246–252, 1981.
28. S.M. Gates. *Surf Sci* 195: 307–329, 1988.
29. A.M. Beers and J. Bloem. *Appl Phys Lett* 41: 153–155, 1982.
30. N. Gonohe, S. Shimizu, K. Tamagawa, T. Hayashi, and H. Yamakawa. *Jpn J Appl Phys* 26: L1189–L1192, 1987.
31. J. Ramm. *III Vs Rev* 15(7): 42–45, 2002.
32. F.W. Smith and G. Ghidini. *J Electrochem Soc* 129: 1300–1306, 1982.
33. G. Ghidini and F.W. Smith. *J Electrochem Soc* 131: 2924–2928, 1984.
34. A.S. Segal, S. Yu. Karpov, A.P. Sid'ko, and Yu. N. Makarov. *J Cryst Growth* 225: 268–273, 2001.
35. D.C. Houghton. *J Appl Phys* 70: 2136–2151, 1991.
36. J.H. Comfort and R. Reif. *J Electrochem Soc* 136: 2386–2398, 1989.
37. C. Li, S. John, E. Quinones, and S. Banerjee. *J Vac Sci Technol A* 14: 170–183, 1996.
38. W.B. De Boer, D. Terpstra, and J.G.M. Van Berkum. *Mater Sci Eng B* 67: 46–52, 1999.
39. B.S. Meyerson, F.K. LeGoues, T.N. Nguyen, and D.L. Harame. *Appl Phys Lett* 50: 113–115, 1987.
40. H. Lafontaine, D.C. Houghton, D. Elliot, N.L. Rowell, J.-M. Baribeau, S. Laframboise, G.I. Sproule, and S.J. Rolfe. *J Vac Sci Technol B* 14: 1675–1681, 1996.
41. S. Bodnar, C. Morin, and J.L. Regolini. *Thin Solid Films* 294: 11–14, 1997.
42. Y. Kiyota, R. Udo, T. Hashimoto, A. Kodama, H. Shimamoto, R. Hayami, E. Ohue, and K. Washio. *IEEE Trans Electron Dev* 49: 739–745, 2002.
43. K. Washio. *IEEE Trans Electron Dev* 50: 656–668, 2003.
44. S. Nayak, D.E. Savage, H.-N. Chu, M.G. Lagally, and T.F. Kuech. *J Cryst Growth* 157: 168–171, 1995.
45. H. Habuka and T. Otsuka. U.S. Patent 6,235,645, 2001.
46. S.R. Stiffler, J.H. Comfort, C.L. Stanis, D.L. Harame, E. de Frésart, and B.S. Meyerson. *J Appl Phys* 70: 1416–1420, 1991.
47. W. Kern and D.A. Puotinen. *RCA Rev* 31: 187–205, 1970.
48. S. Shwartzman. U.S. Patent 4,261,791, 1981.

49. A. Abbadie, J.M. Hartmann, P. Holliger, M.N. Semeria, P. Besson, and P. Gentile. *Appl Surf Sci* 225: 256–266, 2004.
50. M.S. Carrol, J.C. Sturm, and M. Yang. *J Electrochem Soc* 147: 4652–4659, 2000.
51. R. Loo, M. Caymax, I. Peytier, S. Decoutere, N. Collaert, P. Verheyen, W. Vandervorst, and K. De Meyer. *J Electrochem Soc* 150: G638–G647, 2003.
52. M. Caymax, R. Loo, B. Brijs, W. Vandervorst, D.J. Howard, K. Kimura, and K. Nakajima. *Mater Res Soc Symp Proc* 533: 339–344, 1998.
53. J.F. Gibbons, C.M. Gronet, and K.E. Williams. *Appl Phys Lett* 47: 721–723, 1985.
54. S.S. Iyer, M. Arienzo, and E. de Frésart. *Appl Phys Lett* 57: 893–895, 1990.
55. S. Wright and H. Kroemer. *Appl Phys Lett* 36: 210–213, 1980.
56. J.F. Morar, B.S. Meyerson, U.O. Karlson, F.J. Himpsel, F.R. McFreely, D. Rieger, A. Taleb-Ibrahimi, and J.A. Yarnoff. *Appl Phys Lett* 50: 463–465, 1987.
57. K. Nozawa, K. Katayama, Y. Kanzawa, G. Sugahara, T. Saitoh, and M. Kubo. *Mater Res Soc Symp Proc* 569: 83–88, 1999.
58. M. Caymax, S. Decoutere, E. Rohr, W. Vandervorst, M. Heyns, H. Sprey, A. Storm, and J.-W. Maes. *Diffus Def Data pt B* 65–66: 237–240, 1999.
59. H. Ogawa, T. Arai, M. Yanagisawa, T. Ichiki, and Y. Horiike. *Jpn J Appl Phys* 41: 5349–5358, 2002.
60. H.W. Kim, K.S. Kim, and J.H. Lee. *Met Mat Int* 8: 183–190, 2002.
61. T. Ohmi, K. Hashimoto, M. Morita, and T. Shibata. IEDM, 1989, pp. 53–56.
62. J.C. Bean, G.E. Becker, P.M. Petroff, and T.E. Seidel. *J Appl Phys* 48: 907–913, 1977.
63. Y. Ma, M.L. Green, L.C. Feldman, J. Sapjeta, K.J. Hanson, and T.W. Weidman. *J Vac Sci Technol B* 13: 1460–1465, 1995.
64. S.-H. Lim, S. Song, T.-S. Park, and E. Yoon. *J Vac Sci Technol B* 21: 2388–2392, 2003.
65. T.J. Donahue and R. Reif. *J Electrochem Soc* 131: 1691–1697, 1886.
66. J.L. Regolini, D. Bensahel, Y.I. Nissim, J. Mercier, E. Scheid, A. Perio, and E. Andre. *Electron Lett* 24: 408–409, 1988.

8

Defects and Diffusion in SiGe and Strained Si

Anthony R. Peaker and
V.P. Markevich
University of Manchester

8.1 Introduction

Defects play a crucial role in semiconductor processing and in finished devices by mediating diffusion, defining junction leakage, and controlling carrier recombination. As a result, issues as wide ranging as junction depth, noise, and memory retention time are critically dependent on the defect population. By far the most important source of point defects is the ion implantation process, which produces high concentrations of intrinsic defects (self-interstitials and vacancies). The issues regarding diffusion enhancement in silicon by intrinsic defects have been studied in great detail in relation to small devices and although some work exists for the SiGe case, the problem is vastly more complicated because of the interaction of defects with strain, interfaces, and, in some cases, the introduction of dislocations.

In this section of the handbook we consider briefly the principles behind these effects and then review the relevant literature on point defects and defect complexes in silicon, in germanium, and in silicon–germanium. We consider the electronic properties of the defects and in particular their position in the bandgap and as a consequence the likely effects on the electrical behavior of devices. The recent literature on diffusion in SiGe and strained Si is reviewed and the effects of point defects on diffusion are considered along with the impact of strain on diffusivities of the common dopants. Finally we look at the significance of dislocations in relation to device performance.

Much of our understanding of ion implantation defects in silicon and their annealing kinetics is based on studies of electron-irradiated silicon. A high-energy electron will displace an atom in the semiconductor matrix to produce a vacancy and an interstitial as a Frenkel pair. The interstitial is quite mobile and diffuses either to react with a vacancy or substitutional impurities or to add to the surface of the crystal. The remaining vacancies pair with impurities in the immediate vicinity of the displacement or with themselves to form divacancies. The most common reaction products in Si and SiGe are the vacancy–oxygen pair and the divacancy.

However there are very substantial differences between the point defects introduced by electron damage and the defects commonly seen in ion-implanted semiconductors. As the mass and energy of the implanted species increases, a more complex population of defects is created. There is clear evidence that this is due to clusters of interstitials or vacancies depending on the depth below the surface relative to the implant range. Many of the vacancies and interstitials recombine to reconstitute the lattice, others react with each other to form clusters of vacancies or clusters of interstitials. Such clusters result in relatively stable entities, which may dissociate during the annealing process used to activate the implanted dopants. The released point defects can result in substantial changes in the diffusivity of the dopants. By far the best known example is the enhancement of boron diffusivity in silicon by excess of self-interstitials in which the diffusivity is increased by several orders of magnitude.

The generation of intrinsic defects is very nonuniform throughout the ion path, with the majority of defects created near the end of range. Redistribution of the defects at the implant temperature results in a vacancy-rich region near the surface and an interstitial-rich region just beyond the concentration peak of the implanted species. On annealing, particularly under oxidizing conditions where the surface cannot act as a sink for the interstitials, extrinsic stacking faults with their bounding Frank partial dislocation loops can be formed at the end of the implant range. These accommodate the excess of self-interstitials locally as an extra plane of atoms. Some of the defects react with impurities (dopants, carbon, oxygen, and hydrogen) during the annealing process. However the majority of these defects can be removed from the active region of the device provided the device design will withstand a sufficient thermal budget.

In the next section we examine the populations of defect complexes that evolve and their position in the bandgap. This is an issue of crucial importance in relation to device-significant properties. Figure 8.1 shows the conceptual behavior of such defect levels in terms of their behavior as generation, recombination, or trapping centers. Process (A) is the dominant mechanism affecting the *generation* lifetime. It is a very important process in memory applications particularly DRAM but can also be a major contributor to CMOS quiescent current. In Si and SiGe, process (D) is the dominant mechanism in defining the

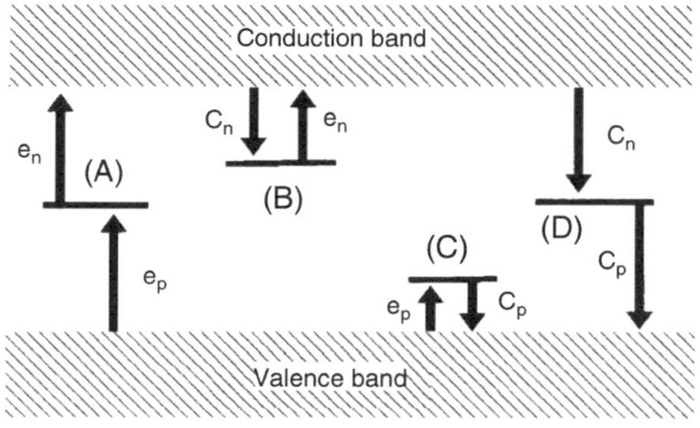

FIGURE 8.1 The classic Shockley–Hall–Read representation of (A) a generation center, (B) an electron trap, (C) a hole trap, and (D) a recombination center.

recombination lifetime; it is important in bipolar devices and can have a defining influence on the base transport factor and hence the gain in some transistor structures.

The trapping processes (B) and (C) are potential sources of noise. In practice the noise frequency is low and in general trapping processes are regarded as contributors to $1/f$ noise and in consequence are of most significance in analog circuits, particularly RF mixers. If such centers are present in very high concentrations, the charge they store can shift the operating point of devices, either slowly with time or after overload conditions.

8.2 Intrinsic Defects and Their Interactions with Impurities in Crystalline Si and Ge

The Lattice Vacancy and Vacancy–Impurity Complexes in Si

The isolated lattice vacancy in Si has been identified by electron paramagnetic resonance (EPR) in both its paramagnetic single positive (V^+) and negative (V^-) charge states after *in situ* electron irradiation at cryogenic temperatures (4.2 to 20.4 K) [1,2]. It was found that the vacancy can take on five different charge states in the silicon band gap (++), (+), (0), (−), and (− −). The information available on the electronic and structural properties of the vacancy is summarized in Table 8.1.

The EPR spectra of the isolated vacancy disappear after a 15-min annealing at ~70 K in n-type Si, ~150 K in p-type, and ~200 K in high-resistivity material [1,2]. The disappearance is argued to be related to the long-range diffusion of the vacancy, and the difference in the annealing temperatures is caused by the different values of the activation energy for the vacancy migration in different charge states. The activation energies for migration were determined as 0.18 ± 0.02 eV for V^{--}, 0.45 ± 0.04 eV for V^0, and 0.32 ± 0.02 eV for V^{++} [1,2].

The disappearance of the EPR spectra of the isolated vacancy is accompanied by the appearance of other EPR signals, which are identified as due to vacancies paired with other defects. The Si vacancy forms pairs with group V donor atoms (P, As, Sb) as well as with oxygen, germanium, tin, and some other impurity atoms [1–5]. The interaction of two vacancies was found to result in the formation of the divacancy. The information available on the electronic properties, symmetry, and thermal stability of some vacancy-related point defects is presented in Table 8.1.

Important issues to be considered in relation to diffusion problems are equilibrium concentrations of intrinsic defects at high temperatures. The equilibrium concentration of a point-defect species X can be expressed as [6]:

$$\frac{C_X}{C_s} = \Theta_X \exp\left(-\frac{\Delta G_X^f}{kT}\right) = \Theta_X \exp\left(\frac{\Delta S_X^f}{k}\right) \exp\left(\frac{-\Delta H_X^f}{kT}\right)$$

where ΔG_X^f, ΔH_X^f, and ΔS_X^f are, respectively, Gibbs free energy, the enthalpy, and the entropy changes related to the formation of one defect X. ΔS_X^f has a configurational contribution and a vibrational contribution that arises from a change in the lattice vibrational frequency spectrum in the vicinity of X:

$$\Delta S_X^f = \Delta S_X^{f,c} + \Delta S_X^{f,v}$$

The term Θ_X is the number of equivalent, distinguishable configurations that X can take (per lattice site); for a given lattice site, at most, one of the Θ_X configurations that are associated with it can be occupied at any time. C_s is the number of available lattice sites in the crystal (i.e., number of substitutional sites, tetrahedral interstitial sites, etc.).

It should be mentioned that there are no experiments where equilibrium concentrations of vacancies or interstitials in Si have been measured definitely. Estimates of the vacancy formation enthalpy based on experimental studies gave the values in the range of 2 to 3.2 eV [4,7], while the values obtained from theoretical calculations are higher, 3.5 to 3.7 eV [8,9].

TABLE 8.1 Charge States in the Gap, Symmetry, Position of Energy Levels, and Temperature Ranges of Annealing Out for the Most Abundant Vacancy-Related Defects in Si

Defect	Charge States	Configuration (Point-Group Symmetry) At Low Temperatures	Energy Levels	Annealing Out Temperature Ranges (°C) (~15–30 min Isochronal)
Vacancy	− −	C_{2v}	$E(- -/-) \leq E_c - 0.17$ eV	(−225) to (−175)
	−	C_{2v}	$E(-/0) = E_c - 0.57$ eV[a] (?)	
	0	D_{2d}	$E(0/+) = E_v + 0.05$ eV	(−75) to (−25)
	+	D_{2d}	$E(+/++) = E_v + 0.13$ eV	
	++	T_d		(−125) to (−75)
Vacancy—oxygen	−	C_{2v}	$E(-/0) = E_c - 0.17$ eV	325–375
	0	C_{2v}		
Vacancy—phosphorus	−	—	$E(-/0) = E_c - 0.43$ eV	100–175
	0	C_{2h}		
Vacancy—antimony	−	—	$E(-/0) = E_c - 0.44$ eV	175–225
	0	C_{2h}		
Vacancy—germanium	− −	—	$E(- -/-) = E_c - 0.29$ eV[b] (?)	(−75) to (−25)
	0	—	—	
	+			
	++			
Divacancy	− −	D_{3d}	$E(- -/-) = E_c - 0.24$ eV	
	−	C_{2h}	$E(-/0) = E_c - 0.43$ eV	
	0	C_{2h}	$E(0/+) = E_v + 0.20$ eV	225–300
	+	C_{2h}		
Vacancy—boron	0	C_{1h}	—	(−25) to 25
	+	—		
Vacancy—aluminum	−	C_{3v}	—	
	0	—		225–275

Source: GD Watkins. *Deep Centers in Semiconductors*, ST Pantelides (ed.). Gordon and Breach, New York, 1986, Chapter 3, pp. 147–183. With permission; GD Watkins. *Properties of Crystalline Silicon* (EMIS Data Review Series No. 20), R Hull (ed.), INSPEC, London, 1999, Chapter 11.1, pp. 643–652. With permission.

[a] From PM Fahey, PB Griffin, and JD Plummer. *Reviews of Modern Physics* 61:289–388, 1989. With permission.

[b] From CV Budtz-Jorgensen, P Kringhoj, A Nylandsted Larsen, and NV Abrosimov. *Physical Review B* 58:1110–1113, 1998. With permission.

Question marks indicate the data that are doubtful.

The Lattice Vacancy and Vacancy–Impurity Complexes in Ge

In contrast to the silicon case, reliable information available on the properties of the lattice vacancy and vacancy–impurity complexes in crystalline Ge is rather limited.

According to recent theoretical calculations [10,11], the structure and electronic properties of the vacancy in Ge are similar to those of the vacancy in Si: the vacancy in Ge is present as five charge states: (++), (+), (0), (−), and (− −). However, the magnitudes and energies of Jahn–Teller distortions were found to be smaller for the vacancy in Ge compared to Si. Comparison with the results for silicon showed that the vacancy formation energies in germanium (from 1.7 to about 2.5 eV) are significantly smaller than in silicon, for all charge states, which makes the vacancy much more important for diffusion in Ge than in Si [10].

Very little information on the positions of energy levels of the isolated vacancy and activation energies for the vacancy migration in Ge is available. However, it was suggested recently on the basis of perturbed angular correlation (PAC) spectroscopy measurements that the (−/0) energy level of the Ge vacancy should be at about $E_v + (0.2 \pm 0.04)$ eV [12]. A PAC signal, which was assigned to a complex of a ^{111}In atom with a vacancy nearby, appeared upon annealing at temperatures around 200 K in Ge samples subjected to electron irradiation at 77 K. It was proposed that this process is related to the long-range migration of neutral vacancies, and the activation energy for the V^0 diffusion was estimated as 0.6 ± 0.1 eV [12]. Regarding the vacancy diffusivity in Ge, there is another study [13], the results of which look inconsistent with the results obtained from PAC measurements. It was found by Whan [13] that in oxygen-rich Ge samples, which were electron-irradiated at 25 K, oxygen-related LVM lines emerged upon annealing in the temperature range 45 to 80 K. These lines were associated with vacancy–oxygen complexes, and the above observation was considered as an evidence for the vacancy diffusion in Ge at low (~50 K) temperatures. It might be possible, however, that the vacancy diffusion occurred in different charge states in PAC [12] and infrared (IR) absorption [13] measurements.

As in the case of silicon, there are some strong indications that the vacancy in Ge can interact with oxygen and group V donor atoms (P, As, Sb, Bi) with the formation of complexes stable at room temperature.

It was shown recently in a combined deep-level transient spectroscopy (DLTS) and IR absorption spectroscopy study that the vacancy–oxygen complex in Ge can exist in three charge states: (− −), (−), and (0), with the corresponding energy levels at about $E(- -/-) = E_c - 0.21$ eV and $E(-/0) = E_v + 0.27$ eV [14,15].

Strong arguments were presented recently that the Sb–V vacancy in Ge introduces two acceptor levels in the gap at about $E(- -/-) = E_c - 0.20$ eV and $E(-/0) = E_v + 0.30$ eV [16,17]. The Sb–V complex in Ge annealed out in the temperature range 140 to 180°C [16,17].

The Silicon Interstitial and Interstitial-Related Interactions in Si

In contrast to the isolated vacancy, the isolated self-interstitial has not been observed directly either in p-type or n-type Si after low-temperature electron irradiation, and the structure and electronic properties of the Si self-interstitial are not known definitely [3]. In p-type Si after low temperature irradiation, EPR signals due to defects which were identified as interstitials trapped at impurities were observed and the same concentration as isolated vacancies. It is argued that the isolated interstitial atoms must be highly mobile in p-type material, even at 4.2 K. There is also strong evidence for low-temperature (<175 K) long-range migration of Si self-interstitials in n-type Si.

Si self-interstitials were found to interact effectively with group III acceptor atoms (B, Al) and substitutional carbon atoms [3]. Information available on electronic properties and thermal stability of the first generation trapped interstitials (Al_i B_i, C_i) is given in Table 8.2. It was found that these defects can, in turn, migrate over long distances and form pairs with other defects. Table 8.2 shows information on the properties of some of these second generation interstitial defects. The values of formation enthalpy of the Si self-interstitial estimated on the basis of some experimental studies are in the range of 3.2–3.8 eV [4,21]. These values are close to those obtained from theoretical calculations, 3.2–3.7 eV [8,9].

The Germanium Interstitial and Interstitial-Related Interactions in Ge

Reliable information available on the properties of the Ge self-interstitial and its interaction with impurities in crystalline germanium is rather limited. It was found in a recent theoretical study that the Ge interstitial could be an amphoteric defect with an acceptor and a donor level in the gap [22]. For all the charge states the ⟨1 1 0⟩-split dumbbell configuration was found to be the energetically favored one. The formation energy for the neutral Ge interstitial (2.3 eV) was found to be significantly lower than the corresponding value for Si (3.2 eV) according to the calculations [22].

It was suggested in a recent study based on PAC measurements that Ge self-interstitial has an energy level at about $E_c - 0.04$ eV and is mobile, either in neutral or in singly positively charged state, at temperatures above 220 K [12]. The activation energy for the diffusion was estimated as 0.6 ± 0.1 eV.

TABLE 8.2 Charge States in the Gap, Position of Energy Levels, and Temperature Ranges of Annealing Out for the Most Abundant Interstitial-Related Defects in Si

Defect	Charge States	Energy Levels	Annealing Out Temperature Ranges (°C) (~15–30 min Isochronal)
Al_i	0	—	—
	+	$E(+/++) = E_v + 0.17$ eV	175–225[a]
	++		—
B_i	−	$E(-/0) = E_c - 0.37$ eV	
	0	$E(0/+) = E_c - 0.13$ eV	
	+		(−25) to 25[a]
C_i	−	$E(-/0) = E_c - 0.10$ eV	
	0	$E(0/+) = E_v + 0.28$ eV	50–100
	+		
B_i–O_i	+	$E(+/++) = E_c - 0.27$ eV[b] (?)	
	++		150–200
C_i–O_i	0	$E(0/+) = E_v + 0.35$ eV	350–400
	+		
C_i–C_s	−	$E(-/0) = E_c - 0.16$ eV	
	0	$E(0/+) = E_v + 0.09$ eV[c]	225–275
	+		

Source: GD Watkins. *Properties of Crystalline Silicon* (EMIS Data Review Series No. 20), R. Hull (ed.), INSPEC, London, 1999, Chapter 11.1, pp. 643–652. With permission.
[a]Annealing out of B_i and Al_i is free-carrier injection dependent, much faster upon injection conditions; see Refs. [3,18].
[b]From PM Money, LI Cheng, M Süli, JD Gerson, and JW Corbett. *Physical Review B* 15: 3836–3843, 1977. With permission.
[c]The C_i–C_s is a bistable defect; see Ref. [20] for the details of its electronic structure.
Question marks indicate the data that are doubtful.

8.3 Effects of SiGe Alloy Composition and Strain on Defect Characteristics

Effect of Alloy Composition on the Formation Energies of Intrinsic Defects in SiGe

The effect of SiGe alloy composition on the formation energy of vacancies has been studied by *ab initio* calculations in Refs. [23–25]. There was a consensus that the formation energy of a vacancy in $Si_{1-x}Ge_x$ alloys drops by about 0.2 to 0.25 eV for each Ge atom surrounding the vacancy. The energy of the system with a vacancy was found to drop by 0.04 eV for every second nearest-neighbor Ge atom to the vacancy [25]. There are no reliable data on the effect of SiGe alloy composition on the self-interstitial formation energy.

Effect of Alloy Composition on Electronic Properties of Defects in SiGe

Electronic properties of defects in SiGe alloys have been considered recently in Ref. [26]. It was argued that there are three types of behavior a defect may exhibit in the $Si_{1-x}Ge_x$ alloys. First, the core of the defect might be occupied with Ge atoms and consequently the properties of the center are quite different from those of the defect in pure Si. Second, the change in the lattice parameter can cause the vibrational modes or energy levels to shift often in a linear way for dilute concentrations of Ge, and third, the presence of Ge atoms affect the conduction and valence bands in a way beyond the change in bond length and this has an influence on the lattice parameter and other properties. It is often difficult to distinguish these three effects in the alloy experimentally, but the recently developed technique of Laplace deep-level transient spectroscopy (LDLTS) [27] has enabled them to be separated in some cases and so used to study the local environment of defects [26,28]. As an example, Figure 8.2 shows results of a study of the siting of Au in SiGe. The populations could be quantified and so it was possible to state that gold has a strong preference to site in the vicinity of germanium [28]. Each of the three above mentioned effects might cause changes in diffusivity of defects and binding energies of complexes.

Effect of Strain on Defect Characteristics

The application of uniaxial or biaxial stress to a semiconductor changes the equilibrium population of interstitials and vacancies. These effects are significant enough to change diffusion coefficients [29] as will be discussed later. Stress modifies other important parameters, such as the formation of interstitial dislocation loops after implantation and the behavior of implantation-induced point defects.

The effect of strain on point defects is to lift the degeneracy so that the electronic behavior is different in different directions. This can be observed as a splitting of the properties in techniques with sufficient resolution; classically this has been done in optical absorption, electron spin resonance, and photoluminescence. Unfortunately, these techniques cannot be used on thin layers, but the application of LDLTS [27] allows the direct measurements of electronic properties and the observation of defect motion in thin epitaxial layers under uniaxial stress.

An example is given in Figure 8.3 for the divacancy [30]. In this experiment, compressive stress is applied to the sample during measurements and the splitting can be observed as a function of stress. Combined with heating, the same technique can be used to study the motion of impurities in strained silicon. In this case, the impurity is aligned at a temperature above that used for the measurement and then cooled to the measurement temperature. The motion can be observed as a function of annealing and activation energies for the process extracted in the usual way.

This motion, on an atomic scale, is at the heart of diffusion mechanisms and these novel techniques for observing and quantifying it have the potential to give us a much deeper understanding of the mechanisms involved than the traditional top–down approach and these mechanisms are normally the basis of diffusion studies in more conventional systems.

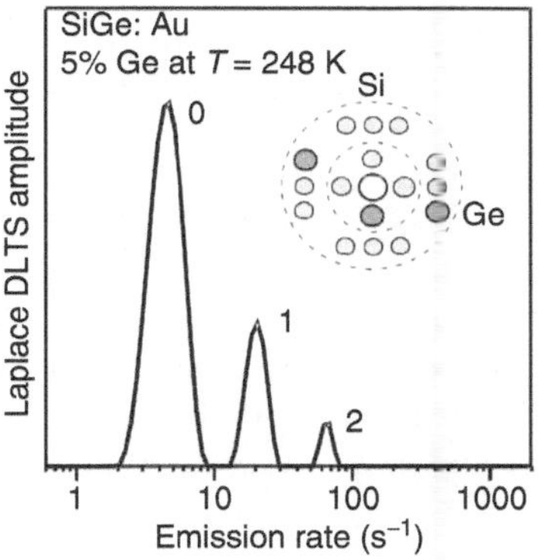

FIGURE 8.2 Electron emission from gold in $Si_{0.95}Ge_{0.05}$ with a flat diagram representation of first and second nearest neighboring atoms. The labelling on the peaks represents emission from gold with 0, 1, or 2 Ge atoms in the first nearest-neighbor position. (Adapted from L Dobaczewski et al. *Physical Review Letters* 83: 4582–4585, 1999. With permission.)

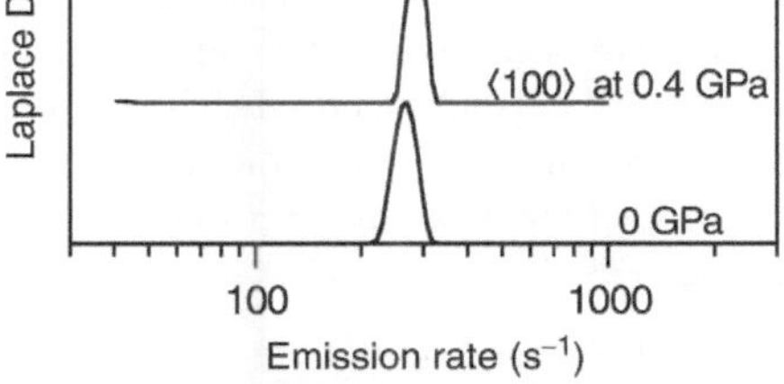

FIGURE 8.3 The effect of uniaxial stress on the dominant ion implantation defect in silicon. The stress lifts degeneracy and shifts in the electron binding energy are observed as line splitting. (Adapted from L Dobaczewski et al. *Physical Review B* 65: 113203-1 to 113203-4, 2002. With permission.)

8.4 Specific Defects in SiGe

Shallow Dopants

Phosphorus donors have been studied in Si-rich $Si_{1-x}Ge_x$ alloys ($0 \leq x \leq 0.27$) by means of IR absorption [31]. A weak shift of the activation energy with germanium content was found: $E(A_1) = (45.5-26x)$ meV. This was explained by multivalley effective mass theory with a phenomenological central cell potential, which does not depend on alloy composition within the investigated range. The experiments showed that P atoms are distributed at random in the alloy and that their positions are uncorrelated to those of the Ge atoms.

Boron–germanium pairs were detected in Si-rich $Si_{1-x}Ge_x$ alloys ($0 \leq x \leq 0.075$) by means of β-nuclear magnetic resonance (NMR) technique [32]. It was found that the binding energy of the B–Ge pair is low and the probability of the formation of the complex matches with statistical expectations.

Radiation (Implantation)-Induced Defects

Vacancy-Related defects

The effect of SiGe alloy composition on the electronic properties and thermal stability of vacancy–donor (P, Sb) atom pairs was studied by means of DLTS in strain-relaxed $Si_{1-x}Ge_x$ samples ($0 \leq x \leq 0.25$), which were grown by the molecular beam epitaxy (MBE) technique [33]. It was found that the activation enthalpy of electron ionization from the single acceptor level of the Sb–V pair increases from 0.41 to 0.51 eV with the increase in Ge content [33]. Recent positron annihilation spectroscopy study showed that Ge atoms are effective sinks for mobile phosphorus–vacancy pairs in strained Si-rich SiGe samples [34]. The V–P–Ge complex was found to anneal out at about 200°C. The 50°C higher annealing temperature of the V–P–Ge complex corresponds to about 0.1 to 0.2 eV larger binding energy than that of the V–P pair [34].

Electronic properties of vacancy–oxygen complexes in SiGe alloys were investigated recently in a combined DLTS and *ab initio* modeling study [26]. Bulk unstrained Czochralski-grown SiGe samples irradiated with electrons were studied. At least three configurations of the VO centers were identified in SiGe alloys. The most stable configuration consists of a Si–O–Si unit and a Ge–Si reconstructed bond in a vacancy. This configuration was found to be about 0.2 eV more stable than separated VO and Ge_s defects and possesses an acceptor level, which is about 25 meV deeper compared to the level of the VO center without a Ge atom in a nearest-neighbor site. The increase of the ionization enthalpy of the VO complexes with the increase in Ge content in $Si_{1-x}Ge_x$ crystals (the rate of the increase was determined to be about $0.55x$ eV) was associated with changes in the average Si–Si bond length [26].

The energy levels of the divacancy have been studied in epitaxially grown, strain-relaxed $Si_{1-x}Ge_x$ as a function of x for $0 \leq x \leq 0.5$, using DLTS on particle-irradiated p^+n and n^+p diodes [35]. The two acceptor levels situated in the upper half of the bandgap in Si (see Table 8.1) were found to move gradually deeper into the gap with increasing x and the single-acceptor level crosses the mid-gap for $x \approx 0.25$, from then on the change of energy changes slope with x. The donor level in the lower half of the bandgap becomes gradually more shallow, pinned to the conduction band. The annealing temperature of the divacancy was found to be independent of composition in the investigated composition range [35].

Interstitial-Related defects

DLTS has been used to study electronic properties and annealing behavior of C_i and C_iC_S defects, created by proton or electron irradiation in strain-relaxed, epitaxial $Si_{1-x}Ge_x$ of $0.005 \leq x \leq 0.5$ [36,37]. It was shown that neither the acceptor levels of C_i and C_iC_S nor the donor levels of the C_i center are pinned to any of the band edges when the composition of the $Si_{1-x}Ge_x$ was varied. The electron ionization enthalpies of the acceptor levels were found to increase with a rate $\delta\Delta H/\delta x \approx 0.3$ eV/x. The two levels of the C_i defect remained at an energy of 0.8 eV separation from each other, independent of composition. The migration enthalpy of C_i was found to be independent of composition ($\Delta H_m \approx 0.75$ eV) in the investigated composition range, $0 \leq x \leq 0.15$. The experimentally observed increased stability of C_i with increasing x was argued to be due to a dramatic decrease of the entropy of migration of the C_i defect. It was argued that the entropy is a crucial parameter when discussing dynamic effects in disordered materials [37].

Electronic properties of the C_i–O_i complex were investigated recently in bulk unstrained Czochralski-grown $Si_{1-x}Ge_x$ samples irradiated with electrons. The enthalpy of hole ionization for the single donor level of the complex relative to the valence band edge, ΔH_p, was found to decrease with a rate $d(\Delta H_p)/dx = -0.96$ eV upon increase in Ge content [38]. The level is not pinned to the conduction band edge or to the valence band edge.

Metallic Impurities

DLTS and high-resolution LDLTS techniques have been also used to study effects of SiGe alloy composition and strain on the electronic properties of 3d transition metal (Fe, Ti, Cr, Ni) and 5d noble metal (Au, Pt) impurities [39–43]. It was found that, as a rule, in strain-relaxed $Si_{1-x}Ge_x$ alloys metal-related energy levels in the lower part of the bandgap move towards the valence band edge with increasing Ge content [39–43]. For the metal-related energy levels in the upper part of the gap both the increase and decrease in activation energies for ionization relative to the conduction band edge were observed upon the increase in Ge content [39–42].

Activation energies of ionization of both electron and hole traps related to metallic impurities were found to increase with the increase of strain in SiGe alloys [41]. The strain dependence was described in terms of the deep state uniaxial and hydrostatic deformation potentials. The difference between the deep state activation energies in Si and $Si_{1-x}Ge_x$ did not agree with the independently determined Si–$Si_{1-x}Ge_x$ band offsets [41,42]. This demonstrated that the concept of 3d state as an independent, internal reference level, well established for III–V and II–VI compounds, fails in the case of group IV semiconductors [41,42].

8.5 Diffusion

The General Picture

Diffusion of the common dopants (boron, phosphorous, arsenic, and antimony) in silicon–germanium alloys is inherently more complex than in the elemental semiconductors. In general, diffusion proceeds via the interaction of the dopant species with point defects. In this case these are predominantly the silicon and germanium interstitials and vacancies. An example of the significance of the importance of intrinsic defects is the diffusion of boron in silicon. There have been numerous studies of the release of interstitials from clusters of silicon atoms formed during ion implantation [44,45]. The dissociation of these clusters during annealing is the basis of enhanced diffusion with diffusivity increases of several orders of magnitude. This has proved to be an immense problem in producing silicon devices with small feature sizes. In contrast species such as antimony, in which diffusion is mediated by vacancies, exhibit a reduced diffusivity in the presence of excess of interstitials. In the alloy the presence of two species results in parallel mechanisms for diffusion. In the general case a classical Arrhenius dependence of the diffusivity D on the temperature T is observed and so we can express the vacancy and interstitial contributions separately:

$$D = D_0^{\mathrm{I}} \exp\left[\frac{-E_a^{\mathrm{I}}}{kT}\right] + D_0^{\mathrm{V}} \exp\left[\frac{-E_a^{\mathrm{V}}}{kT}\right]$$

The activation energies are each derived from two components namely the energy required to form the intrinsic defect and the migration energy. In principle both of these components might be expected to depend on composition of the alloy. For example, the energy required to form a vacancy will depend on the local environment, i.e., how many germanium atoms occupy the four first nearest-neighbor sites. According to the dominant mechanism the diffusion of a particular species is referred to as being interstitially mediated or vacancy mediated.

However, another major difference is the presence of strain, which, among other things, alters the equilibrium concentration of intrinsic defects. It is now generally accepted that the formation energy of the self-interstitial increases with increasing compressive strain and decreases with increasing tensile strain. Strain is present in most SiGe systems of technological interest and in general the strained layers are rather thin bounded by heterojunctions or surfaces. The surfaces and interfaces perturb the vacancy and interstitial populations. In the case of the surface the presence of oxide introduces additional interstitials while a nitrided surface will introduce excess vacancies. This is well documented for the case of silicon but has not been studied in the same detail for germanium or germanium-rich alloys.

In consequence it is very important to deconvolute the effects of alloy composition from that of strain and the effects of interface proximity. Attempts have been made to do this by modeling [46]. Experimentally the simultaneous presence of strain and alloy effects makes quantifying diffusivity in SiGe extremely difficult with many experimental pitfalls. In consequence there are considerable differences in the literature in absolute values. However, it is possible to measure diffusivity in unstrained SiGe although it is important that the unstrained state is not achieved through relaxation via dislocation formation. This is because the presence of these extended defects at diffusion temperatures modulates the intrinsic defect population and introduces localized strain. In this review, we have focussed on studies of single-crystal material with low dislocation density, ignoring some of the older work on polycrystalline or highly dislocated material.

For dilute alloys of $Si_{1-x}Ge_x$ ($x > 0.9$ or $x < 0.1$) bulk Czochralski-grown material can be grown [47] but for $0.1 \leq x \leq 0.9$ virtual substrate layers must be used [48]. Separation of the effects on diffusivity of strain from other parameters must be achieved if SiGe process models are to have general applicability. A further complication is self-diffusion. Interdiffusion is enhanced by strain and in thin layers it may be the dominant mechanism for strain relaxation rather than dislocation formation. Such diffusion tends to reduce the germanium concentration in the germanium-rich region of the heterostructure and vice versa, so introducing uncertainty into both the strain and composition in such experiments. However, self-diffusion also gives us insight into the mechanisms of diffusion in SiGe.

Self-Diffusion in Unstrained $Si_{1-x}Ge_x$

It is generally accepted that the diffusion of most species in Ge is mediated by vacancies and this is true of self-diffusion. However in silicon Ge diffusion depends on both vacancies and interstitials. In the alloy the interstitially mediated diffusion of Ge in $Si_{1-x}Ge_x$ is significant in the range $0 < x < 0.25$ and although there are far fewer data this is probably true of Si diffusion as well. Most recent self-diffusion experiments have been carried out by the introduction of an isotopically enriched region either during growth or by implantation. Considerable care needs to be taken when using implantation as this will perturb the equilibrium concentration of intrinsic defects.

The behavior of Ge diffusion in relaxed material is seen clearly in Figure 8.4 taken from Laitinen et al. [49]. The diffusivity D at a specific temperature is given from the data by

$$D = D_0 \exp\left[\frac{-E_a}{kT}\right]$$

In the diagram the activation energy E_a is presented as an enthalpy H. The diagram includes some data from Strohm et al. [50] and Hettich et al. [51]. Strohm's and Zangenberg's [52] data are quantified in Table 8.3 which also includes a result ($x = 0.8$) from Laitinen. The change of behavior at $x \approx 0.25$ is very pronounced and reflects the transition from vacancy-mediated diffusion in the germanium-rich

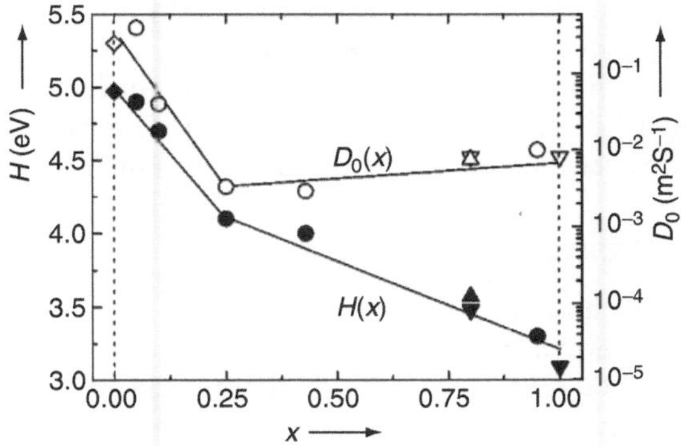

FIGURE 8.4 Self-diffusion enthalpies H (filled symbols) and pre-exponential factors D_0 (open symbols) of diffusion coefficients in SiGe as functions of the Ge content x (circles: ^{71}Ge in SiGe epilayers [50]; diamonds: ^{71}Ge in bulk Si [51]; downtriangles: ^{71}Ge in $Si_{0.20}Ge_{0.80}$ epilayers and in bulk Ge; uptriangles: ^{31}Si in $Si_{0.20}Ge_{0.80}$ epilayers). (Reprinted with permission from Laitinen et al. *Physical Review Letters* 89, 085902 (2002). Copyright 2002 by the American Physical Society.)

TABLE 8.3 Activation Energies (E_a) and Pre-Exponential Factors (D_0) for Self-Diffusion of Ge in $Si_{1-x}Ge_x$ Alloys of Composition x

$Si_{1-x}Ge_x$ Composition x	VPE SiGe[a,b] Implanted ^{71}Ge		MBE SiGe[c] Grown in ^{72}Ge Spike	
	E_a (eV)	D_0 (cm^2/sec)	E_a (eV)	D_0 (cm^2/sec)
0 (Si)	4.97	2500	4.65	310
0.05	4.9	4000		
0.10	4.7	400	4.66	870
0.2			4.00	6.6
0.25	4.1	33		
0.3			3.82	4.7
0.4			3.72	4.2
0.43	4.0	29		
0.50			3.23	0.11
0.8	3.48	81		
0.95	3.3	100		
1 (Ge)	3.09	13.6		

[a]From A Strohm, T Voss, W Frank, J Räisänen, and M Dietrich. *Physica B* 308–310: 542–545, 2001. With permission.
[b]From P Laitinen, A Strohm, J Huikari, A Nieminen, T Voss, C Grodon, I Riihimäki, M Kummer, J Äystö, P Dendooven, J. Räisänen, W Frank, and ISOLDE Collaboration. *Physical Review Letters* 89: 085902-1 to 085902-4, 2002. With permission.
[c]From N Zangenberg, J Lundsgaard Hansen, J Fage-Pedersen, and A Nylandsted Larsen. *Physical Review Letters* 87:125901, 2001. With permission.

material to interstitially dominated mediation in pure silicon. A more detailed study by Strohm et al. [53] reports a very similar trend but puts the transition from interstitial dominance to vacancy dominance at $x \approx 0.35$ and details a comprehensive analysis of the relative contributions. However, in terms of absolute values, there is a clear systematic behavior reflected in all the published data showing increased self-diffusivity of Ge with increasing Ge content. The data on self-diffusion of Si in SiGe are very limited but values very similar to those of Ge have been reported in relaxed $Si_{0.2}Ge_{0.8}$ and silicon [49,53].

The enthalpy values are the sum of the energy of formation of the vacancy and the energy required for its migration. Venezuela et al. have calculated the formation energies of vacancies in $Si_{0.5}Ge_{0.5}$ using first-principle methods [24]. They find a strong dependence of the formation energy on the configuration of the first nearest-neighbor atoms. This varies linearly from 2 to 3 eV as the number of nearest-neighbor Ge atoms varies from 4 to 0. Ramanarayanan et al. Have undertaken *ab initio* calculations of the vacancy migration energy [25]. However as yet these calculations are quite a long way from providing a comprehensive explanation of the experimental observations in relaxed material.

Dopant Diffusion in Unstrained SiGe

Table 8.4 provides data on the diffusion mechanism of a number of species in silicon and germanium by indicating whether diffusion is mediated by interstitials or vacancies. This is important as it enables us to decide whether diffusivity will be enhanced or retarded under implant conditions and to predict the effects of strain [54]. We will review the published data on the common dopant species.

Arsenic

A systematic study of arsenic diffusion in relaxed SiGe across the entire composition range has been undertaken by Laitinen et al. [55] as detailed in Table 8.5. Radioactive isotopes (^{73}As and ^{72}As) were introduced by implantation into 1 μm-thick relaxed $Si_{1-x}Ge_x$ epilayers grown by low-energy plasma enhanced chemical vapor deposition (LEPECVD) on graded buffer layers. The density of the threading dislocations was estimated to be about 10^6 cm^{-2}. As stated previously self-diffusion in $Si_{1-x}Ge_x$ for $x > 0.35$ is vacancy dominated. Although as shown in Table 8.4 diffusion of As in Si is mediated predominantly by interstitials in germanium-rich alloys it is considered to proceed via vacancies. When the arsenic activation energies are compared with Ge self-diffusion values for SiGe, their dependence on the SiGe composition is very similar and it seems likely that the decrease in activation energy as a function of x results from the decrease in vacancy formation energy. The activation energy values for arsenic diffusion are ~0.5 eV lower than for self-diffusion. It is postulated that this may be due to the arsenic migrating via a vacancy–arsenic complex [55].

TABLE 8.4 Role of Vacancies and Interstitials in Promoting Diffusion in Silicon and Germanium

Migrating Species	Silicon Matrix	Germanium Matrix
Boron	Interstitial	Vacancy?
Phosphorous	Interstitial	Vacancy
Arsenic	Interstitial	Vacancy
Antimony	Vacancy	Vacancy
Silicon	Mixed	Vacancy
Germanium	Mixed	Vacancy

TABLE 8.5 Activation Energies (E_a) and Pre-Exponential Factors (D_0) of the As Diffusion Coefficients as a Function of $Si_{1-x}Ge_x$ Composition x

$Si_{1-x}Ge_x$ Composition x	Temperature Range (°C)	E_a (eV)	D_0 (cm^2/sec)	$D_{850°C}$ (cm^2/sec)
0	882–1120	3.81	4.3	*2.2×10^{-19}*
0.20	882–1120	3.83	30	*1.9×10^{-16}*
0.35	812–1050	3.68	23	7×10^{-16}
0.50	750–1020	3.47	18	4.8×10^{-15}
0.65	675–870	3.16	16	4.3×10^{-15}
0.80	696–925	2.97	11	5.2×10^{-13}
1.00	490–600	2.42	5.8	*8×10^{-11}*

Source: P Laitinen, I Riihimäki, J Räisänen, and ISOLDE Collaboration. *Physical Review B* 68: 155209-1 to 155209-6, 2003. With permission.

Note: Values of diffusion coefficient at 850°C are calculated from E_a and D_0 (extrapolated data are shown in italics).

Eguchi et al. [56] have separated the transient diffusion effects following implantation from the equilibrium behavior in relaxed SiGe. They find, as above, that increased germanium content increases the equilibrium diffusivity but observe a decrease in transient diffusion effects with increasing germanium content.

Antimony

Antimony is the only dopant which diffuses in both Si and Ge via a mechanism that is mediated predominantly by vacancies. In silicon the diffusion mechanism is via the vacancy Sb pair (the E center) discussed previously in Section 8.2. Larsen and Kringhøj [57] have studied the diffusion of antimony in relaxed $Si_{1-x}Ge_x$ layers grown by molecular beam epitaxy for $0 < x < 0.5$. The diffusivity of antimony was found to increase with the Ge alloy content. The activation energies for diffusion were found to decrease with increasing germanium content. A summary of their results is given in Table 8.6. Antimony exhibits a particularly strong dependence of diffusivity on concentration. In germanium Boltaks [58] reports a change of E_a from 1.69 to 2.13 eV with an increase in Sb concentration from 10^{15} to $10^{18}\,cm^{-3}$ and a corresponding increase in D_0 from 2.7×10^{-3} to $0.35\,cm^2/sec$. In Table 8.6 Chui's [59] value for $[Sb_{max}] = 6\times10^{19}\,cm^{-3}$ is included for pure Ge which enables comparison with Larsen et al., who have used quite high concentrations of $[Sb_{max}]$ in their experiments ($10^{19}\,cm^{-3}$).

These data represent much lower diffusivities than have been reported by Dan et al. [60] who have used point-defect injection from surface oxidation or nitridation to introduce excess of interstitials or vacancies, respectively. This has confirmed that Sb diffusion is vacancy-mediated in Si and $Si_{0.9}Ge_{0.1}$. However, lower Sb concentrations were used $[Sb_{max}] = 3\times10^{18}\,cm^{-3}$ and the SiGe layer used by Dan was compressively strained, which would reduce the formation energy of the vacancy as will be discussed in Section 8.5. The combination of these two effects probably accounts for the differences in the reported diffusivities.

Phosphorus

Table 8.7 presents the results of Zangenberg et al. [61] and data on pure germanium [59,62]. Here the activation energy of the diffusivity in Si is rather low compared to the generally accepted values. The authors attribute this to non-Arrhenius behavior as their measurements have been made at relatively low temperatures. However, the diffusion coefficient (as distinct from the individual values of E_a and D_o) calculated from the data agrees with the literature below 900°C. The conclusion is similar to that of Eguchi et al. [63], i.e., that the presence of Ge enhances the diffusion coefficient of phosphorus.

Boron

The case of boron is quite complicated. As already discussed the diffusivity of boron is very sensitive to the presence of excess of interstitials and so issues of ion implantation damage and surface generation of interstitials (e.g., by annealing in an oxidizing atmosphere) create variables which confuse both researchers and process engineers. However, there are other variables. The boron interstitial exists in both neutral and positive charge state. This is believed to have an effect on the diffusivity at low temperatures where the intrinsic carrier concentration is less than the extrinsic hole concentration

TABLE 8.6 Activation Energies (E_a) and Pre-Exponential Factors (D_0) of the Sb Diffusion Coefficients as a Function of $Si_{1-x}Ge_x$ Composition x

$Si_{1-x}Ge_x$ Composition x	E_a (eV)	D_0 (cm^2/sec)	$D_{850°C}$ (cm^2/sec)
0 (Si)	4.06	0.2	1.2×10^{-19}
0.1	4.07	0.4	2.2×10^{-19}
0.2	4.07	1.3	7.1×10^{-19}
0.3	3.89	0.8	2.8×10^{-18}
0.5	3.63	4.2	2.1×10^{-16}
1 (Ge)	2.28	0.12	7.1×10^{-12}

Source: Ref. [57] with the exception of the Ge data, which are from Ref. [59].

TABLE 8.7 Activation Energies (E_a) and Pre-Exponential Factors (D_0) of the Phosphorus Diffusion Coefficients as a Function of $Si_{1-x}Ge_x$ Composition x

$Si_{1-x}Ge_x$ Composition x	Temperature Range (°C)	D_0 (cm^2/sec)	E_a (eV)	$D_{850°C}$ (cm^2/sec)
0	825–900	2.0×10^{-3}	2.80	5.5×10^{-16}
0.07	825–900	1.8×10^{-1}	3.24	5.2×10^{-16}
0.12	800–900	1.1×10^{-1}	3.11	1.1×10^{-15}
0.24	800–900	1.7×10^{3}	4.01	1.7×10^{-15}
0.4	800–900	1.7×10^{2}	3.83	1.1×10^{-15}
1 (Ref. [63])	560–850	2.5	2.48	1.8×10^{-11}
1 (Ref. [60])	650 + model	4.4×10^{-2}	2.07	*2.2×10^{-11}*

Source: Ref. [61] with the exception of the Ge data, which are from Refs. [59,62].
Note: Values of diffusion coefficient at 850°C are calculated from E_a and D_0 (an extrapolated figure is shown in italics).

($n_i < p$). The small size of the boron atom compared to silicon and germanium introduces lattice relaxation effects, which have an impact on transit paths in SiGe and favor the formation of complexes. Cowern et al. [64] have proposed a mechanism by which oxidation-enhanced diffusion occurs via the formation of boron–interstitial complexes.

There is a substantial literature on boron diffusion in Si and SiGe but unfortunately there are significant discrepancies in values of activation energies and prefactors. However when this is translated into diffusion coefficients around the temperatures at which the measurements were undertaken the discrepancies are much less pronounced. It is not really clear, as a generality, whether this is due to fitting Arrhenius behavior over a limited temperature range or more fundamental issues associated with the entropy term as exemplified by the Meyer–Neldel rule [65].

Zangenberg et al. [61] have measured diffusion in $Si_{1-x}Ge_x$ for $0 < x < 0.4$ in relaxed MBE material at temperatures in the range 800°C to 900°C. These results are shown in Table 8.8 together with a diffusivity result for pure Ge from Uppal et al. [66]. Perhaps the most noteable result is that the diffusion coefficient shows almost no change across the entire composition range of SiGe. Uppal's result has led to speculation that the longstanding assumption that boron diffusion in Ge is vacancy mediated may not be correct. However Uppal et al. used experimental conditions that injected large concentrations of interstitials, which would be expected to suppress vacancy mediated diffusion. This discrepancy may result from conditions that were very different from previous experiments, which gave much higher diffusivities [62]. Recent first-principle calculations [67] appear to support Uppal's interpretation but further experimental work is needed to resolve this important issue particularly in the Ge-rich SiGe alloys.

The Impact of Strain and Interfaces on Diffusivity

The application of hydrostatic pressure or stress (uniaxial or biaxial) to a semiconductor changes the equilibrium population of interstitials and vacancies. Free energy calculations for the hydrostatic pressure case indicate that the formation energy of the self-interstitial increases with increasing compressive strain and decreases with increasing tensile strain. The formation energy of the vacancy decreases with increasing compressive strain and is almost independent of tensile strain [68]. Aziz [53] and Zhao et al. [69] have considered the hydrostatic case theoretically and experimentally and the relationship between hydrostatic, biaxial, and uniaxial stress in relation to point-defect generation. In the strained Si and SiGe cases, it is obviously biaxial stress that is technologically important and is significant enough to change self-diffusion coefficients and the diffusivity of common impurities under normal process conditions.

Zangenberg et al. [52] have measured the effect of strain on self-diffusion (Ge in $Si_{0.9}Ge_{0.1}$). Their work showed a decrease in diffusion coefficient and an increase in activation energy on going from compressive to relaxed to tensile strain. A compressive strain of 0.21% (corresponding to an

TABLE 8.8 Activation Energies (E_a) and Pre-Exponential Factors (D_0) of the Boron Diffusion Coefficients as a Function of $Si_{1-x}Ge_x$ Composition x

$Si_{1-x}Ge_x$ Composition x	Temperature Range (°C)	D_0 (cm^2/sec)	E_a (eV)	$D_{850°C}$ (cm^2/sec)
0	800–900	3.4×10^{-4}	2.68	3.2×10^{-16}
0.01	800–925	3.4×10^{-2}	3.13	3.1×10^{-16}
0.12	800–925	2.4×10^{-1}	3.30	3.8×10^{-16}
0.24	800–925	5.7×10^{-2}	3.18	3.1×10^{-16}
1	850	—	—	1.3×10^{-16}

Source: Ref. [61] with the exception of the Ge data, which are from Ref. [64].
Note: Values of diffusion coefficient at 850°C are calculated from E_a and D_0.

in-plane stress of 0.4 GPa) produced an increase in diffusivity by a factor of ~1.8 at 1000°C while the same tensile stress reduced the diffusivity by about 20%. Earlier results from Cowern et al. [70] show changes in the same sense but of considerably greater magnitude. However, there is a fundamental difference between the experiments in that Cowern et al. measured the diffusion from a $Si_{1-x}Ge_x$ layer with $0.1< x <0.3$ into Si, while Zangenberg et al. used a substitution of ^{72}Ge for the natural Ge in the epilayer. In this latter experiment the diffusion of the germanium within the SiGe produced no change in composition and hence no stress relief. In Cowern's case the interdiffusion is driven by strain relaxation, which is probably favored kinetically over dislocation growth as the relaxation mechanism.

Work by Zangenberg et al. on boron in strained $Si_{0.88}Ge_{0.12}$ showed a reduction (by a factor of 2) in the diffusion coefficient with compressive strain of 0.21% and a very slight reduction with a similar value of tensile strain. In the case of silicon a factor of 2 increase of boron diffusivity was seen in tensile strained material compared to unstrained silicon [60]. Earlier work by Moriya et al. [71] and by Kuo et al. [72] also showed a retarded B diffusion in compressively stressed SiGe but the separation of the strain effects from the role of germanium was not clear. However, subsequent work discussed above [61] seems to indicate that the chemical effect is small in the case of boron and the dominant effect is strain. Paine et al. [73] have measured the diffusion of boron in compressively strained $Si_{1-x}Ge_x$ layers over the entire composition range but again their results convolute the effects of strain and composition. Over the range $0 < x < 0.5$ the diffusivity decreases by almost an order of magnitude with increasing germanium content but then in the germanium-rich region the diffusivity was observed to increase with increasing Ge concentration.

Unlike boron in which diffusion is mediated by interstitials the diffusion of antimony is mediated by vacancies; as a consequence the expectation is that compressive stress will enhance the diffusivity due to the reduced energy of formation of the vacancy while tensile stress will have little effect. Kringhøj et al. [29] have studied Sb diffusion under both compressive and tensile stress in Si and $Si_{0.91}Ge_{0.09}$ and indeed diffusivity is enhanced by compressive stress and retarded by tensile stress.

In real devices the issue of composition change is almost always convoluted with strain but in addition interfaces are present. We have already referred to the studies of Cowern et al. [70] where the interdiffusion across a heterojunction was driven by strain relaxation. More recently Lever et al. [74] have studied diffusion across interfaces. The test structures used were grown on Si substrates by LPCVD and consisted of a boron-doped silicon layer sandwiched between two dilute undoped $Si_{1-x}Ge_x$ layers with $x = 0.1$ or 0.03. After annealing at 850°C the boron profile was determined by SIMS. In all cases there was a clear accumulation of boron in the SiGe layers, which was attributed to a preferential siting of the boron with Ge as a first nearest neighbor. This confirms earlier work by Hu et al. [75] in which experimental results were obtained on the segregation coefficients of boron, phosphorus, and arsenic in Si–(Ge_xSi_{1-x}) heterostructures. It was found that boron tended to segregate into the germanium-rich region, while phosphorus and arsenic tend to segregate away from it. Within experimental error, the diffusivities were not appreciably affected.

8.6 Dislocations

The only issue we consider here is the electrical behavior of dislocations although of course they can have a significant effect in device processing as diffusion pipes with resultant shorts across junctions.

It is well established that dislocations in Si and in SiGe have levels in the bandgap [76,77] and so it is apparent there are two key questions to address; firstly the position of the levels in the gap and secondly whether the concentration is sufficient to affect device performance. Dislocations introduce an ensemble of electronic levels into the gap, which can appear as a continuum or mini band of states. The electronic behavior is distinctly different to that of point defects, which, in general, have well-defined discrete energy levels. A very important issue in assessing the significance of dislocations is the role of decoration.

The energy state associated with the undecorated dislocation is relatively shallow, typically 120 meV, but as impurities are drawn into the region of the strain field the level moves towards mid-gap [78]. This is very significant as the mid-gap states are extremely effective generation centers while the shallower 'clean' state is ineffective in this role. This work has been done mostly on Frank partial dislocations bounding stacking faults but dislocations with larger Burgers vectors behave similarly. In these experiments extremely small levels of 'decoration' with transition metals were used. These are well below those detectable in transmission electron microscopy (TEM) and probably typical of levels found in 'adequate' fabs in the late 1980s (a few ppb averaged over the layer volume). The behavior with different metals is remarkably similar provided they can diffuse to the defect. However, as the concentration of impurity is increased, metals which precipitate (e.g., copper) produce electrically inert objects which draw the impurity away from dislocations with small Burgers vectors and the state moves away from mid-gap, making it less effective as a generation center. There is also evidence from scanning electron microscopy studies using electron beam induced currents and from luminescence mapping that decorated dislocations are also much more important as recombination centers than clean dislocations [79].

In the limit a dislocation can trap one charge at each atomic plane it threads through but experimental observations indicate that on average the concentration of charge traps is much lower. This gives us a basis for calculating the effects of dislocations on devices assuming a worst-case scenario in which the decoration is such as to move the level to mid-gap. A generation state is only active in the depletion region so if we know the depletion width we can calculate the leakage current produced by one dislocation. For the purpose of this calculation we will assume a device operating temperature of 70°C and transistors typical of the 90 nm node in the International Technology Roadmap for Semiconductors (ITRS). At this node the physical gate length is ~37 nm, the junction depth ~33 nm and V_{dd} 1 V. The worst-case leakage current associated with a single dislocation is 200 pA. The value is essentially the same for a threading dislocation or a misfit dislocation. This is dramatically different from measurements on devices with wide depletion regions (e.g., power devices where leakage currents as high as 1 μA/dislocation have been reported) [80]. The reason is due to the size of the device and the length of dislocation which can be accommodated in the depletion region. The probability of finding more than one dislocation in a 35-nm device is extremely remote so the question is will our technology accept a leakage current of 200 pA? The design rules for the 90-nm CMOS mode accommodate 20 nA/μm and so it seems that even this worst-case scenario is acceptable for conventional CMOS although the requirements for some memory devices are more stringent.

The case of bipolar transistors can be considered very simply. The leakage considerations are very similar to those discussed above. In relation to the effects of dislocations on gain (recombination), the effects are negligible. Beyond the 180-nm node the base width in both hetero- and homojunction transistors is sufficiently small for us to be able to neglect the role of base recombination in the transistor gain.

8.7 Summary

In this section of the handbook we have tried to give an overview of defects and diffusion in SiGe. It is clear that in the last few years very substantial progress has been made towards resolving one of the major problems that plagued early work, namely, separating the issues associated with strain from the chemical effects associated with the presence of two species in the host matrix. There is now a very substantial and reliable literature on defects in silicon and in dilute $Si_{1-x}Ge_x$ (i.e., where $x < 0.3$). However recent work has caused us to doubt the validity of some of the previously accepted work on germanium. Very little experimental data exist on defects or on diffusion in the germanium-rich alloys ($x > 0.5$) and extrapolation from $x < 0.3$ is not feasible because of changes in intrinsic defect mechanisms at around $x = 0.25$. Interpolation between $x = 0.3$ and $x = 1$ is dubious partly for the same reason but also because of doubts regarding the pure germanium data.

It is evident that much more work is required if we are to have a sufficiently rigorous understanding of defects and defect reactions to develop reliable process models for devices containing pure germanium or germanium-rich layers.

Acknowledgments

We are grateful to our colleagues for valuable discussions and their contribution to this work. In particular we thank A.N. Larsen, K. Bonde Nielsen, L. Dobaczeweski, G. Davies, A.F.W. Willoughby, J.M. Bonar, L.I. Murin, J.L. Lindström, A. Mesli, and N.V. Abrosimov. We would also like to thank W. Frank and A. Strohm for providing data on self-diffusion.

References

1. GD Watkins. EPR studies of the lattice vacancy and low-temperature damage processes in silicon. *Lattice Defects in Semiconductors* 1974, edited by FA Huntley. Inst. Phys. Conf. Ser. No 23, IOP, Bristol, 1975, pp 1–22.
2. GD Watkins. The lattice vacancy in silicon. *Deep Centers in Semiconductors*, ST Pantelides (ed.). Gordon and Breach, New York, 1986, chapter 3, pp 147–183.
3. GD Watkins. Vacancies and interstitials and their interactions with impurities in c-Si. *Properties of Crystalline Silicon* (EMIS Data Review Series No. 20), R. Hull (ed.), INSPEC, London, 1999, chapter 11.1, pp. 643–652.
4. PM Fahey, PB Griffin, and JD Plummer. Point defects and dopant diffusion in silicon. *Reviews of Modern Physics* 61: 289–388, 1989.
5. CV Budtz-Jorgensen, P Kringhøj, A Nylandsted Larsen, and NV Abrosimov. Deep-level transient spectroscopy of the Ge-vacancy pair in Ge-doped n-type silicon. *Physical Review B* 58: 1110–1113, 1998.
6. M Lannoo and J Bourgoin. *Point Defects in Semiconductors I: Theoretical Aspects.* Springer Series in Solid State Sciences No 22, Springer, New York, 1981.
7. S Dannefaer, P Mascher, and D Kerr. Monovacancy formation enthalpy in silicon. *Physical Review Letters* 56: 2195–2198, 1986.
8. JS Nelson, PA Schultz, and AF Wright. Valence and atomic size-dependent exchange barriers in vacancy-mediated dopant diffusion. *Applied Physics Letters* 73: 247–249, 1998.
9. J Xie and SP Chen. Diffusion and clustering in heavily arsenic-doped silicon: discrepancies and explanation. *Physical Review Letters* 83: 1795–1798, 1999.
10. A Fazzio, A Janotti, AJR da Silva, and R Mota. Microscopic picture of the single vacancy in germanium. *Physical Review B* 61: R2401–R2404, 2000.
11. S Ögüt and R Chelikowsky. *Ab initio* investigation of point defects in bulk Si and Ge using a cluster method. *Physical Review B* 64: 245206-1 to 245206-11, 2001.
12. H Haesslein, R Sielemann, and C Zistl. Vacancies and self-interstitials in germanium observed by perturbed angular correlation spectroscopy. *Physical Review Letters* 80: 2626–2629, 1998.
13. RE Whan. Investigations of oxygen-defect interactions between 25 and 700 K in irradiated germanium. *Physical Review* 140: A690–A698, 1965.

14. VP Markevich, ID Hawkins, AR Peaker, VV Litvinov, LI Murin, L. Dobaczewski, and JL Lindström. Electronic properties of vacancy–oxygen complex in Ge crystals. *Applied Physics Letters* 81: 1821–1823, 2002.
15. VP Markevich, VV Litvinov, l Dobaczewski, JL Lindström, LI Murin, and AR Peaker. Radiation-induced defects and their transformations in oxygen-rich germanium crystals. *Physica Status Solidi (c)* 0: 702–706, 2003.
16. J Fage-Pedersen, A Mesli, and A Nylandsted Larsen. Irradiation-induced defects in Ge studied by transient spectroscopies. *Physical Review B* 62: 10116–10125, 2000.
17. VP Markevich, AR Peaker, VV Litvinov, VV Emtsev, and LI Murin. Electronic properties of antimony–vacancy complex in Ge crystals. *Journal of Applied Physics* 95: 4078–4083, 2004.
18. JR Troxell, AP Chatterjee, GD Watkins, and LC Kimerling. Recombination-enhanced migration of interstitial aluminum in silicon. *Physical Review B* 19: 5336–5348, 1979.
19. PM Money, LI Cheng, M Süli, JD Gerson, and JW Corbett. Defect energy levels in boron-doped silicon irradiated with 1-MeV electrons. *Physical Review B* 15: 3836–3843, 1977.
20. LW Song, XD Zhan, BW Benson, and GD Watkins. Bistable interstitial-carbon–substitutional-carbon pair in silicon. *Physical Review B* 42: 5765–5783, 1990.
21. H Bracht, NA Stolwijk, and H Mehrer. Properties of intrinsic point defects in silicon determined by zinc diffusion experiments under nonequilibrium conditions. *Physical Review B* 52: 16542–16560, 1995.
22. AJR da Silva, A Janotti, A Fazzio, RJ Baierle, and R Mota. Self-interstitial defect in germanium. *Physical Review B* 62: 9903–9906, 2000.
23. J Lento, M Pesola, J-L Moros, and RM Nieminen. Vacancies in SiGe: Jahn–Teller distortion and spin effects. *Applied Physics Letters* 77: 232–234, 2000.
24. P Venezuela, GM Dalpian, AJR da Silva, and A Fazzio. Vacancy-mediated diffusion in disordered alloys: Ge self-diffusion in $Si_{1-x}Ge_x$. *Physical Review B* 65: 193306-1 to 193306-4, 2002.
25. P Ramanarayanan, K. Cho, and BM Clemens. Effect of composition on vacancy mediated diffusion in random binary alloys: first principles study of the $Si_{1-x}Ge_x$ system. *Journal of Applied Physics* 94: 174–185, 2003.
26. VP Markevich, AR Peaker, J Coutinho, R Jones, VJB Torres, S Öberg, PR Briddon, LI Murin, L Dobachewski, and NV Abrosimov. Structure and properties of vacancy–oxygen complexes in $Si_{1-x}Ge_x$ alloys. *Physical Review B* 69: 125218-1 to 125218-11, 2004.
27. L Dobaczewski, P Kaczor, ID Hawkins, and AR Peaker. Laplace transform deep-level transient spectroscopic studies of defects in semiconductors. *Journal of Applied Physics* 76: 194–198, 1994.
28. L Dobaczewski, K Goscinski, K Bonde Nielsen, A Nylandsted Larsen, J Lundsgaard Hansen, and AR Peaker. Alloy splitting of gold and platinum acceptor levels in SiGes. *Physical Review Letters* 83: 4582–4585, 1999.
29. P Kringhøj, A Nylandsted Larsen, and Yu Shiryaev. Diffusion of Sb in strained and relaxed Si and SiGe. *Physical Review Letters* 76: 3372–3375, 1996.
30. L Dobaczewski, K Goscinski, ZR Zytkiewicz, K Bonde Nielsen, L Rubaldo, O Andersen, and AR Peaker. Piezoscopic deep-level transient spectroscopy studies of the silicon divacancy. *Physical Review B* 65: 113203-1 to 113203-4, 2002.
31. R Krüssmann, H Vollmer, and R Labusch. The activation energy of phosphorus donors in silicon-rich silicon–germanium alloys. *Physica Status Solidi (b)* 118: 275–282, 1983.
32. J Hattendorf, W-D Zeitz, W Schröder, and NV Abrosimov. On the formation of boron–germanium pairs in silicon–germanium mixed crystals. *Physica B* 340–342: 858–862, 2003.
33. P Kringhøj and A Nylandsted Larsen. Irradiation-induced defect states in epitaxial n-type $Si_{1-x}Ge_x$ alloy layers. *Physical Review B* 52: 16333–16336, 1995.
34. S-L Sihto, J Slotte, J Lento, K Saarinen, EV Monakhov, AYu Kuznetsov, and BG Svensson. Vacancy–phosphorus complexes in strained $Si_{1-x}Ge_x$: structure and stability. *Physical Review B* 68: 115307-1 to 115307-10, 2003.

35. H av Skardi, A Bro Hansen, A Mesli, and A Nylandsted Larsen, The divacancy in particle-irradiated, strain-relaxed SiGe. *Nuclear Instruments and Methods in Physics Research B* 186: 195–200, 2002.
36. TP Leevard Pedersen, A Nylandsted Larsen, and A Mesli. Carbon-related defects in proton-irradiated, n-type epitaxial $Si_{1-x}Ge_x$. *Applied Physics Letters* 75: 4085–4087, 1999.
37. A Nylandsted Larsen, A Bro Hansen, D Reitze, J-J Goubet, and J Fage-Pedersen. Interstitial-carbon defects in $Si_{1-x}Ge_x$. *Physical Review B* 64: 233202-1 to 233202-4, 2001.
38. VP Markevich, AR Peaker, LI Murin, and NV Abrosimov. Oxygen-related radiation-induced defects in SiGe alloys. *Journal of Physics: Condensed Matter* 15: S2835–S2842, 2003.
39. F Nikolajsen, P Kringhoj, and A Nylandsted Larsen. Substitutional gold in $Si_{1-x}Ge_x$. *Applied Physics Letters* 69: 1743–1745, 1996.
40. A Mesli, P Kringhoj, and A Nylandsted Larsen. Pinning behaviour of gold-related levels in Si using $Si_{1-x}Ge_x$ alloy layers. *Physical Review B* 56: 13202–13217, 1997.
41. K Nauka and TI Kamins. Deep state defects in strained and relaxed epitaxial $Si_{1-x}Ge_x$ on Si introduced by 3d transition metal and 5d noble metal impurities. *Physica B* 273–274: 603–607, 1999.
42. K Goscinski, L Dobaczewski, K Bonde Nielsen, A Nylandsted Larsen, and AR Peaker. High-resolution deep-level transient spectroscopy studies of gold and platinum acceptor states in diluted SiGe alloys. *Physical Review B* 63: 235309-1 to 235309-10, 2001.
43. A Mesli, B Vileno, C Eckert, A Slaoui, C Pedersen, A Nylandsted Larsen, and NV Abrosimov. Iron in relaxed $Si_{1-x}Ge_x$ alloy: band gap related levels, diffusion, and alloying effects. *Physical Review B* 66: 045206-1 to 045206-12, 2002.
44. NEB Cowern, GFA van de Walle, PC Zalm, and DJ Oostra. Reactions of point defects and dopant atoms in silicon. *Physical Review Letters* 69: 116–119, 1992.
45. DJ Eaglesham, PA Stolk, H-J Gossmann, and JM Poate. Implantation and transient B diffusion in Si: the source of the interstitials. *Applied Physics Letters* 65: 2305, 1994.
46. A Pakfar. Dopant diffusion in SiGe: modeling stress and Ge chemical effects. *Materials Science and Engineering* B89: 225–228, 2002.
47. NV Abrosimov, SN Rossolenko, V Alex, A Gerhardt, and W Shröder. Single crystal growth of $Si_{1-x}Ge_x$ by the Czochralski technique. *Journal of Crystal Growth* 166: 657, 1996.
48. A Nylandsted Larsen. Growth and process-induced defects in SiGe-based heterostructures. *Diffusion and Defect Data Part B (Solid State Phenomena)* 69–70: 43–45, 1999.
49. P Laitinen, A Strohm, J Huikari, A Nieminen, T Voss, C Grodon, I Riihimäki, M Kummer, J Äystö, P Dendooven, J. Räisänen, W Frank, and ISOLDE Collaboration. Self-diffusion of ^{31}Si and ^{71}Ge in relaxed $Si_{0:20}Ge_{0:80}$ layers. *Physical Review Letters* 89: 085902-1 to 085902-4, 2002.
50. A Strohm, T Voss, W Frank, J Räisänen, and M Dietrich. Self-diffusion of ^{71}Ge in Si–Ge, *Physica B* 308–310: 542–545, 2001.
51. G Hettich, H Mehrer, and K Maier. Tracer diffusion of ^{71}Ge and ^{31}Si in intrinsic and doped silicon. *Institute of Physics Conference Series* 46: 500–504, 1979.
52. N Zangenberg, J Lundsgaard Hansen, J Fage-Pedersen, and A Nylandsted Larsen. Ge self-diffusion in epitaxial $Si_{1-x}Ge_x$ layers. *Physical Review Letters* 87: 125901, 2001.
53. A Strohm, T Voss, W Frank, P Laitinen and J Räisänen. Self-diffusion of ^{71}Ge and ^{31}Si in Si–Ge. *Zeitschrift für Metallkunde* 93: 737–744, 2002.
54. MJ Aziz. Thermodynamics of diffusion under pressure and stress: relation to point defect mechanisms. *Applied Physics Letters* 70: 2810–2812, 1997.
55. P Laitinen, I Riihimäki, J Räisänen, and ISOLDE Collaboration. Arsenic diffusion in relaxed $Si_{1-x}Ge_x$. *Physical Review B* 68: 155209-1 to 155209-6, 2003.
56. S Eguchi, CN Chleirigh, OO Olubuyide, and JL Hoyt. Germanium-concentration dependence of arsenic diffusion in silicon germanium alloys. *Applied Physics Letters* 84: 368–370, 2004.
57. A Nylandsted Larsen and P Kringhøj. Diffusion of Sb in relaxed $Si_{1-x}Ge_x$. *Applied Physics Letters* 68: 2684–2686, 1996.

58. BI Boltaks. *Diffusion in Semiconductors.* Infosearch Ltd., London, 1963, p. 190.
59. Chi On Chui, K Gopalakrishnan, PB Griffin, JD Plummer and KC Saraswat. Activation and diffusion studies of ion-implanted p and n dopants in germanium. *Applied Physics Letters* 83: 3275–3277, 2003.
60. Aihua Dan, AFW Willoughby, JM Bonar, BM Mcgregor, MG Dowsett, and RJH Morris. Antimony diffusion in silicon gemanium alloys under point defect injection. *International Journal of Modern Physics B* 16: 4195–4198, 2002.
61. N Zangenberg, J Fage-Pedersen, J Lundsgaard Hansen, and A Nylandsted Larsen. Boron and phosphorus diffusion in strained and relaxed Si and SiGe. *Journal of Applied Physics* 94: 3883–3890, 2003.
62. WC Dunlap Jr. Diffusion of impurities in germanium. *Physical Review* 94: 1531–1540, 1954.
63. S Eguchi, JL Hoyt, CW Leitz, and EA Fitzgerald. Comparison of arsenic and phosphorus diffusion behavior in silicon–germanium alloys. *Applied Physics Letters* 80: 1743–1745, 2002.
64. NEB Cowern, KTF Janssen, GFA van de Walle, and DJ Gravesteijn. Impurity diffusion via an intermediate species: the B–Si system. *Physical Review Letters* 65: 2434–2437, 1990.
65. NR Zangenberg and A Nylandsted Larsen, The Meyer–Neldel rule for diffusion in Si and SiGe. *Physica B* 340–342: 780–783, 2003.
66. S Uppal, AFW Willoughby, JM Bonar, AGR Evans, NEB Cowern, R Morris, and MG Dowsett. Diffusion of ion-implanted boron in germanium. *Journal of Applied Physics* 90: 4293–4295, 2001.
67. P Delugas and V Fiorentini. Energetics of transient enhanced diffusion of boron in Ge and SiGe. *Physical Review B* 69: 085203-1 to 085203-5, 2004.
68. A Antonelli and J Bernholc. Pressure effects on self-diffusion in silicon. *Physical Review B* 40: 10643–10646, 1989.
69. Y Zhao, MJ Aziza, H-J Gossmann, S Mitha, and D Schiferl. Activation volume for boron diffusion in silicon and implications for strained films. *Applied Physics Letters* 74: 31–33, 1999.
70. NEB Cowern, PC Zalm, P van der Sluis, DJ Gravesteijn, and WB de Boer. Diffusion in strained Si(Ge). *Physical Review Letters* 72: 2585–2588, 1994.
71. N Moriya, LC Feldman, HS Luftman, CA King, J Bevk, and B Freer. Boron diffusion in strained $Si_{1-x}Ge_x$ epitaxial layers. *Physical Review Letters* 71: 883–886, 1993.
72. P Kuo, JL Hoyt, JF Gibbons, JE Turner, RD Jacowitz, and TI Kamins. Comparison of boron diffusion in Si and strained $Si_{1-x}Ge_x$ epitaxial layers. *Applied Physics Letters* 62: 612–614, 1993.
73. ADN Paine, AFW Willoughby, and JM Bonar. Point defect redistribution in $Si_{1-x}Ge_x$ alloys. *Journal of Material Science: Materials in Electronics* 10: 339–343, 1999.
74. RF Lever, JM Bonar, and AFW Willoughby. Boron diffusion across silicon–silicon germanium boundaries. *Journal of Applied Physics* 83: 1988–1994, 1998.
75. SM Hu, DC Ahlgren, PA Ronsheim, JO Chu. Experimental study of diffusion and segregation in a Si–(Ge_xSi_{1-x}) heterostructure. *Physical Review Letters* 67: 1450–1453, 1991.
76. P Omling, ER Weber, L Montelius, HM Alexander, and J Michel. Electrical properties of dislocations and point defects in plastically deformed silicon. *Physical Review B* 32: 6571–6581, 1988.
77. PN Grillot, SA Ringel, EA Fitzgerald, GP Watson, and YH Xie. Electron trapping kinetics at dislocations in relaxed $Ge_{0.3}Si_{0.7}$/Si heterostructures. *Journal of Applied Physics* 77: 3248–3256, 1995.
78. A Berg, I Brough, JH Evans, G Lorimer, and AR Peaker. Recombination–generation behaviour of decorated defects in silicon. *Semiconductor Science and Technology* 7: 263–268, 1992.
79. V Higgs and M Kittler. Investigation of the recombination activity of misfit dislocations in Si/SiGe epilayers by cathodoluminescence imaging and the electron beam induced current technique. *Applied Physics Letters* 63: 2085–2087, 1993.
80. BO Kolbesen, W Bergholz, H Cerva, F Gelschorf, H Wendt, and G Zoth. Effects of extended lattice defects on silicon semiconductor devices. Structure and Properties of Dislocations in Semiconductors 1989. Proceedings of the Sixth International Symposium, 1989, *Inst. Phys. Conf. Ser.* 104: 421–430, 1990.

9

Stability Constraints in SiGe Epitaxy

Armin Fischer
Innovations for High Performance Microelectronics (IHP)

9.1 Introduction

The growth of coherent thin layers on rigid crystalline substrates is possible when biaxial compressive or tensile strain in the layer accommodates the lattice mismatch between the film and substrate material. When the stored strain energy exceeds a certain threshold, the heterostructure becomes metastable and the film strain may give way to misfit dislocations. The basic energetic and kinetic parameters describing mismatch accommodation by elastic strain and misfit dislocation in metastable heterostructures appear to be well described by the framework of Matthews and Blakeslee [1–3] and Dodson and Tsao [4,5]. However, it is evident that they cannot adequately explain the point of strain relief onset via plastic flow and the work hardening behavior of strained layers at the end of thermal relaxation process. This stems from ignoring the effects of elastic surface relaxation on the film lattice cells and the elastic interaction between straight misfit dislocations within the film–substrate interface. The first includes the problem of developing a relationship between the equilibrium critical thickness at which dislocations form and the bulk lattice mismatch [6]. The latter involves balancing the force required to move misfit dislocations against the internal elastic stress field due to dislocation–dislocation interactions [7]. Finally, suffice it to say that classical equilibrium and kinetic models for strained layer case do not imply rigorously the conditions of equilibrium at the boundary.

In this chapter we present a modified Volterra approach in equilibrium theory for strain relaxation in metastable heteroepitaxial semiconductor structures, which includes the surface effects on mismatch accommodation by tetragonal distortion of the cubic lattice cells and the elastic interaction between straight misfit dislocations. Because of the mathematical complexities involved in a proper atomistic description of the competing forces, the subject is not treated here in a Frenkel–Kontorowa model or its approximations. The principle of our theoretical method is straightforward. The free-surface boundary conditions are satisfied by placing an image dislocation outside the crystal so that its stress field cancels that of the real misfit dislocation at the surface. We will discuss the effect of the Airy stress function that removes the fictitious shear and normal stresses at the crystal surface and thus makes the crystal stress-free. We show that this image-force method provides an equilibrium theory that correctly predicts experimentally observed values of critical layer thickness. To demonstrate the physical significance of the present approach in equilibrium theory for strained layer relaxation, we have calculated the equilibrium critical thickness and Ge content via coherency strain of SiGe/Si strained layer structures and compared our results to those predicted by using relaxation models based on the absence of surface relaxation effects and elastically noninteracting dislocations. Furthermore, by considering the exact solution for the elastic interaction of real and image dislocations, our equilibrium model can completely describe the strain relief via lattice distortion, plastic flow, and work hardening in lattice-mismatched epilayers. We show that our Volterra dislocation model is also appropriate when the film thickness is smaller than the misfit dislocation spacing. It is pointed out that the elastic coherency stress of the strained material is really affected by a large surface relaxation stress. This is essential for experimental determination of the Ge content of extremely thin films as a function of the tetragonal distortion of the cubic lattice cells. The equilibrium theory is also used to define the degree of strain relaxation and to predict the incomplete strain relief at the end of thermal relaxation process of metastable SiGe/Si heterostructures. We consider here only the thermodynamic equilibrium and homogeneous deformations of strained layer structures, recognizing that kinematic factors can influence the attainment of the equilibrium state of strain. In the final portion of the chapter we treat the case of Si-capped SiGe epilayer on Si substrate in some detail to expand the basic equilibrium theory for strained layer relaxation to such a Si/SiGe/Si geometry with two adjacent interfaces and a free surface at the top of the heteroepitaxial stack.

9.2 Image-Force Method for Strained Layer Relaxation

In thermodynamic equilibrium, misfit dislocations appear at the interface of a strained layer heterostructure when the strained layer is thick enough that it is energetically favorable for the mismatch to be accommodated by a combination of elastic strain and interfacial misfit dislocations rather than by elastic strain alone [8]. This equilibrium critical thickness h_{crit} has been calculated and discussed by many authors in the continuum picture as well as through phenomenological description of dislocation dynamics [1,3,5,9–14]. However, there have been many reports (e.g., in Refs. [11,14–16]) of experimental determinations of h_{crit}, indicating that coherence apparently persists to thicknesses much greater than that predicted by classical equilibrium theories. The semiempirical kinetic model of Dodson and Tsao [4,5] is more appropriate to describe the latter stages of relaxation, where the effective stress is decreasing due to a reduction in misfit strain produced by the high dislocation density. For the case of sufficiently low dislocation content in the strained layer near the point of strain relief onset, this model reduces to the equilibrium form of Matthews and Blakeslee [1,3]. Thus the challenge remains to develop a predictive model appropriate for strained layer relaxation [17].

To introduce an appropriate continuum model, we begin by analyzing the conditions under which the strained layer relaxation should occur in metastable heterostructures, and then we modify the governing models to account for the discrepancies mentioned above. In a finite body, boundary conditions at the surface must be satisfied. For example, no forces can act on a free surface. The image-force method provides a powerful tool to solve such problems in the continuum theory of elasticity [18]. Let us now consider the schematic illustration in Figure 9.1. For the strained layer case, the free-surface boundary conditions are satisfied by placing an image dislocation outside the crystal such that its stress field cancels that of the real interfacial misfit dislocation at the surface. The condition is fulfilled if the self-stress of an image misfit dislocation of equal strength and opposite sign at a position $2h$ along the strained interface normal is superposed on the self-stress of the real primary dislocation. In general, the complete stress distribution for a mixed dislocation is given by the superposition of the stress fields of the real dislocation, the image dislocation, and a stress term derived from the Airy stress function that makes the surface traction-free. In our case of a mixed dislocation in a coaxial cylinder with its line

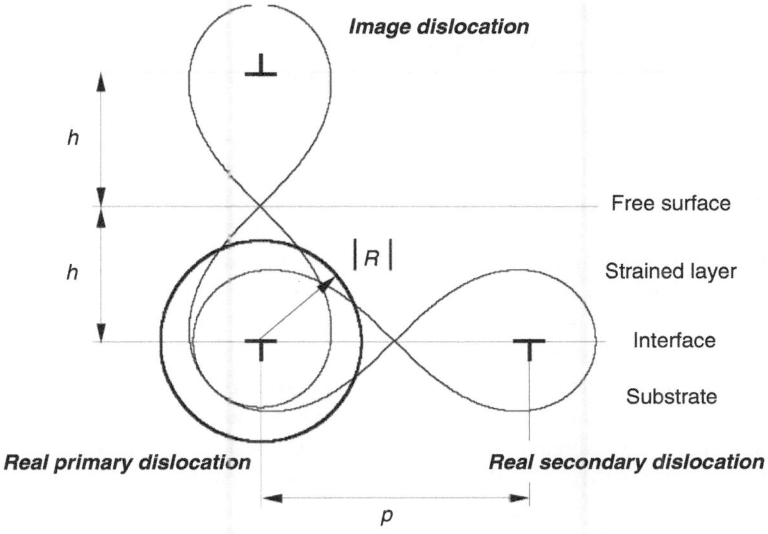

FIGURE 9.1 Schematic illustration of the configuration of real and image misfit dislocations in a strained heteroepitaxial structure. The modulus of complex dislocation semispacing $|\boldsymbol{R}|$ is indicated. (From A Fischer, H Kühne, M Eichler, F Holländer, and H Richter. *Physical Review B* 54:8761–8768, 1996. With permission.)

parallel to the surface, the image construction gives the dominant part of the shear stress component. Since the corresponding stress function term exerts no force component along the strained interface normal, the shear stress of the dislocation construction is approximately obtained from the shear stress of the image dislocation alone. Notice that only shear stresses in the slip system produce glide forces on a dislocation.

In the continuum picture, the presence of dislocations causes strains around the line and, as a response to these, stresses as known from conventional elasticity theory. These stresses are defined by the contact forces transmitted through internal area elements. We speak of self-stresses to distinguish them from the applied misfit stresses. In linear approximation, the Volterra expression for the shear self-stress σ_s of a mixed straight dislocation due to its line tension in a region bounded by a coaxial cylinder of radius R is

$$\sigma_s = \frac{Gb(1 - \nu\cos^2\theta)}{4\pi(1-\nu)R\cos\phi}\left(\ln\frac{\alpha R}{b} - 1\right), \tag{9.1}$$

where G is the anisotropic shear modulus in the $\langle 1\ 1\ 0\rangle$ direction of the $(0\ 0\ 1)$ plane of the epilayer material, ν is its Poisson ratio, b is the magnitude of the Burgers vector, θ is the angle between the dislocation Burgers vector and its line direction, ϕ is the angle between the slip plane and the strained interface normal, and α is a factor that accounts for the energy in the dislocation core where linear elasticity does not apply. α is generally taken to be in the range from 1 to 4 for covalently bonded semiconductor materials [18]. Because of the logarithmic dependence and the $R \gg b$ Volterra regime, the elastic self-stress is insensitive to the precise value of α. We set $\alpha/2.7 = 1$. Here σ_s is considered to act on the plane containing the dislocation line direction and the interface normal. So, referred to the slip plane surface, the shear component of the self-stress of a straight dislocation τ_s is given by $\tau_s = \cos\phi\,\sigma_s$. For our case, an interfacial 60°-type misfit dislocation on the slip plane causes resolved shear stress, i.e.,

$$\tau_s = \frac{Gb\left(1 - \frac{\nu}{4}\right)}{4\pi(1-\nu)R\cos\phi}\ln\frac{R}{b}, \tag{9.2}$$

where the ratio $G/\cos\phi$ is the isotropic shear modulus in the {1 1 1} slip planes. Thus, in this case the quantity $b\tau_s$ is an image force given by the simple image construction. Under these conditions the attraction of the real primary dislocation toward the surface is obtained from the stress of the image dislocation alone.

So far we have considered the image-force and shear self-stress problems for a misfit dislocation in a metastable heterostructure. As shown in Figure 9.1, imagine now a real secondary dislocation of equal strength and sign lying parallel to the real primary dislocation at a distance of p and moving continuously towards the primary dislocation. For further deformation under the driving force produced by the external misfit stress, it is necessary that the moving misfit dislocation overcomes the resistance caused by superposed shear self-stress field of the image dislocation and the real primary dislocation bounded by a virtual cylinder of radius $R = h$. Note that at larger distances from the real primary dislocation the image stresses largely cancel the dislocation stresses. Now for the elastic shear stress field extension of the real interfacial misfit dislocations lying parallel to another, a reasonable approximation would be to take roughly one-half the distance p between dislocations for R in Equation 9.2 (see Ref. [18]). However, in this case the quantity $b\tau_s$ is a real repulsive force that drives the real primary dislocation in its slip plane toward the surface. According to the Green function method for the elastic displacements $u(R) \propto b/R^2$ and the principle of superposition of displacement fields [18], we now combine the imaginary, i.e., fictitious, and the real free-surface term of the shear self-stress of a misfit dislocation, i.e., h and $p/2$, respectively, and get a complex dislocation semispacing $\boldsymbol{R}$ as the first-order solution. Its modulus $|\boldsymbol{R}| = R_{h,p}$ is given by

$$\frac{1}{R_{h,p}^2} = \frac{1}{h^2} + \frac{4}{p^2}, \tag{9.3}$$

where h is the thickness of the epilayer and the subscripts h and p stand for the fictitious and the real component, respectively. We can say that the modulus of complex dislocation semispacing $R_{h,p}$ is an approximate solution for the stress-free boundary associated with the presence of two free surfaces. The relationship represented by Equation 9.3 is plotted in Figure 9.2 for different layer thicknesses h. The pursual of Figure 9.2 shows some interesting relationships. For dislocation spacings p greater than $5h$, the modulus of complex dislocation semispacing as a measure of the extension of the elastically strained continuum about a misfit dislocation is dominated by its fictitious term, i.e., by the layer thickness h. In this case, Equation 9.3 reduces to $R_{h,p} \sim h$. If p diminishes continuously, then the effect of the real term on $R_{h,p}$ increases slowly. When the dislocation spacing reaches the value $p \sim 2h$, the fictitious and the real free-surface terms make the same contribution to the modulus of complex dislocation semispacing. Below $p \sim h/5$, as the real misfit dislocations approach one another, the effect of the fictitious component vanishes and $R_{h,p}$ does not depend on h. Replacing R with $R_{h,p}$ in Equation 9.2, the shear component of the total self-stress created by present dislocation content in a finite body becomes finally

$$\tau_s = \frac{Gb\left(1 - \frac{\nu}{4}\right)}{4\pi(1-\nu)R_{h,p}\cos\phi}\ln\frac{R_{h,p}}{b}. \tag{9.4}$$

We notice that, up until now, we have considered a type of plane strain deformation under shear stresses, in which the body undergoes only changes of shape, but no changes of volume. Now let us consider the effect of the Airy stress function, which removes the fictitious tangential and radial normal stresses at the cylinder surface and thus changes the volume of the epilayer.

9.3 Surface Relaxation Stress

In the case of the normal self-stress components of 60° misfit dislocations the above solution is inadequate and large additional stresses must be superposed. For such dislocations the normal self-stress components given by the simple image construction do not vanish at the free surface. The

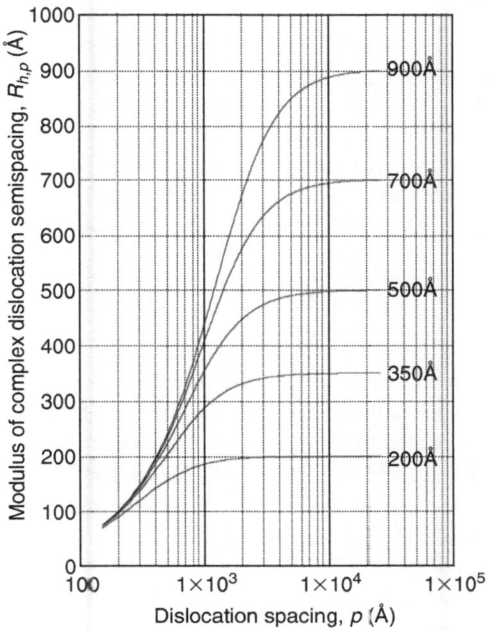

FIGURE 9.2 Plot of the calculated modulus of complex dislocation semispacing $|\boldsymbol{R}| = R_{h,p}$ as a function of misfit dislocation spacing p for strained layer thicknesses $h = 200$, 350, 500, 700, and 900 Å. (From A Fischer, H Kühne, M Eichler, F Holländer, and H Richter. *Physical Review B* 54:8761–8768, 1996. With permission.)

procedure for solving this case is the same as that outlined above: first to superpose the simple image and then to devise a stress function that cancels the remaining forces acting at the free surface and thus satisfy the boundary conditions. Hirth and Lothe [18] have given the explicit solution of the Airy stress function for a pure edge dislocation in a cylinder parallel to a free surface. In analogy to the edge dislocation case, the first-order solution for a straight 60° misfit dislocation gives the normal self-stress components $\sigma_{f,t}$ and $\sigma_{f,r}$ that remove the fictitious plane strain and make the crystal stress-free:

$$\sigma_{f,t} = \frac{Gb\left(1 - \frac{\nu}{4}\right)}{2\pi(1-\nu)\cos\phi}\left(\frac{3\eta}{h^2} - \frac{b^2}{\eta^3}\right), \tag{9.5a}$$

$$\sigma_{f,r} = \frac{Gb\left(1 - \frac{\nu}{4}\right)}{2\pi(1-\nu)\cos\phi}\left(\frac{\eta}{h^2} + \frac{b^2}{\eta^3}\right), \tag{9.5b}$$

where η is the distance from the interface in direction to the crystal surface. These tangential and radial stress terms are in fact the external state of stress that gives rise to the image dislocation construction, which in turn creates the state of tension within the epilayer. Then, according to the generalized Hooke's law, the normal stress $\sigma_{f,n}$ acting perpendicular to the slip plane is given by the combined effect of the tangential $\sigma_{f,t}$ and radial $\sigma_{f,r}$ stress components on the slip plane. Therefore, from Equations 9.5a and 9.5b we can write the relation

$$\sigma_{f,n} = \frac{1+2\nu}{(1-2\nu)(1+\nu)}2G(1+\nu)\frac{b\left(1 - \frac{\nu}{4}\right)}{2\pi(1-\nu)\cos\phi}\frac{\eta}{h^2}. \tag{9.6}$$

Here $2G(1+\nu)$ is Young's modulus of the epilayer. $\sigma_{f,n}$ creates a state of tension between the slip planes of the epilayer. We call the normal stress $\sigma_{f,n}$ the surface relaxation stress. A transformation of this normal stress into the interface plane with the tensile axis normal to the dislocation line gives the state of in-plane tension within the epilayer. In these coordinates, Equation 9.6 yields the in-plane surface relaxation stress σ_f. We get $\sigma_f = 2\cos\lambda\sigma_{f,n}\cos\phi/\cos^2\phi$ where λ is the angle between the Burgers vector and the direction in the interface, normal to the dislocation line. For typical values of ν from 0.26 to 0.28, the quantity $(1 + 2\nu)(1 - \nu/4)/\pi(1 - 2\nu)(1 + \nu)\cos^2\phi$ varies from 1.1 to 1.2, close to unity. Thus the result for the in-plane surface relaxation stress becomes approximately

$$\sigma_f = \frac{2G(1+\nu)}{1-\nu} b\cos\lambda \frac{\eta}{h^2}. \tag{9.7}$$

In words, the in-plane surface relaxation stress decreases linearly from its maximum value at the crystal surface $\eta = h$ to zero at the interface $\eta = 0$, whereas the strength of surface relaxation effects on the strained layer structure is related to the reciprocal of the square of the layer thickness h. Obviously, it is a remarkable result with regard to our considerations of the lattice-mismatch accommodation. When these effects are ignored and a coherency strain calculation is carried out, one obtains a spurious result, leading to an erroneous strain that acts on a misfit dislocation—image dislocation pair for its generation. In addition, these effects should play a role in the lattice resistance to plastic flow of the material. For practical purposes, a reasonable approximation would be to take an average value σ_f^{av} for $\sigma_f(\eta)$, i.e.,

$$\sigma_f^{av} = \frac{1}{h}\int_0^h \sigma_f(\eta)\,d\eta = \frac{2G(1+\nu)}{1-\nu}\frac{b\cos\lambda}{2h}. \tag{9.8}$$

Notice once again that G is the anisotropic shear modulus in the $\langle 1\ 1\ 0\rangle$ direction of the $(0\ 0\ 1)$ plane of the epilayer. Furthermore, we have measured the effect of the elastic surface relaxation on the strained layer heterostructure directly. The results are presented in Section 9.5.

9.4 Interaction between Internal Stress and External Stress

For a complete description of the equilibrium state of stress in lattice-mismatched epilayers, we should now consider the resolved shear stress τ that acts on the slip system as a consequence of an externally applied misfit stress. In an initially misfit dislocation-free substrate–epilayer system, the in-plane misfit strain is given by $(a_l - a_s)/a_s$, where a denotes the bulk lattice parameter and the subscripts 's' and 'l' refer to the substrate and the layer, respectively. When the elastic strain is partially relieved by a single array of misfit dislocations created at the interface, the residual in-plane strain becomes $\varepsilon = [(a_l - a_s)/a_s] - (b\cos\lambda/p)$, where the term $b\cos\lambda/p$ represents the strain relief via plastic flow [19]. Here p is the average distance between the dislocations. Hence it follows that the lattice-mismatch accommodation would occur without in-plane tension within the epilayer caused by the surface relaxation stress. In other words, as explained above, for the case where $p > h/5$, the heterostructure is not traction-free, because normal stress components acting on the crystal surface are not taken into account. To overcome this difficulty we will now substitute the modulus of complex dislocation semispacing $R_{h,p}$ for $p/2$ in the residual in-plane strain expression above. So the true residual in-plane strain becomes $\varepsilon = [(a_l - a_s)/a_s] - [b\cos\lambda/(2R_{h,p})]$, where the h term accounts for the purely elastic surface relaxation and the p term for plastic flow. In this case, the resolved shear stress τ acting on the slip system on a misfit dislocation is given by

$$\tau = \cos\lambda\cos\phi\frac{2G(1+\nu)}{(1-\nu)\cos\phi}\left(\frac{a_l - a_s}{a_s} - \frac{b\cos\lambda}{2R_{h,p}}\right). \tag{9.9}$$

Thus the surface relaxation stress term is implied in the expression for the resolved shear stress that produced the driving force for plastic flow of the crystal. Note that the ratio $G/\cos\phi$ is the isotropic shear modulus in the {1 1 1} slip planes.

As we have previously argued, a finite heterostructure is subjected to point and lattice-mismatch forces. The stresses and strains caused by such internal and external sources of stress can be superposed. We should now consider the resolved shear stress that acts on the slip system as a consequence of the externally applied stress and the elastic shear stress field due to dislocation–dislocation interactions. As shown above, during misfit strain relief via plastic flow, an intrinsic elastic stress field is built up. For further deformation, the moving misfit dislocations have to overcome the resistance caused by this stress field. Consequently, the dislocation shear self-stress field is in the direction opposite the applied misfit stress. The excess resolved shear stress required to produce plastic flow will then be given by the difference between the two stress components. Equations 9.4 and 9.9 yield $\tau_{exc} = \tau - \tau_s$, where the second term also accounts for work hardening of the material. Combining the two terms, we obtain an expression for the excess resolved shear stress:

$$\tau_{exc} = \cos\lambda\cos\phi\frac{2G(1+\nu)}{(1-\nu)\cos\phi}\left(\frac{a_l - a_s}{a_s} - \frac{b\cos\lambda}{2R_{h,p}}(1+\beta)\right), \tag{9.10}$$

with

$$\beta = \frac{1-(\nu/4)}{4\pi\cos^2\lambda\cos\phi(1+\nu)}\ln\frac{R_{h,p}}{b}.$$

Here the quantity $[b\cos\lambda/(2R_{h,p})]\beta$ corresponds to the decrease in active shear stress through elastic interaction between dislocations depending on the current misfit dislocation density of the crystal.

As a final topic, a few words on the expression for the excess resolved shear stress are appropriate. This is a convenient expression. The applications are numerous: problems of elastic strain retained by the epilayer, equilibrium critical thickness of a SiGe layer on Si substrate, initial and residual in-plane strain, film stress, work hardening of the material, degree of elastic and plastic strain relaxation, and misfit dislocation density reached at the equilibrium strain state, for example. Some of the applications are discussed in the following sections.

9.5 Critical Thickness and Film Stress of SiGe/Si Layers

To demonstrate the physical significance of the present approach in equilibrium theory for strained layer relaxation proposed here, we have calculated the equilibrium critical thickness of $Si_{1-x}Ge_x$/Si strained layer structures as a function of the fractional atomic Ge content x. We have compared our results with those predicted by classical relaxation models, which do not include surface relaxation effects and dislocation–dislocation interactions. Taking the in-plane misfit strain $[(a_l - a_s)/a_s] = 0.0418x$, Equations 9.3 and 9.10 and the equilibrium conditions for the point of strain relief onset via plastic flow, i.e., $\tau_{exc} = 0$ and $p \to \infty$, lead to the following expression for the critical strained layer thickness h_{crit}:

$$x = \frac{b\cos\lambda}{0.0836h_{crit}}\left[1 + \frac{1-\frac{\nu}{4}}{4\pi\cos^2\lambda\cos\phi(1+\nu)}\ln\frac{h_{crit}}{b}\right]. \tag{9.11}$$

In the absence of surface relaxation forces, the equilibrium critical thickness of a single strained epilayer upon a substrate of different lattice parameter according to Matthews and Blakeslee is given by the formulation reported in Ref. [1]. Inserting appropriate material parameters in Equation 9.11 and in the Matthews–Blakeslee formulation, $\cos\lambda = 0.5$, $\cos\phi = 0.816$, $\nu = 0.36$ (in the (1 1 1) plane) [21], and $b = 3.84$ Å for growth on the (0 0 1) surface, the equilibrium critical thickness calculated for the two different models is plotted in Figure 9.3. It is seen, at first glance, that our values of h_{crit} represented by Equation 9.11 are much larger than the values calculated using equation of Matthews and Blakeslee. Moreover, for a fractional atomic Ge content x greater than 0.5, the Matthews–Blakeslee formulation does not provide any value for h_{crit}. Additionally, we have compared our theoretical results with the published experimental data [14,15] of h_{crit} obtained from $Si_{1-x}Ge_x$/Si structures, grown by molecular beam epitaxy (MBE) at a growth temperature of 750°C. For each composition, there is good agreement between our theoretical results and experimental data reported. For example, for $x = 0.25$, h_{crit} was found to be approximately 200 Å. As seen in Figure 9.3, for this case, our analysis yields the value of 200 Å. The Matthews–Blakeslee model predicts only an equilibrium critical thickness value of 60 Å.

So far, we have been considering the equilibrium critical thickness of a coherently strained structure. Let us now consider film stresses for elastically strained $Si_{1-x}Ge_x$/Si layers. For the pseudomorphically strained layer case, where $p \rightarrow \infty$, according to the previous Equation 9.9, the in-plane film stress σ becomes

$$\sigma = \frac{2G(1+\nu)}{(1-\nu)}\left(0.0418x - \frac{b\cos\lambda}{2h}\right). \tag{9.12}$$

This equation can then be used to predict the stress response to a strain increment, which is related to the volume change of the layer. The first term on the right-hand side corresponds to a volume expansion due to Ge incorporation and the second one to a volume contraction via surface relaxation. Thus the residual in-plane strain is given by $\varepsilon = 0.0418x - (b\cos\lambda/2h)$. Figure 9.4 shows a plot of σ versus h for $x = 0.3$, $\nu = 0.28$ (in the (1 0 0) plane) [21], and $G = 51$ GPa. This relationship between coherency stress and surface relaxation stress has also been studied experimentally. The in-plane epitaxial film stress σ as a function of layer thickness was measured by a thin-film stress measuring apparatus, which

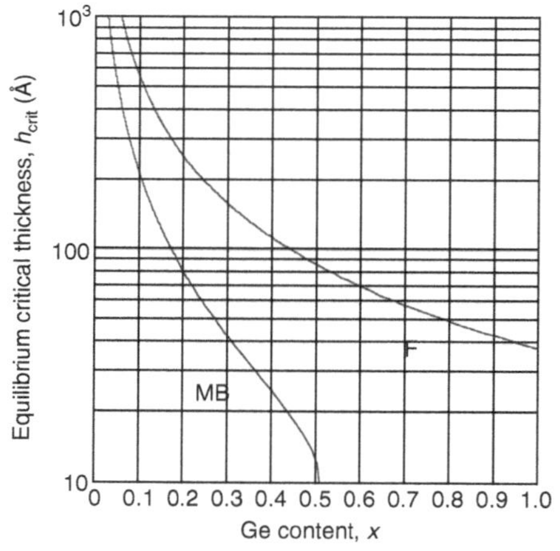

FIGURE 9.3 Comparison of predicted equilibrium critical thickness h_{crit} for relaxation models based on a nonstress-free body (MB, Matthews–Blakeslee) and a stress-free body (F, this study) as a function of Ge content x. (From A Fischer, H Kühne, M Eichler, F Holländer, and H Richter. *Physical Review B* 54:8761–8768, 1996. With permission.)

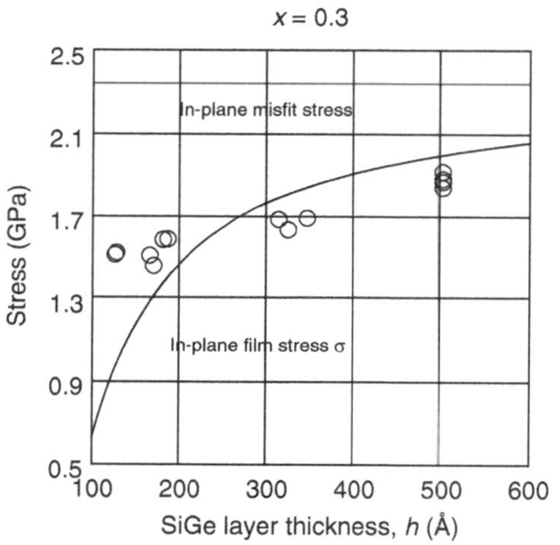

FIGURE 9.4 Theoretical (solid line) and experimental (○) in-plane epitaxial film stress σ as a function of thickness h for as-grown SiGe/Si. The in-plane misfit stress (elastic coherency stress) corresponding to a strain of $0.0418x$ is also shown. (From A Fischer, H Kühne, M Eichler, F Holländer, and H Richter. *Physical Review B* 54: 8761–8768, 1996. With permission.)

measures the changes in the radius of curvature of a substrate created by deposition of a stressed thin film on its surface. Our experimental data are also shown in Figure 9.4. To our knowledge, such measurements of surface relaxation stresses have not been reported previously. It is seen that there is good agreement between our theoretical results and experimental data. This tells us that the elastic coherency stress of the strained material in the plane of the surface is really affected by a large surface relaxation stress.

9.6 Plastic Deformation and Work Hardening in SiGe/Si

SiGe/Si structures may be grown to substantially greater thickness than our equilibrium critical thickness predictions, before misfit dislocation nucleation and propagation are observed. The heterostructure becomes metastable during growth because $\tau_{exc} > 0$. This metastable growth regime is concerned with the initial period of creep of diamond structure materials, in which marked plastic deformation under static load does not start abruptly. In our case of strained layer relaxation, such an initial period is characterized by an incubation time encouraged by lower growth temperature and lattice mismatch. The initial creep period is finally terminated by the onset of in-plane strain relief via plastic flow.

The ramifications of our model for plastic flow and work hardening in $Si_{1-x}Ge_x$/Si strained layer structures will now be discussed. According to Equation (9.9), the in-plane film stress σ becomes

$$\sigma = \frac{2G(1+\nu)}{(1-\nu)}\left(0.0418x - \frac{b\cos\lambda}{2R_{h,p}}\right). \tag{9.13}$$

Putting the values of the material parameters into Equations 9.4, 9.10, and 9.13, assuming a metastable epilayer with $x = 0.3$ and $h = 500$ Å, and recalling Equation 9.3, we can evaluate the variation in excess resolved shear stress τ_{exc} acting on the slip system and in-plane epitaxial film stress σ for the complete thermal relaxation process. At the beginning of deformation, where $p \to \infty$, $\tau_{exc} = 0.7$ GPa and $\sigma = 1.9$ GPa, as illustrated in Figure 9.5. During plastic flow via misfit dislocation generation and propagation, τ_{exc} and σ diminish continuously and then remain unchanged at zero and 1.15 GPa, respectively, whereas the shear self-stress component τ_s rises to its maximum value. At this equilibrium deformation stage, the externally applied misfit stress and the shear component of elastic stress field due to dislocation interaction compensate one another and strain relief via plastic flow comes to rest. Thus, as a result of work hardening, the lattice-mismatched epilayer will remain in a certain state of strain at the end of thermal relaxation process, i.e., the strained layer is in a stable state.

It should be mentioned here that investigations by Gillard et al. [20] suggest that dislocation blocking plays a certain role in limiting strain relaxation. We believe, as shown in the present and the previous sections, that the creep process in all the stages of strain relaxation in these structures is essentially determined by a purely elastic interaction between dislocations, which leads ultimately to work hardening of the material. Further experiments would be required to assess the relative contribution from blocking mechanism.

For homogenous deformation, our equilibrium theory for strained layer relaxation is able to predict the degree of in-plane misfit strain relaxation both at the point of onset of plastic flow and at the point of equilibrium strain state where the plastic flow comes to rest. Returning to Equation 9.13, which is the general in-plane film stress equation, we define the degree of strain relaxation γ as the relative lattice-mismatch accommodation by elastic surface relaxation and interfacial misfit dislocations compared to mismatch accommodation by the elastic in-plane strain alone. It can be written as the ratio of the in-plane strain relief to the in-plane misfit strain (coherency strain). For $Si_{1-x}Ge_x$/Si heterostructures, we get

$$\gamma = \frac{b\cos\lambda}{0.0836xR_{h,p}}. \tag{9.14}$$

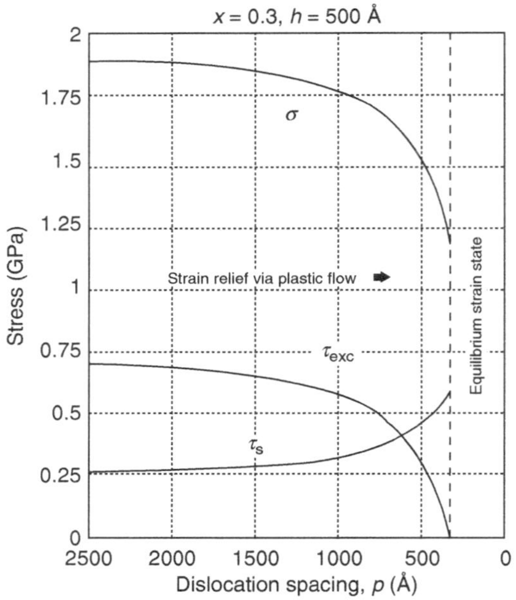

FIGURE 9.5 Decrease in the in-plane epitaxial film stress σ and excess resolved shear stress τ_{exc} during strain relief via plastic flow as a function of the dislocation spacing p. The shear self-stress component τ_s, the direction of strain relief via plastic flow, and the equilibrium strain state at the end of the thermal relaxation process are indicated. (From A Fischer, H Kühne, M Eichler, F Holländer, and H Richter. *Physical Review B* 54:8761–8768, 1996. With permission.)

For the point of strain relief onset via plastic flow, where $p \rightarrow \infty$, Equation 9.14 reduces to $\gamma_{el} = b\cos\lambda/0.0836xh$, which can be a general expression for the degree of elastic strain relief via surface relaxation. γ_{el} is a constant for a given metastable or stable SiGe/Si strained layer system. Again, according to Equation 9.14, the degree of total strain relaxation $\gamma_{t(0)}$ for the point of equilibrium strain state at the end of thermal relaxation process is given by $\gamma_{t(0)} = b\cos\lambda/0.0836xR_{h,p(0)}$, where the subscript $p(0)$ stands for the misfit dislocation spacing reached at the equilibrium deformation stage. In our case, for $Si_{0.7}Ge_{0.3}/Si$, $h = 500$ Å, we get $\gamma_{el} = 0.15$ and $\gamma_{t(0)} = 0.49$ with $p(0) = 330$ Å. As the corresponding degree of plastic strain relaxation is $\gamma_{pl(0)} = \gamma_{t(0)} - \gamma_{el}$, we thus have $\gamma_{pl(0)} = 0.34$. Thus, as stated previously, the lattice-mismatched epilayer remains in a certain state of strain at the end of thermal relaxation process. For the given heterosystem, as a result of work hardening, the residual in-plane strain is approximately 51%. Again, an elastic strain of 15% is retained in the epilayer by surface relaxation effects and about 34% of the in-plane strain due to lattice mismatch is relieved via plastic flow. It should be noted that, up to now, neither equilibrium theory was able to predict the correct relaxed strain in heteroepitaxial films.

9.7 Force Balance Model for Buried SiGe Strained Layers

Up to now, both the classical force balance [1,3–5,11,22,23)], and energy minimization [15,24] approaches for the determination of the critical thickness h_{crit} of a coherently strained layer structure at which dislocations form cannot predict this value for a Si-capped SiGe epilayer on Si substrate. The main difficulties to determine the exact strain relief onset via plastic flow in Si/SiGe/Si multilayer structures frequently arise because the classical theories are insensitive both to the effect of the free surface and the ratio of the stored elastic strain energy in the Si cap layer to that in the buried SiGe layer. Within the classical theoretical framework, it is impossible to decide at which cap layer thickness the plastic relaxation may take place through single misfit dislocation injection at the lower SiGe/Si interface or through the generation of misfit dislocation dipoles with segments located at both interfaces. Detailed experimental results of h_{crit} for different cap layer thicknesses have also not been available.

Here we extend the basic image-force equilibrium model for strained layer relaxation explained above to Si/SiGe/Si layers with a free surface at the top of the heteroepitaxial stack. The solution of the force balance problem will be carried out in plane stress [25] assuming equal elastic moduli. Far from any lateral free surface and referred to the slip plane {1 1 1} and slip directions ⟨1 1 0⟩, the state of stress is uniformly biaxial of magnitude τ, which is related to the in-plane strain ε through

$$\tau = \cos\lambda \cos\phi \frac{2G(1+\nu)}{(1-\nu)\cos\phi}\varepsilon. \tag{9.15}$$

The misfit strain arising from the lattice parameter difference between silicon and germanium in a SiGe epilayer on Si substrate is $0.0418x$. In a Si/SiGe/Si heterostructure, the strain that acts on a misfit dislocation–image dislocation pair for its generation is given by [26]

$$\varepsilon = 0.0418x - \frac{b\cos\lambda}{2(h+H)}, \tag{9.16}$$

where h and H are the thicknesses of the buried SiGe layer and the Si cap layer, respectively. This relation includes the surface relaxation effects on the strained layer stack and makes the crystal stress-free at the surface to satisfy the boundary conditions [17,18].

Counteracting the shear stress component τ are the dislocation self-stress terms [11,26]. These are associated with the strain energy required to create straight misfit dislocation segments at interfaces and the dislocation–dislocation interaction energy between the two parallel segments. The self-stress of a strain-relaxing 60° misfit dislocation on the slip plane of the lower SiGe/Si interface is

$$\tau_{SiGe} = \frac{Gb\left(1-\frac{\nu}{4}\right)}{4\pi(1-\nu)h\cos\phi}\ln\frac{h+H}{b}, \tag{9.17.1}$$

where $h + H$ is the outer cutoff distance of the interfacial dislocation, a measure of the energy stored in the elastic stress field of the misfit dislocation. Note that the strain energy can be regarded as either the work done on a radial cylinder cut surface in creating the dislocation or, equivalently, by Stoke's theorem, the stored elastic strain energy in a coaxial cylinder containing the dislocation. Then, the self-stress of a misfit dislocation on the slip plane of the upper Si/SiGe interface is of the form

$$\tau_{Si} = \frac{Gb\left(1-\frac{\nu}{4}\right)}{4\pi(1-\nu)h\cos\phi}\ln\frac{H}{b}, \tag{9.17.2}$$

where, in this case, H is the outer cutoff distance of the strain-relaxing interfacial dislocation, identical to the cap layer thickness. The attraction of the dipole segments of equal strength and opposite sign lying on adjacent interface planes is given by a stress term connected with a negative dislocation–dislocation interaction energy between the two parallel segments [18,27]:

$$\tau_{Si/SiGe} = -\frac{Gb\left(1-\frac{\nu}{4}\right)}{4\pi(1-\nu)h\cos\phi}\ln\frac{h+H}{h}. \tag{9.17.3}$$

where the negative interaction energy represents the decrease in free energy of the crystal when the dipole segments of opposite sign are brought together from a distance of $h + H$ to a separation h. Since the free energy change is a state function, it is independent of the path taken.

Now consider the ratio of the stored misfit strain in the Si cap layer to that in the buried SiGe layer of a coherent Si/SiGe/Si system to find a measure for the stored elastic strain energy in the layers to drive misfit dislocation generation. This ratio is also a measure for the different frequency of the occurrence of misfit dislocations at the heterointerfaces. The total misfit strain ε_t in a Si-capped SiGe epilayer with an average value of the fractional atomic Ge content $xh/(h+H)$ (see Equation 9.11) is given by

$$\varepsilon_t = 0.042x\frac{h}{h+H}. \tag{9.18.1}$$

Taking only the misfit strain ε_b in the buried SiGe layer, Equation 9.18.1 leads for $H = 0$ to the well-known expression

$$\varepsilon_b = 0.042x. \tag{9.18.2}$$

The misfit strain ε_c in the Si cap layer will then be given by the difference between the two strain components ε_t and ε_b. Thus we have finally

$$\varepsilon_c = -0.042x\frac{H}{h+H}. \tag{9.18.3}$$

This shows that, if the buried SiGe layer incorporates the nucleation energy for one misfit dislocation, the Si cap layer holds energy only for $H/(h+H)$ misfit dislocations. This energy density factor, denoted by δ can vary from zero for a capless structure up to 1 for an infinitely thick Si cap layer. In the latter case, only misfit dislocation dipoles occur during strained layer relaxation. Note that, up to now, no equilibrium or energy minimization model takes into account the stored elastic strain energy in the Si cap layer.

The excess resolved shear stress τ_{exc} driving the bending of threading dislocations to form single and/or double misfit dislocation segments will then be given by the difference between the external stress τ [Equations 9.15 and 9.16] and the internal stress components τ_{SiGe}, τ_{Si}, τ_{Si-Ge} [Equations 9.17.1–9.17.3]. Including the energy density factor δ (from Equations 9.18.1–9.18.3), we have $\tau_{exc} = \tau - \tau_{SiGe} - \delta\tau_{Si} - \delta\tau_{Si-SiGe}$. Taking the equilibrium condition for the point of strain relief onset via plastic flow, i.e., $\tau_{exc} = 0$, this force balance approach leads to the following implicit expression for the critical strained layer thickness h_{crit} as a function of the Ge content x and the cap layer thickness H:

$$x = \frac{b\cos\lambda}{0.0836h_{crit}}\left[\frac{h_{crit}}{h_{crit}+H} + \frac{1-\frac{\nu}{4}}{4\pi\cos^2\lambda\cos\phi(1+\nu)}\left(\ln\frac{h_{crit}+H}{b} + \delta\ln\frac{H}{b} - \delta\ln\frac{h_{crit}+H}{h_{crit}}\right)\right] \tag{9.19}$$

with $\delta = \dfrac{H}{h_{crit}+H}$.

This formulation gives a generalized statement of the balance between external and internal forces acting at planar Si/SiGe/Si structures. It is a measure of the transition from the thermodynamically stable to the metastable state of the strained layer system. Finally, it is worth considering the transition thickness for loss of coherency in uncapped SiGe heterostructures. For $H = 0$, we easily obtain the solution for the critical thickness of a SiGe/Si layer substrate system given by Equation 9.11. Inserting appropriate material parameters given above in Equation 9.19 the equilibrium critical thickness h_{crit} calculated for the Si cap layer thicknesses $H = 0$, 500 Å, and 1000 Å is plotted versus Ge content of the SiGe layer in Figure 9.6. Our experimental data obtained from MBE [26,28] and chemical vapor deposition (CVD) epitaxy [26,29] are also shown. The agreement between calculated and measured values is evident.

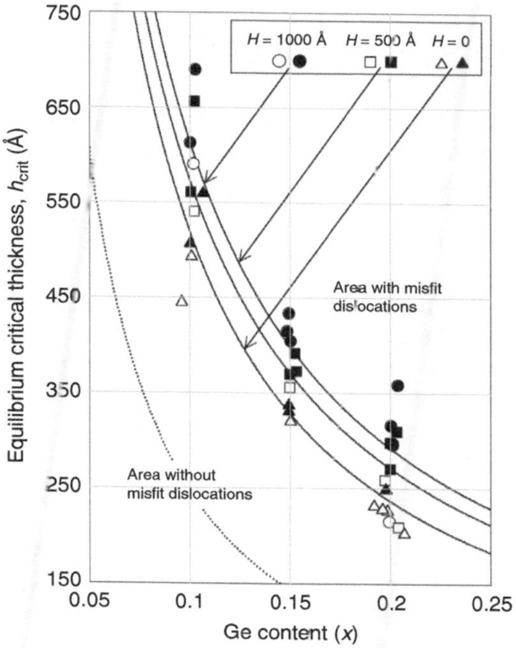

FIGURE 9.6 Theoretical values (solid lines) and experimental measurements (symbols) of the equilibrium critical thickness h_{crit} as a function of fractional atomic Ge content x for different Si cap layer thicknesses H. The open data points correspond to structures that maintained coherency; the filled data points correspond to structures that became partially incoherent. The x-ray measuring accuracies are Δh, $\Delta H = 10$ Å, and $\Delta x = 0.005$. For comparison, the predicted equilibrium critical thickness for relaxation models based on a nonstress-free body (From JW Matthews and AE Blakeslee. *Journal of Crystal Growth* 27:118–125, 1974; JW Matthews and AE Blakeslee. *Journal of Crystal Growth* 32:265–273, 1976) is shown by the dotted line for uncapped SiGe/Si heterostructures. (From A Fischer, H-J Osten, and H Richter. *Solid-State Electronics* 44:869–873, 2000. With permission.)

9.8 Summary

In this chapter I have attempted to provide a physical basis for a different approach in equilibrium theory for strain relaxation in metastable heteroepitaxial semiconductor structures. This approach includes the surface relaxation effects on the strained layer structure and the elastic interaction between straight misfit dislocations. The main purpose of this work was to develop a straightforward treatment, valid in linear elasticity, for the relatively complicated case of deformation of a finite body containing internal stresses and subjected to lattice-mismatch forces. The model we have outlined yields an expression in terms of the excess resolved shear stress for strained layer relaxation, so that the elastic strain retained by the epilayer, the equilibrium critical thickness of the strained layer, the initial and residual in-plane strain, the film stress, work hardening effects on the material, the degree of strain relaxation, and the misfit dislocation density, reached at the equilibrium strain state, can be predicted.

I believe that the image-force method proposed here provides an equilibrium theory that correctly predicts the coherency and relaxation behavior of SiGe/Si and Si/SiGe/Si heterostructures. Finally, this more refined model yields better agreement between computed and measured values, and provides a consistent picture of the complex mechanism for strain relief and defect propagation in a strained layer on a lattice-mismatched substrate.

References

1. JW Matthews and AE Blakeslee. Defects in epitaxial multilayers: I. Misfit dislocations. *Journal of Crystal Growth* 27:118–125, 1974.
2. JW Matthews and AE Blakeslee. Defects in epitaxial multilayers: II. Dislocation pile-ups, threading dislocations, slip lines and cracks. *Journal of Crystal Growth* 29:273–280, 1975.
3. JW Matthews and AE Blakeslee. Defects in epitaxial multilayers: III. Preparation of almost perfect multilayers. *Journal of Crystal Growth* 32:265–273, 1976.
4. BW Dodson and JY Tsao. Relaxation of strained-layer semiconductor structures via plastic flow. *Applied Physics Letters* 51:1325–1327, 1987.
5. BW Dodson and JY Tsao. Erratum: relaxation of strained-layer semiconductor structures via plastic flow. *Applied Physics Letters* 52:852, 1988.
6. E Kasper and HJ Herzog. Elastic strain and misfit dislocation density in $Si_{0.92}Ge_{0.08}$ films on silicon substrates. *Thin Solid Films* 44:357–370, 1977.
7. A Fischer and H Richter. On plastic flow and work hardening in strained layer heterostructures. *Applied Physics Letters* 64:1218–1220, 1994.
8. GC Osbourn. Strained-layer superlattices: a brief review. *IEEE Journal Quantum Electronics* QE-22:1677–1681, 1986.
9. DC Houghton. Misfit dislocation dynamics in $Si_{1-x}Ge_x$/(100)Si: Uncapped alloy layers, buried strained layers, and multiple quantum wells. *Applied Physics Letters* 57:1434–1436, 1990.
10. DC Houghton. Nucleation rate and glide velocity of misfit dislocations in $Si_{1-x}Ge_x$/(100) Si heterostructures. *Applied Physics Letters* 57:2124–2126, 1990.
11. DC Houghton. Strain relaxation kinetics in $Si_{1-x}Ge_x$/Si heterostructures. *Journal of Applied Physics* 70:2136–2151, 1991.
12. R Hull, JC Bean, DJ Werder, and RE Leibenguth. Activation barriers to strain relaxation in lattice-mismatched epitaxy. *Physical Review B* 40:1681–1684, 1989.
13. R Hull. Equilibrium theories of misfit dislocation networks in the SiGe/Si system. In: E Kasper (ed.), *Properties of Strained and Relaxed Silicon Germanium.* London: Inspec, 1995.
14. E Kasper. Strained layer superlattices. In: EE Mendez and K von Klitzing (eds), *Physics and Applications of Quantum Wells and Superlattices.* New York: Plenum Press, 1987.
15. R People and JC Bean. Calculation of critical layer thickness versus lattice mismatch for Ge_xSi_{1-x}/Si strained layer heterostructures. *Applied Physics Letters* 47: 322–324, 1985.
16. BA Fox and WA Jesser. The effect of frictional stress on the calculation of critical thickness in epitaxy. *Journal Applied Physics* 68:2801–2808, 1990.
17. A Fischer, H Kühne, M Eichler, F Holländer, and H Richter. Strain and surface phenomena in SiGe structures. *Physical Review B* 54:8761–8768, 1996.
18. JP Hirth and J Lothe. *Theory of Dislocations.* New York: McGraw-Hill, 1968.
19. JR Willis, SC Jain, and R Bullough. Work hardening and strain relaxation in strained-layer buffers. *Applied Physics Letters* 59:920–922, 1991.
20. VT Gillard, WD Nix, and LB Freund. Role of dislocation blocking in limiting strain relaxation in heteroepitaxial films. *Journal of Applied Physics* 76:7280–7287, 1994.
21. JJ Wortmann and RA Evans. Young's modulus, shear modulus, and Poisson's ratio in silicon and germanium. *Journal of Applied Physics* 36:153–156, 1965.
22. CG Tuppen and CJ Gibbings. A quantitative analysis of strain relaxation by misfit dislocation glide in $Si_{1-x}Ge_x$/Si heterostructures. *Journal of Applied Physics* 68:1526–1534, 1990.
23. R Hull, JC Bean, D Bahnck, LJ Peticolas Jr, and KT Short. Interpretation of dislocation propagation velocities in strained Ge_xSi_{1-x}/Si(100) heterostructures by the diffusive kink pair model. *Journal of Applied Physics* 70:2052–2065, 1991.
24. JH van der Merwe. Structure of epitaxial crystal interfaces. *Surface Science* 31:198–228, 1972.
25. SP Timoshenko and NJ Goodier. *Theory of Elasticity.* New York: McGraw-Hill, 1982.

26. A Fischer, H-J Osten, and H Richter. An equilibrium model for buried SiGe strained layers. *Solid-State Electronics* 44:869–873, 2000.
27. FRN Nabarro. The mathematical theory of stationary dislocations. *Advances in Physics* 1:269–394, 1952.
28. H-J Osten, HP Zeindl, and E Bugiel. Considerations about the critical thickness for pseudomorphic $Si_{1-x}Ge_x$ growth on Si(001). *Journal of Crystal Growth* 143:194–199, 1994.
29. B Tillack, HH Richter, G Ritter, A Wolff, G Morgenstern, and C Eggs. Monitoring of deposition and dry etching of Si/SiGe multiple stacks. *Journal Vacuum Science Technology B* 14:102–105, 1996.

10

Electronic Properties of Strained Si/SiGe and $Si_{1-y}C_y$ Alloys

Judy L. Hoyt
Massachusetts Institute of Technology

10.1 Introduction

Knowledge of the basic electronic properties of silicon-based heterostructures is essential to understand their applications. This chapter is a review of the electronic properties of strained SiGe, strained Si on relaxed SiGe, and a brief introduction to strained $Si_{1-y}C_y/Si$ heterostructures. Because the energy band lineups are critical in determining the types of applications of these materials, the emphasis here is on reviewing theoretical and experimental determinations of the energy band lineups (Sections 10.2 and 10.3). Section 10.4 introduces transport properties in such materials, with an eye towards field-effect transistor (FET) applications, though readers are referred to other chapters of this volume for details on transport and FET applications.

10.2 Strain and its Impact on Energy Bands

Figure 10.1a schematically illustrates the relaxed, or equilibrium lattice structures for Si, random $Si_{1-x}Ge_x$ alloys, and Ge. The equilibrium lattice parameters a_0 for Si and Ge are 5.431 and 5.646 Å, respectively [1]. This corresponds to a lattice mismatch to Si

$$m = 100 \times \frac{a_0(\text{alloy}) - a_0(\text{Si})}{a_0(\text{Si})}$$

equal to approximately 4.1% (for pure Ge). When an alloy is formed of Si and Ge the equilibrium lattice parameter can roughly be linearly interpolated between a_0 for Si and Ge, though there is a slight negative deviation from Vegard's law [1]. Figure 10.1b illustrates the case of pseudomorphic epitaxial growth in which the in-plane lattice parameter of the epitaxial layer compresses or stretches in order to match that of the substrate, and the out-of-plane parameter is elongated or compressed accordingly. Thus, when strained SiGe is grown pseudomorphically on relaxed Si (left-hand side of Figure 10.1b), the layer experiences in-plane biaxial compression. Conversely, when a thin layer of strained Si is grown on

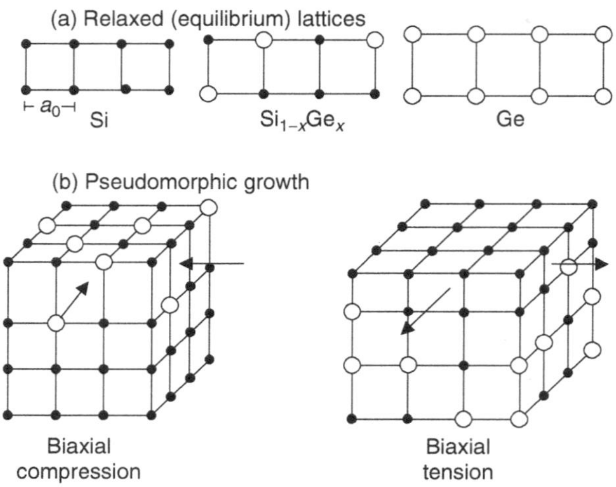

FIGURE 10.1 Schematic illustration of the lattice structures for (a) relaxed Si, $Si_{1-x}Ge_x$, and Ge, and (b) pseudomorphic epitaxial layers of strained SiGe on Si in biaxial compression (left-hand side), and of strained Si on relaxed SiGe in biaxial tension (right-hand side). The symbol a_0 refers to the equilibrium lattice parameter of the material.

relaxed SiGe (right-hand side of the figure), the Si experiences biaxial tensile stress. As will be discussed below, the energy band lineups are quite different in these two cases, and determine, to a large extent, the device applications relevant for the two materials structures.

Perfect pseudomorphic growth is only possible up to a certain strained layer thickness, above which misfit dislocations form and the strain begins to relax, as discussed elsewhere in this handbook. Si and Ge are completely miscible and the full range of alloy compositions can be grown in single crystal form. This is not true of Si and C, though it is possible to grow alloys of $Si_{1-y}C_y$ with up to ~2 at.% C in substitutional lattice sites. The lattice mismatch between Si and $Si_{0.5}C_{0.5}$ is −20%, much larger than that between Si and $Si_{0.5}Ge_{0.5}$ (+2%). $Si_{1-y}C_y$ layers containing up to a few percent C can be grown pseudomorphically on Si, with the $Si_{1-y}C_y$ layer experiencing biaxial tensile stress, analogous to the case of strained Si on relaxed SiGe. Growth of $Si_{1-y}C_y$ is discussed elsewhere in this handbook.

Perhaps the most well-known electronic property of SiGe alloys is the reduction in the indirect energy bandgap compared to that of Si as Ge is added to the alloy. Figure 10.2 shows the energy bandgap difference, $\Delta E_g = E_g(\text{SiGe}) - E_g(\text{Si})$ between SiGe alloys and Si as a function of the Ge percentage in the alloy. The upper curve corresponds to the data for unstrained or relaxed alloys, measured by Braunstein et al. by optical absorption at ~300 K [2]. The conduction band minimum is along the Δ direction in the Brillion zone for alloy compositions up to roughly 85% Ge [2]. For higher Ge contents, where the magnitude of the bandgap difference increases rapidly with Ge composition, the conduction band minimum appears along the L direction, similar to that for pure Ge. The remainder of the data in Figure 10.2 corresponds to ΔE_g for strained SiGe grown on relaxed Si. As shown in the figure, ΔE_g as determined from photocurrent measurements by Lang et al. [3], and extracted from heterojunction bipolar transistors (HBTs) by King et al. [4], is in reasonable agreement with theoretical predictions by People [5]. It is clear that strain makes a large contribution to the bandgap reduction in SiGe. This is true in general in silicon-based heterostructures, where the strain splitting of degeneracies at the band edges tends to close the bandgap relative to the unstrained state, as discussed below. Note that the electron affinity of relaxed SiGe alloys is not well documented in the literature. Morar et al. employed electron-energy-loss

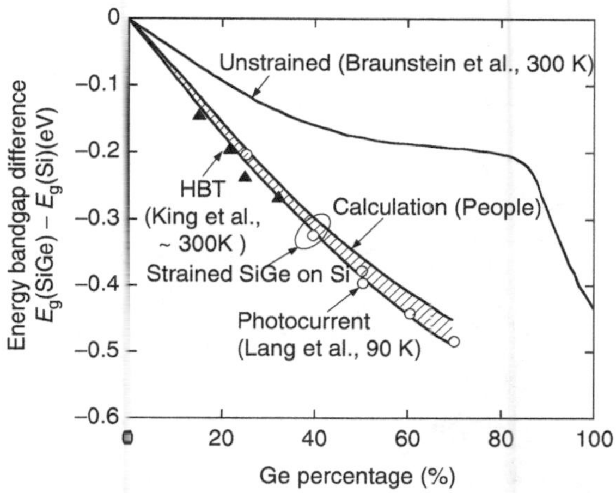

FIGURE 10.2 The energy bandgap difference, $\Delta E_g = E_g$(SiGe) − E_g(Si) between SiGe alloys and Si as a function of the Ge percentage in the alloy. The upper curve is for unstrained alloys, measured by Braunstein et al. by optical absorption at ~ 300 K [2]. The remainder of the data corresponds to ΔE_g for strained SiGe grown on relaxed Si, including photocurrent measurements by Lang et al. (○) [3], and heterojunction bipolar transistors (HBTs) by King et al. (▲) [4]. The hatched region represents the range of theoretical predictions due to uncertainties in the values of the deformation potentials [5]. (From C.A. King, J.L. Hoyt, and J.F. Gibbons. *IEEE Trans. Elec. Dev.* 36:2093–2104, 1989. With permission.)

spectroscopy to extract the variation in the conduction band minimum for a range of relaxed SiGe alloy compositions [6], which gives some indication of electron affinity variation. In many applications, it is assumed that the conduction bands of relaxed Si and relaxed SiGe are closely aligned.

Figure 10.3 illustrates the energy band lineups for some important silicon heterostructures, including (a) compressively strained SiGe on relaxed Si and (b) strained Si grown on relaxed SiGe. The band alignment for strained SiGe on Si (a) is such that almost all of the bandgap difference appears in the valence band [7]. This material is thus ideally suited for the fabrication of npn Si/SiGe/Si heterojunction bipolar transistors, as well as for applications such as p-channel FETs where confinement of holes in the SiGe layer is desired. Strained Si on relaxed SiGe results in a Type II alignment, with electrons preferentially populating the strained Si layers and holes the SiGe layers (Figure 10.3b). In the figure, ΔE_s represents the strain-induced energy splitting of the conduction band minimum, where the sixfold degenerate conduction band at the Δ point splits into a set of twofold (Δ_2) and fourfold (Δ_4) degenerate valleys. The constant energy surfaces corresponding to these valleys are shown in Figure 10.3c. ΔE_s is directly proportional to the lattice distortion, and can be calculated from deformation potential theory [8,9]. It can be shown that ΔE_s corresponds to roughly 67 meV/10% Ge in the substrate, for strained Si on relaxed SiGe, and approximately 66 meV/1% C for strained $Si_{1-y}C_y$ on cubic Si [10], as illustrated in Figure 10.3d. The strain-induced splitting of the conduction band degeneracy for strained Si on relaxed SiGe leads to a configuration that is particularly favorable to in-plane electron transport [11,12]. In this case, electrons preferentially occupy the lower-energy Δ_2 conduction band, which has a smaller in-plane effective transport mass than for the case of unstrained Si, where the in-plane effective mass is calculated by averaging over all six valleys. In addition to this effect, intervalley scattering is suppressed by the conduction band energy splitting. These effects contribute to the enhanced electron mobility in strained Si that is being leveraged in CMOS circuits, as discussed in Chapter 18 of the book *Silicon Heterostructure Devices.*

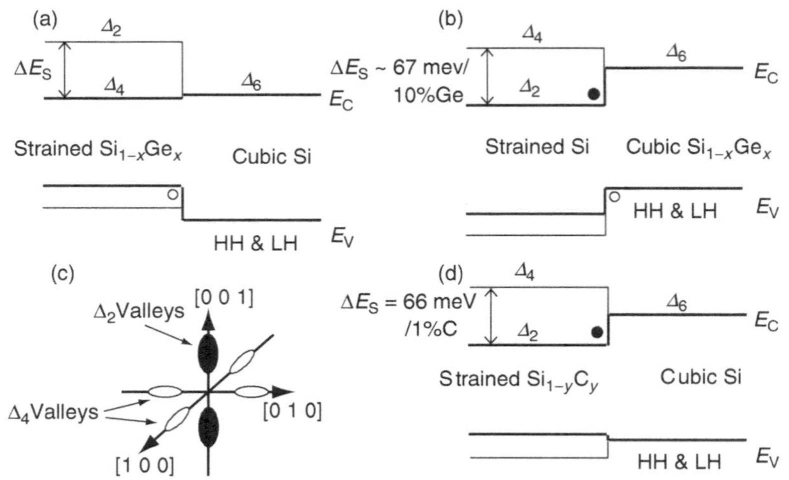

FIGURE 10.3 Energy band lineups for various silicon heterostructures, including (a) compressively strained SiGe on relaxed Si and (b) strained Si grown on relaxed SiGe. The heavy solid lines indicate the band edges. ΔE_s represents the strain-induced energy splitting of the conduction band minimum. The conduction band constant energy surfaces are shown in (c), and the band lineup for strained $Si_{1-y}C_y$ on cubic (relaxed) Si is illustrated in (d).

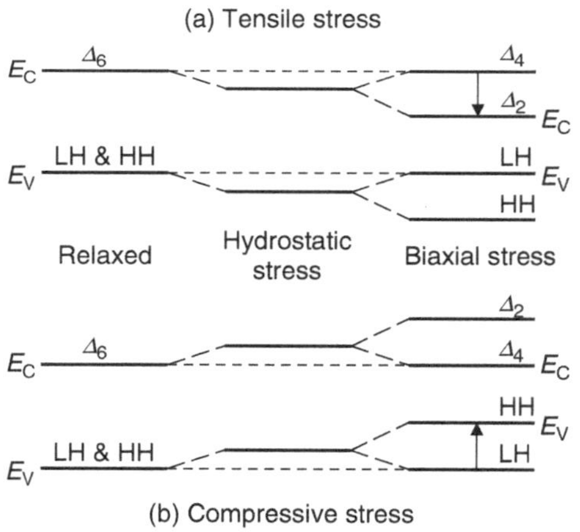

FIGURE 10.4 Schematic illustration of the effects of hydrostatic and biaxial stress on the energy bands in Si for (a) tensile and (b) compressive stress.

Biaxial strain can be decomposed into a hydrostatic contribution (equal dilation in all three directions) plus uniaxial strain in the direction normal to epitaxial growth. The hydrostatic component leads to an energy band shift (and change of the bandgap) while the uniaxial component results in a splitting of the conduction and valence band edge degeneracies. Figure 10.4 shows schematically the effects of hydrostatic and biaxial stress on the energy bands in Si for both (a) tensile and (b) compressive stress. For example, biaxial tensile stress (a) corresponds to hydrostatic tension combined with uniaxial compression in the *z*-direction. For this case, a uniform hydrostatic stretching of the lattice shifts the conduction and valence band edges to lower energies relative to their equilibrium positions, while the uniaxial contribution induces a splitting that does not affect the average band energy. The Δ_2 and Δ_4 bands move in opposite directions and the Δ_2 level moves by twice the amount as the Δ_4 level. One can also see from the figure that in the absence of other (e.g., chemical) effects, biaxial tension favors the introduction of a conduction band offset while biaxial compression tends to produce a valence band offset. This is due to the particular way in which hydrostatic and uniaxial stress affect the bands.

Strain effects are only part of the story of the energy band lineup at a heterointerface. Another effect comes from the chemical difference between the two semiconductor materials. Van de Walle and Martin were the first to discuss both the chemical and strain effects on the band alignment, and presented a methodology for calculating the valence band offsets in the Si/Ge system for various strain states [13]. Figure 10.5 shows the energy bands as calculated by Van de Walle for both strained SiGe on cubic Si (biaxial compression) and strained SiGe on cubic Ge (biaxial tension). For thorough theoretical reviews of the calculation of energy band lineups and the impact of stress on the band offsets, the reader is referred to Refs. [13–16]. Rieger and Vogl [15] and Schaffler [16] calculated and plotted the theoretical conduction and valence band offsets for a $Si_{1-y}Ge_y$ strained layer on a $Si_{1-x}Ge_x$ substrate.

10.3 Measurements of Energy Band Lineups

For compressively strained SiGe on cubic Si with Ge contents below 50% it is reasonably well established that the conduction band offset is less than 20 meV. In fact, rigorous studies using photoluminescence suggest that the Si and strained SiGe conduction bands are very close to each other for a Ge composition of 30% [17]. In the case of strained Si/strained SiGe/relaxed SiGe heterostructures, the uncertainty in the theoretical values of the band lineups can be as high as ± 100 meV, particularly for higher Ge contents, which is too large for device applications. For example, in heterojunction bipolar transistor applications, the collector current varies exponentially with the band offset $\Delta E_g/kT$, and in FETs the threshold voltage is often proportional to an energy band offset in the heterostructure. Hence, accurate determination of

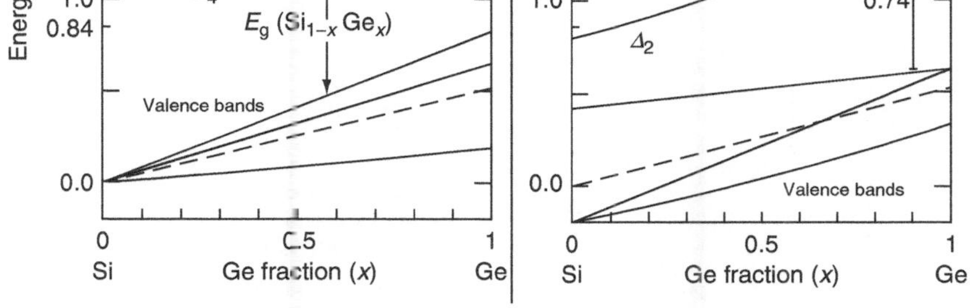

FIGURE 10.5 Energy bands as calculated by Van de Walle and Martin [13] for both strained SiGe on cubic Si (biaxial compression) and strained SiGe on cubic Ge (biaxial tension). The three strain-split valence bands represent the heavy hole, light hole, and split-off hole bands. The dashed lines represent the weighted average positions of the conduction and valence bands before the uniaxial strain splitting is taken into account. It should be noted that the data of Braunstein et al. [2] were used to place the weighted average conduction band relative to the valence bands. (From G.G. Van de Walle and R.M. Martin. *Phys. Rev. B* 34: 5621–5634, 1986. With permission.)

such quantities is required. The following sections discuss measurements of the energy band lineups in silicon heterostructures.

The Strained Si/Strained $Si_{1-y}Ge_y$/Relaxed $Si_{1-x}Ge_x$ System

There are a number of methods used for estimating energy band offsets at semiconductor heterojunctions, including methods such as photoluminescence and x-ray photoelectron spectroscopy (XPS). XPS generally has error bars on the order of 0.1 eV, too large for electronic device applications. Electrical techniques such as admittance spectroscopy and capacitance voltage profiling of reverse-biased Schottky diodes are popular methods [18,19], though care must be taken as both techniques can be influenced by the presence of deep levels within the depletion region or interface states that can impact the magnitude of the extracted energy band offset.

A variation on the Schottky diode methods that is particularly useful for thin heterostructures of interest in FETs is the MOS capacitance–voltage (*C–V*) method, as applied to MOS capacitors or MOSFETs [20–22]. By simulating and fitting the measured gate-to-channel capacitance characteristics of strained Si MOSFETs, Welser was able to extract values for the conduction and valance band offsets, ΔE_c and ΔE_v, for strained Si/relaxed SiGe heterojunctions [21]. These results are illustrated in Figure 10.6. As shown in the figure, there is generally good agreement between the magnitude of the extracted band offsets (symbols) and the theoretical calculations (lines). The bandgap of strained Si can be estimated by

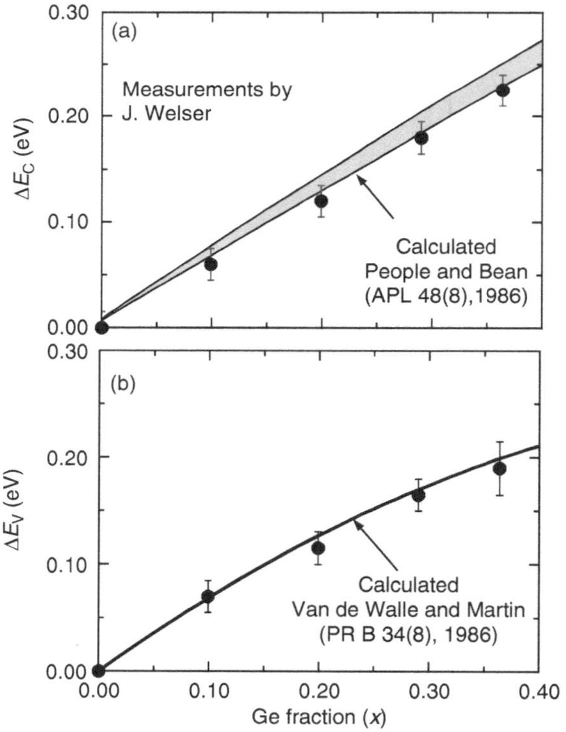

FIGURE 10.6 Conduction (a) and valence (b) band offsets for strained Si/relaxed SiGe heterojunctions determined by Welser [21]. Solid symbols are experimentally extracted from modeling of *C–V* characteristics measured for strained Si MOSFETs. The band alignment is as shown in Figure 10.3(b). (From J.J. Welser. Ph.D. Thesis. Stanford, 1994. With permission.)

combining these data with the measured data for the bandgap of relaxed SiGe from Braunstein, according to E_g(strained Si) = E_g(relaxed SiGe) − ΔE_c + ΔE_v. Figure 10.7 illustrates the resulting estimate of the strained Si bandgap as a function of the Ge fraction of the SiGe substrate. The results are in good agreement with the calculations from People and Bean [7]. Note that the error bars in the figure represent the uncertainty in the fitting as opposed to an absolute error. The impact of uncertainties in the strain-induced density of states and quantum corrections on such analysis is discussed in more detail below in the analysis of dual-channel structures.

A particularly exciting heterostructure for MOSFET applications is the dual-channel structure [23,24] shown schematically in Figure 10.8 This structure has a top strained Si layer that forms a high-mobility electron channel and a slightly buried, strained high-Ge-content SiGe channel that serves as a high-hole mobility channel. The high Ge content in the buried layer is made possible in part by the relaxed SiGe layer that reduces the overall mismatch. It has been demonstrated that both high electron and hole mobility enhancements can be achieved simultaneously in dual-channel structures [25].

It is critical to know the energy band offsets in such structures in order to predict hole confinement, mobility, and device properties such as threshold voltage and subthreshold slope. The valence band offset, ΔE_v, between the strained Si and strained SiGe layers is particularly important. For the purpose of extracting this energy band offset, p-type heterostructures shown schematically in Figure 10.8 have been fabricated [26]. Figure 10.9a illustrates the energy band diagram for these MOS capacitors at flatband (solid lines) and negative bias (dashed lines). Under increasing negative bias it is clear that holes first accumulate in the buried SiGe well and then at the oxide-strained Si interface. The resulting $C-V$ curves of these MOS capacitors are illustrated in Figure 10.9b. The plot is divided into three regions: inversion on the surface oxide–silicon interface (gate voltages greater than zero) with electrons in the strained Si layer, accumulation of holes in the SiGe well, and accumulation of holes at the oxide–silicon interface. The width of the characteristic plateau region is indicative of the valence band offset, ΔE_v, as shown in the figure. This feature is utilized in order to extract the valence band offset.

One-dimensional simulations have been performed to fit the experimental data using the Dessis device simulator [27]. Quantum effects are taken into account in the simulation using the density

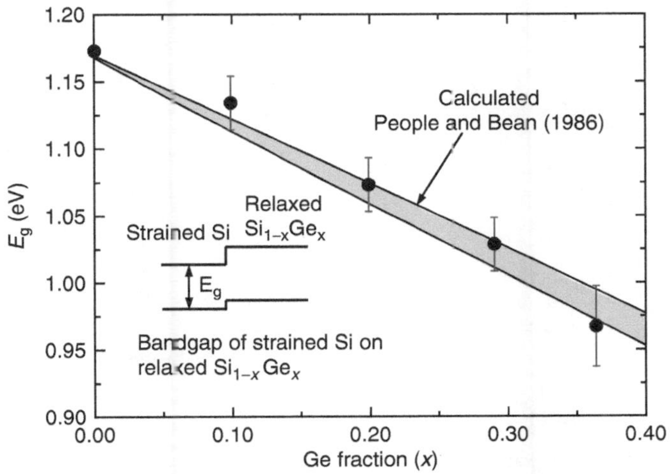

FIGURE 10.7 The bandgap energy of strained Si on relaxed SiGe as determined by Welser [21], by combining the data in Figure 10.6 with the data for the bandgap of relaxed SiGe from Braunstein et al. [2], according to E_g(strained Si) = E_g(relaxed SiGe) − $\Delta E_c + \Delta E_v$. (From J.J. Welser. Ph.D. Thesis. Stanford, 1994. With permission.)

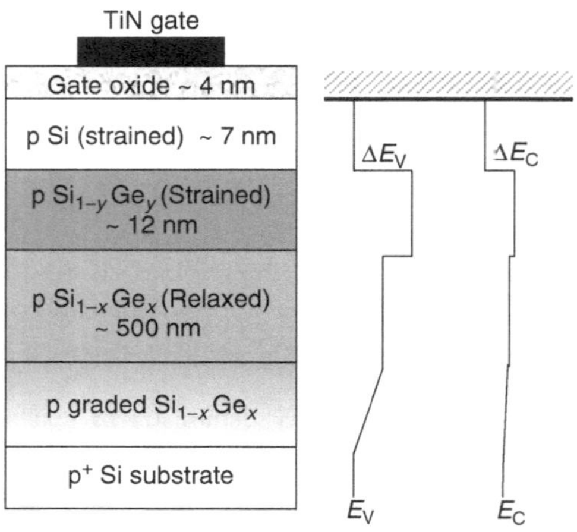

FIGURE 10.8 Schematic illustration of an MOS capacitor formed on a dual-channel heterostructure, and associated energy band diagram. This structure has a top strained Si layer that serves as a high-mobility electron channel and a slightly buried, strained high-Ge-content SiGe layer that serves as an enhanced hole mobility channel.

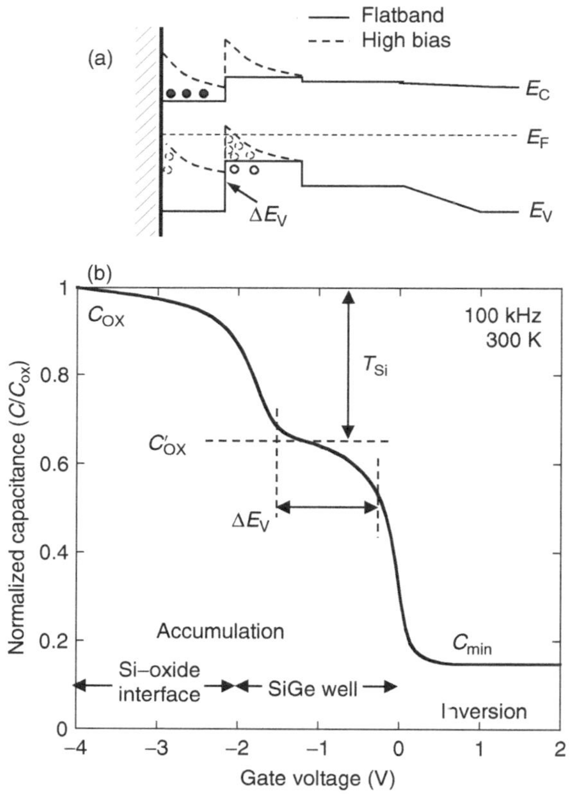

FIGURE 10.9 (a) Energy band diagram at flatband (solid lines) and negative gate bias (dashed lines), and (b) C–V data for a dual-channel MOS capacitor shown schematically in Figure 10.8, with $y = 0.6$ and $x = 0.3$. The plateau feature in the C–V characteristics, observed when holes accumulate in the SiGe well, is related to the magnitude of the valence band offset ΔE_V. (From C. Ni Chleirigh, C. Jungemann, J. Jung, O.O. Olubuyide, and J.L. Hoyt. Proceedings of the ECS Symposium on SiGe: Materials, Processing and Devices, October 2004, PV 2004-7, pp. 99–109, 2004. Pennington, N.J. With permission.)

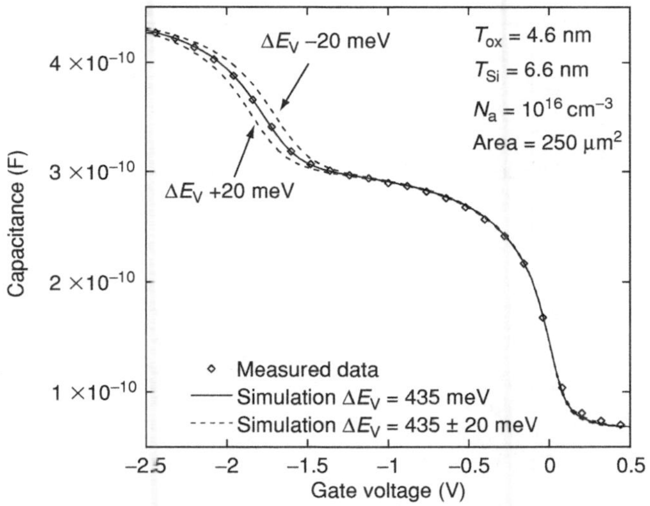

FIGURE 10.10 Measured (symbols) and simulated (lines) *C–V* characteristics for an MOS structure containing strained Si/strained $Si_{0.4}Ge_{0.6}$ on a relaxed $Si_{0.7}Ge_{0.3}$ substrate. The dashed lines show simulations for $\Delta E_v = 435 \pm 20$ meV. (From C. Ni Chleirigh, C. Jungemann, J. Jung, O.O. Olubuyide, and J.L. Hoyt. Proceedings of the ECS Symposium on SiGe: Materials, Processing and Devices, October 2004, PV 2004-7, pp. 99–109, 2004. Pennington, N.J. With permission.)

gradient quantum correction model. Figure 10.10 compares measured (symbols) and simulated (lines) *C–V* characteristics for a structure consisting of strained Si/strained $Si_{0.4}Ge_{0.6}$ on a relaxed $Si_{0.7}Ge_{0.3}$ substrate. The dashed lines show simulations for $\Delta E_v = 435 \pm 20$ meV, illustrating that the quality of the fit is very sensitive to small changes in the valence band offset.

It has been found that the *C–V* simulations are sensitive to the values used for the effective valence band density of states N_v in the strained Si. Note that this density of states decreases with increasing strain and Ge content due to the splitting and deformation of the valence bands. Use of the theoretical values for N_v for strained Si produces a much better fit to the *C–V* data in the portion of the curve corresponding to the population of the strained Si valence bands [26]. Analysis indicates that for this sample, a 15 meV difference in the extracted value of ΔE_v can result from the use of the unstrained values of the density of states in place of the strained values. It is thus important to use a self-consistent parameter set including both N_v and ΔE_v when modeling dual-channel MOSFETs.

Figure 10.11 illustrates the extracted ΔE_v for strained Si/strained $Si_{1-y}Ge_y$ on relaxed $Si_{1-x}Ge_x$ heterostructures as a function of the compositions x and y. The valence band offset increases with both composition and strain in the $Si_{1-y}Ge_y$ layer. Linear fits to the data yield an average slope of roughly 100 meV/10% Ge in the strained $Si_{1-y}Ge_y$ layer. This follows the theoretical trend, which displays a slope of ~ 75 meV/10% Ge.

The Si/$Si_{1-y}C_y$ System

Figure 10.3d illustrates the band lineup for strained $Si_{1-y}C_y$ grown on Si, for C compositions less than ~2 at.%. Such $Si_{1-y}C_y$ layers are strained in biaxial tension. The situation is complementary to that of strained SiGe grown on Si, and the $Si_{1-y}C_y$/Si band lineup, with a conduction band offset and little valence band offset is ideal for the fabrication of pnp Si/$Si_{1-y}C_y$/Si HBTs [28]. Theoretical predictions of conduction band offsets have been reported to accompany the addition of C to Si [29], and there has been some controversy concerning the presence of large conduction band offsets in the SiGeC/Si system [22,30–32]. The estimated conduction band offset between strained $Si_{1-y}C_y$ and Si is approximately 65 meV/at.% C. A combination of MOS *C–V* analysis to extract ΔE_c, similar to that employed above, and

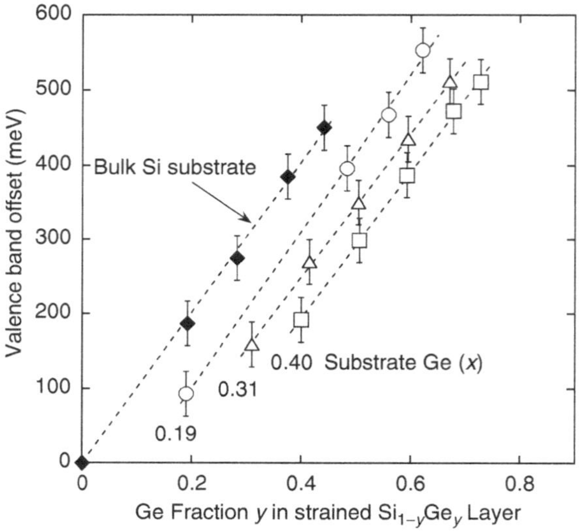

FIGURE 10.11 Extracted ΔE_v for strained Si/strained $Si_{1-y}Ge_y$ on relaxed $Si_{1-x}Ge_x$ heterostructures as determined by fitting MOS C–V analysis. Theoretical values for the effective density of states in strained Si and strained $Si_{1-y}Ge_y$ were used in the extractions. (From C. Ni Chleirigh, C. Jungemann, J. Jung, O.O. Olubuyide, and J.L. Hoyt. Proceedings of the ECS Symposium on SiGe: Materials, Processing and Devices, October 2004, PV 2004–7, pp. 99–109, 2004. Pennington, N.J. With permission.)

temperature-dependent measurements of the collector current in pnp $Si/Si_{1-y}C_y/Si$ HBTs to extract ΔE_g, indicates that for strained $Si_{1-y}C_y$ on Si, most of the band offset is in the conduction band [33]. These results are summarized in Figure 10.12. Note that when plotted as a function of the magnitude of the lattice mismatch to silicon (upper x-axis in the figure), the magnitude of the bandgap offset per percent mismatch for strained $Si_{1-y}C_y$ on Si is close to that for strained SiGe on Si.

10.4 Introduction to Transport Properties

Although minority carrier diffusion in SiGe has been studied for modeling of npn Si/SiGe HBTs [35–37], the greatest excitement exists today for understanding transport in strained Si and strained SiGe layers for high-speed FET applications, including digital CMOS. For the Si conduction band, it has already been mentioned that the biaxial tensile strain-induced splitting of the degeneracies at the Δ point is beneficial to in-plane electron transport. There is little deformation of the Si conduction band ellipsoids themselves, but energy shifting and repopulation of the valleys in k-space [15]. The resulting enhancement in electron mobility by a factor of $\sim 1.8\times$ has been the subject of investigation for CMOS applications [38] as discussed elsewhere in this volume. Although models have been used to explain the measured strain dependence of peak electron mobility enhancements in strained Si/relaxed SiGe n-MOSFETs [12], *ab initio* calculations have not been able to account for the enhanced electron mobilities in strained Si n-MOSFETs measured at high vertical electric fields [39], and further work aimed at understanding these effects is required.

In addition to the shifts in band edge positions, the shape of the energy bands can change under strain and alloying effects. The valence band is considerably more complex than the conduction band, undergoing warping, splitting, and deformation with strain. Reviews of calculations of the valence band structure in strained SiGe can be found in Refs. [40–43]. To first order as the valence bands split the mass of the uppermost band near $k = 0$ is reduced, and this is associated with an enhanced in-plane hole mobility in biaxial tensile-strained Si on relaxed SiGe [44]. Recent three-dimensional calculations

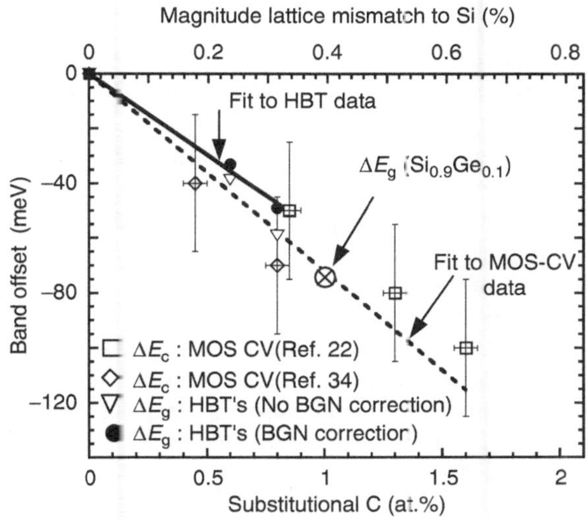

FIGURE 10.12 Summary of band offsets for strained $Si_{1-y}C_y$ on Si heterojunctions. The solid line is a fit to the ΔE_g values extracted from pnp HBTs. (From D.V. Singh, J.L. Hoyt, and J.F. Gibbons. Technical Digest of International Electron Devices Meeting 2000, pp. 749–752. With permission.) The dashed line is a fit to the ΔE_c values extracted from MOS $C-V$ analysis. The estimated conduction band offset between strained $Si_{1-y}C_y$ and Si is approximately 65 meV/at.% C. The bandgap offset ΔE_g for $Si_{0.9}Ge_{0.1}$/Si (mismatch = 0.4%) is shown (⊗) for reference. (From D.V. Singh, J.L. Hoyt, and J.F. Gibbons *IEEE Trans. Elec. Dev.* 50:425–432, 2003. With permission.)

of the Si band structure under "uniaxial" compressive stress indicate a complex behavior that enhances the hole mobility for transport in the direction of compression [45]. The addition of Ge to SiGe leads to a more Ge-like valence band with a higher curvature and reduced effective mass, as well as to strain-enhanced mobilities for pseudomorphic growth. Hole mobilities increase dramatically for strained SiGe with Ge contents above 50% The highest mobilities in the silicon heterostructure system have been obtained for high-Ge-content strained $Si_{1-y}Ge_y$ or strained Ge grown on relaxed $Si_{1-x}Ge_x$ layers with $y > x$ [46].

10.5 Summary

There has been considerable progress in the understanding of the band structure and basic electronic properties of silicon heterostructures over the years. Strain plays a critical role in determining the energy band structure and the lineup of the bands at strained SiGe/Si and strained Si/SiGe heterojunctions. Research continues in this area as interest in the use of strained Si and SiGe in FET applications grows. More experimental and theoretical work is needed to formulate an accurate picture of transport in strained Si and strained SiGe heterostructures, particularly in inversion layers where electric field quantization is also important, and this is currently an exciting area of investigation.

Acknowledgments

I am indebted to a number of talented people with whom I have worked in the area of silicon heterostructures, and whose efforts are represented in this chapter. In particular I would like to thank past collaborators including James Gibbons, Clifford King, Jeff Welser, Kern Rim, Shinichi Takagi, and Dinkar Singh, as well as present collaborators, especially Cait Ni Chleirigh, Oluwamuyiwa Olubuyide, Dimitri Antoniadis, and Eugene Fitzgerald.

References

1. R.W. Olesinski and G.J. Abbaschian. *Bull. Alloy Phase Diagrams* 5:180, 1984.
2. R. Braunstein, A.R. Moore, and F. Herman. Intrinsic optical absorption in germanium–silicon alloys. *Phys. Rev.* 109:695–710, 1958.
3. D.V. Lang, R. People, J.C. Bean, and A.M. Sergent. Measurement of the band gap of $Si_{1-x}Ge_x$/Si strained-layer heterostructures. *Appl. Phys. Lett.* 47:1333–1335, 1985.
4. C.A. King, J.L. Hoyt, and J.F. Gibbons. Bandgap and transport properties of $Si_{1-x}Ge_x$ by analysis of nearly ideal Si/$Si_{1-x}Ge_x$/Si heterojunction bipolar transistors. *IEEE Trans. Elec. Dev.* 36:2093–2104, 1989.
5. R. People. Indirect band gap of coherently strained Ge_xSi_{1-x} bulk alloys on (0 0 1) silicon substrates. *Phys. Rev. B* 32:1405–1408, 1985.
6. J.F. Morar, P.E. Batson, and J. Tersoff. Heterojunction band lineups in Si–Ge alloys using spatially resolved electron-energy-loss spectroscopy. *Phys. Rev. B* 47:4107–4110, 1993.
7. R. People and J.C. Bean. Band alignments of coherently strained Ge_xSi_{1-x}/Si heterostructures on (0 0 1) Ge_ySi_{1-y} substrates. *Appl. Phys. Lett.* 48:538–540, 1986.
8. C. Herring and E. Vogt. *Phys. Rev.* 101:944, 1956.
9. I. Balslev. *Phys. Rev.* 143:636, 1966.
10. J.L. Hoyt. Substitutional carbon incorporation and electronic characterization of $Si_{1-y}C_y$/Si and SiGeC/Si heterojunctions, in *Silicon–Germanium Carbon Alloys: Growth, Properties and Applications*, Taylor and Francis, New York, 2002, pp. 59–89.
11. T. Vogelsang and K.R. Hofmann. Electron transport in strained Si layers on $Si_{1-x}Ge_x$ substrates. *Appl. Phys. Lett.* 63:185–187, 1993.
12. S. Takagi, J.L. Hoyt, J. Welser, and J.F. Gibbons. Comparative study of phonon-limited mobility of two-dimensional electrons in strained and unstrained Si metal-oxide-semiconductor field-effect transistors. *J. Appl. Phys.* 80:1567–1577, 1996.
13. G.G. Van de Walle and R.M. Martin. Theoretical calculations of heterojunction discontinuities in the Si/Ge system. *Phys. Rev. B* 34:5621–5634, 1986.
14. R. People. Physics and applications of Ge_xSi_{1-x}/Si strained layer heterostructures. *IEEE J. Quantum Electronics* 22:1696–1710, 1986.
15. M.M. Rieger and P. Vogl, Electronic-band parameters in strained $Si_{1-x}Ge_x$ alloys on $Si_{1-y}Ge_y$ substrates. *Phys. Rev. B* 48:14276–14287, 1993.
16. F. Schaffler, High-mobility Si and Ge structures. *Semicond. Sci. Technol.* 12:1515–1549, 1997.
17. M.L.W. Thewalt, D.A. Harrison, C.F. Reinhart, J.A. Wolk, and H. Lafontaine. Type II band alignment in $Si_{1-x}Ge_x$/Si(0 0 1) quantum wells: the ubiquitous type I luminescence results from band bending. *Phys. Rev. Lett.* 79:269–272, 1997.
18. R.E. Cavicchi, D.V. Lang, D. Gershoni, A.M. Sergent, J.M. Vandenberg, S.N.G. Chu, and M.B. Panish. Admittance spectroscopy measurement of band offsets in strained layers of $In_xGa_{1-x}As$ grown on InP. *Appl. Phys. Lett.* 54:739–741, 1989.
19. H. Kroemer, W.-Y. Chen, J.S. Harris Jr. and D.D. Edwall. Measurement of isotype heterojunction barriers by *C–V* profiling. *Appl. Phys. Lett.* 36:295–297, 1979.
20. S.P. Voinigescu, K. Iniewski, R. Lisak, C.A.T. Salama, J.P. Noel, and D.C. Houghton. New technique for characterization of Si/SiGe layers using heterostructure MOS capacitors. *Solid-State Electron.* 37:1491, 1994.
21. J.J. Welser. The application of strained Si/relaxed-silicon germanium heterostructures to metal-oxide-semiconductor field-effect transistors. Ph.D. Thesis. Stanford, 1994.
22. D.V. Singh, K. Rim, T.O. Mitchell, J.L. Hoyt, and J.F. Gibbons. Measurement of the conduction band offsets in Si/$Si_{1-x-y}Ge_xC_y$ and Si/$Si_{1-y}C_y$ heterostructures using metal-oxide-semiconductor capacitors. *J. Appl. Phys.* 85:978–984, 1999.
23. C.W. Leitz, M.T. Currie, M.L. Lee, Z.-Y. Cheng, D.A. Antoniadis, and E.A. Fitzgerald. Hole mobility enhancements in strained Si/$Si_{1-y}Ge_y$ p-type metal-oxide-semiconductor field-effect transistors grown on relaxed $Si_{1-x}Ge_x$ ($x < y$). *Appl. Phys. Lett.* 79:4246–4248, 2001.

24. C.W. Leitz, M.T. Currie, M.L. Lee, Z.-Y. Cheng, D.A. Antoniadis, and E.A. Fitzgerald. Channel engineering of SiGe-based heterostructures for high mobility MOSFETs, in Materials Issues in Novel Si-Based Technology. Symposium (Materials Research Society Symposium Proceedings) Vol. 686:113–118, 2002.
25. J. Jung, M.L. Lee, S. Yu, E.A. Fitzgerald, and D.A. Antoniadis. Implementation of both high-hole and electron mobility in strained Si/strained $Si_{1-y}Ge_y$ on relaxed $Si_{1-x}Ge_x$/ $(x < y)$. *IEEE Electron Dev. Lett.* 24:460–462, 2003.
26. C. Ni Chleirigh, C. Jungemann, J. Jung, O.O. Olubuyide, and J.L. Hoyt. Extraction of band offsets in strained Si/strained $Si_{1-y}Ge_y$ on relaxed $Si_{1-x}Ge_x$ dual-channel enhanced mobility structures. Proceedings of the ECS Symposium on SiGe: Materials, Processing and Devices, October 2004, PV 2004–7, pp. 99–109, 2004. Pennington, N.J.
27. Dessis: Comprehensive semiconductor device simulator, Integrated Systems Engineering AG, Switzerland, http://www/ise.com.
28. D.V. Singh, J.L. Hoyt, and J.F. Gibbons. Novel epitaxial p-Si/n-$Si_{1-y}C_y$/p-Si heterojunction bipolar transistors. Technical Digest of International Electron Devices Meeting 2000, pp. 749–752.
29. A.A. Demkov and O.F. Sankey. Theoretical investigation of random Si–C alloys. *Phys. Rev. B* 48: 2207–2214, 1993.
30. B.L. Stein, E.T. Yu, E.T. Croke, A.T. Hunger, B. Laursen, A.E. Bair, J.W. Mayer, and C.C. Ahn. Band offsets in Si/$Si_{1-x-y}Ge_xC_y$ heterojunctions measured by admittance spectroscopy. *Appl. Phys. Lett.* 70:3413–3415, 1997.
31. B.L. Stein, E.T. Yu, E.T. Croke, A.T. Hunger, B. Laursen, A.E. Bair, J.W. Mayer, and C.C. Ahn. Measurement of band offsets in Si/$Si_{1-x}Ge_x$ and Si/$Si_{1-x-y}Ge_xC_y$ heterojunctions. *J. Vac. Sci. Technol.* B15:1108–1111, 1997
32. D.V. Singh, K. Rim, T.O. Mitchell, J.L. Hoyt, and J.F. Gibbons. Admittance spectroscopy analysis of the conduction band offsets in Si/$Si_{1-x-y}Ge_xC_y$ and Si/$Si_{1-y}C_y$ heterostructures. *J. Appl. Phys.* 85:985–993, 1999.
33. D.V. Singh, J.L. Hoyt, and J.F. Gibbons. Effect of band alignment and density of states on the collector current in p-Si/n-$Si_{1-y}C_y$/p-Si HBTs. *IEEE Trans. Elec. Dev.* 50:425–432, 2003.
34. K. Rim, T.O. Mitchell, D.V. Singh, J.L. Hoyt, J.F. Gibbons, and G. Fountain. Metal-oxide semiconductor-capacitance–voltage characteristics and band offsets for Si/$Si_{1-y}C_y$/Si heterostructures. *Appl. Phys. Lett.* 72: 2286–2288, 1998.
35. J. Poortmans, M. Caymax, A. Van Ammel, M. Libezny, and J. Nijs. Study of the transversal electron mobility in heterojunction bipolar transistors with strained $Si_{1-x}Ge_x$-base. Semiconductor Heterostructures for Photonic and Electronic Applications Symposium, 1993, pp. 409–414.
36. A. Gruhle, A. Schuppen, H. Kibbel, and U. Konig. Minority carrier mobility and density of states in SiGe HBTs determined by a new method. Compound Semiconductors 1994. Proceedings of the Twenty-First International Symposium, 1995, pp. 253–258.
37. T. Ghani. Characterization of Minority Carrier Transport and Heavy Doping Effects in n-Si/p-SiGe/n-Si Heterojunction Bipolar Transistors, Ph.D. Thesis, Stanford University, 1994.
38. J.L. Hoyt, H.M. Nayfeh, S. Eguchi, I. Aberg, G. Xia, T. Drake, E.A. Fitzgerald, and D.A. Antoniadis. Strained Silicon MOSFET Technology. Tech. Digest IEEE International Electron Devices Meeting, Dec. 2002, pp. 23–26, and references therein.
39. M.V. Fischetti, F. Gamiz, and W. Hansch. On the enhanced electron mobility in strained-silicon inversion layers. *J. Appl. Phys.* 92:7320–7324, 2002.
40. J.M. Hinckley and J. Singh. Hole transport theory in pseudomorphic SiGe alloys grown on Si(001) substrates. *Phys. Rev. B* 41:2912–2926, 1990.
41. M.V. Fischetti and S.E. Laux. Band structure, deformation potentials, and carrier mobility in strained Si, Ge, and SiGe alloys. *J. Appl. Phys.* 80:2234–2252, 1996.
42. T. Manku and A. Nathan, Energy-band structure for strained p-type SiGe. *Phys. Rev. B* 43: 12634–12637, 1991.

43. T. Manku and A. Nathan. Valence energy-band structure for strained group-IV semiconductors. *J. Appl. Phys.* 73:1205–1213, 1993.
44. D.K. Nayak and S.K. Chun. Low-field hole mobility of strained Si on (001) SiGe substrate. *Appl. Phys. Lett.* 64:2514–2516, 1994.
45. M.D. Giles, M. Armstrong, C. Auth, S.M. Cea, T. Ghani, T. Hoffmann, R. Kotlyar, P. Matagne, K. Mistry, R. Nagisetty, B. Obradovic, R. Shaheed, L. Shifren, M. Stettler, S. Tyagi, X. Wang, C. Weber, and K. Zawadzki. Understanding Stress Enhanced Performance in Intel 90 nm CMOS Technology, Proceedings of IEEE Symposium on VLSI Technology, 2004, pp. 118–119.
46. M.L. Lee and E.A. Fitzgerald. Optimized strained Si/strained Ge dual-channel heterostructures for high mobility p- and n-MOSFETs. Technical Digest of the IEEE International Electron Devices Meeting 2003, pp. 433–436.

11

Carbon Doping of SiGe

H. Jörg Osten
University of Hanover

11.1 Introduction

Silicon is the most important semiconductor material in the microelectronic industry. There have been several attempts to increase the material variety that is compatible with Si technologies. Silicon–germanium alloys and heterojunctions will extend the performance of future Si-based devices. However, the $Si_{1-x}Ge_x$ on Si(0 0 1) system exhibits severe limitations. Examples include the existence of a critical thickness for perfect pseudomorphic growth, which depends on the amount of germanium (the introduced strain). In addition, the main band offset between Si and strained SiGe is located in the valence band, that is, this system is much better suited for hole channel than for electron channel devices.

Recently, it was shown that $Si_{1-y}C_y$ and $Si_{1-x-y}Ge_xC_y$ layers can also be grown pseudomorphically on Si(0 0 1). This new material might alleviate some of the limitations of SiGe on Si(0 0 1). Research on highly supersaturated, C-containing alloys on silicon substrates started only a few years ago. Meanwhile, knowledge has been accumulated on growth, strain manipulation, thermal stability, and carbon effects on band structure and charge carrier transport [1–3]. Low carbon concentrations can also be used to control diffusion of dopants without affecting strain and band alignment. This unique property has already found its first device application, the SiGe:C heterojunction bipolar transistor (HBT) [4].

11.2 Basic Considerations

Epitaxial growth of C-containing $Si_{1-y}C_y$ or $Si_{1-x-y}Ge_xC_y$ alloys is more complex than SiGe growth. Several additional problems have to be addressed [5,6]: Unlike to the Si/Ge system, silicon and carbon are not miscible. According to the binary Si—C phase diagram, stoichiometric SiC (silicon carbide) is the only stable compound. Any alloy with a smaller C concentration is thermodynamically metastable. Such alloy layers can be achieved by kinetically dominated growth methods, like molecular beam epitaxy (MBE) [5,6], various kinds of chemical vapor deposition (CVD) at relatively low temperatures [7], or solid-phase epitaxy (SPE). The SPE approach is based on recrystallization of amorphous mixtures of Si/Ge/C obtained mainly by ion implantation with well-defined temperature profiles [8,9]. All of these methods generally work under far from thermodynamic equilibrium conditions, allowing a kinetic stabilization of metastable phases. Although the bulk solubility of carbon in silicon is small (3×10^{17} atoms/cm^3 at the melting point of Si [10]), epitaxial layers with more than 1% C can be fabricated. Once deposited as a random alloy at low temperatures (500–700°C), the $Si_{1-y}C_y$ appears to be able to withstand anneals up to significantly higher temperatures (up to 900°C) [11,12].

Contrary to the Si/Ge alloy, carbon in silicon can easily form interstitial defect complexes. Thus, one of the most crucial questions is the relationship between substitutional and interstitial carbon incorporation, which has a large impact on the electrical properties of these layers. Substitutionally incorporated C atoms allow strain manipulation, including the growth of strain-compensated or inversely strained $Si_{1-x-y}Ge_xC_y$ layers. The striking feature of the Si—C system is the extreme mismatch of the Si—Si and Si—C bond lengths. Disordered alloys will exhibit large local strain. Local ordering will occur due to atomic size differences.

11.3 Growth of C-Containing Alloys

Any alloy with a substitutional C concentration above the bulk solubility limit is thermodynamically metastable. In equilibrium, the solid solubility of a substitutional C impurity (i.e., the value of y for $Si_{1-y}C_y$) is given by: $y = \exp[-(E - \mu)/kT]$, where E is the energy of a substitutional C atom in Si, μ the chemical potential, and k the Boltzmann constant. The concentration of C is limited by the formation of SiC precipitates, which prevents μ from rising higher than the value corresponding to a SiC reservoir [13]. However, during growth it is not the equilibrium bulk solubility that is important but rather a "surface solubility." The increased solubility results from two factors [13]: First, the presence of a surface breaks the bulk symmetry and therefore creates sites for adatoms that are energetically nonequivalent to bulk sites. The stress associated with the atomic size mismatch between C and Si is partially relieved and the energy of the substitutional C decreases smoothly as it approaches the surface. Second, there is a stress field near the surface, associated with the atomic reconstruction there [14]. The coupling of the carbon stress to this spatially varying surface stress substantially lowers the carbon energy on certain sites. For a (2×1) reconstructed Si(0 0 1) surface there are two nonequivalent sites: One lies beneath the surface dimer (α site), and the other site lies between the dimers (β site). As discussed by Kelires and Tersoff [14], the α sites are under compressive stress. Thus, they are favorable sites for a smaller atom such as C. The adatoms are rapidly buried on subsurface sites, which serve like a sink, and thus immobilized. The high carbon density near the surface is frozen in as the crystal grows, permitting the growth of a highly supersaturated solid-solution, e.g., high carbon concentrations in silicon or silicon–germanium. Tersoff used an empirical many-body potential to model the interaction between atoms in the $Si_{1-y}C_y$ system [13]. Including also the surface reconstruction, he found an enhancement of 10^4 over equilibrium bulk solubility. Thus, the surface solubility of carbon can easily be in the percentage range.

The lowest-energy position of carbon in Si is found to be the substitutional site. In addition, the observed activation energy for the diffusion of substitutional carbon (~3 eV) is quite large [7]. This is the origin for the achieved stability of these metastable alloys. Carbon precipitation is a rare process that has been reported to occur only in the presence of a supersaturation of silicon self-interstitials or oxygen [15]. The mechanism described here has another possible consequence, in addition to enhanced solubility. If the C atoms have sufficient mobility within the first layers, they will lie only on sites beneath the dimers, with negligible occupancy of the other sites. This ordering should not be lost as the layers are buried deeper, if the C is immobile in deeper layers. First indications of local ordering in $Si_{1-y}C_y$ layers have been observed [16,17].

The carbon substitutionality (n_S/n_0) is strongly influenced by the growth conditions, such as temperature and the Si growth rate. Similar trends were observed for MBE and CVD growth [18–21]. All authors found a significant decrease of the substitutional carbon concentration with increasing growth temperature or decreasing growth rate. Osten et al. [18] proposed the following simple model: If the temperature is high enough for Si epitaxy, one can assume that all arriving C adatoms are rapidly incorporated in positions beneath the Si dimers. When another Si adatom arrives above a surface position filled with a C atom, either (i) it migrates further on the surface, or (ii) it binds to the carbon ("freezing in" of the C atom on lattice position), or (iii) it kicks the C atom out of a lattice position, forming a Si—C interstitial complex or a C—C dimer on the surface. The substitutional versus interstitial carbon concentration is therefore primarily controlled by process (iii), i.e., by all parameters affecting the formation of the interstitial defect complexes [22]. Desorption processes are likely excluded for the used growth temperatures. Recently, Kelires and Kaxiras [23] reported atomistic calculations of the surface structure of Si with substitutional carbon atoms. One of their important results is that the lowest energy configuration involves the C atoms in the very top layer of the Si(0 0 1) surface, forming dimers in the (2 × 1) reconstruction. The authors suggest that the strong preference of C to remain on the surface is due to the bond between the two threefold coordinated C atoms (similar to the Si surface atoms forming dimers). The ability of C to form double bonds makes this a low-energy structure. This has important consequences. It implies that if C atoms have a chance to form C—C dimers upon arrival on the Si(0 0 1) surface, their subsequent incorporation in the growing film will be limited. The formation of such dimers depends on several factors, including the frequency of surface hops (i.e., the growth temperature), the time before further deposition immobilizes the atoms (i.e., the growth rate), and the probability of finding another C atom (i.e., the total carbon concentration). In addition, some carbon evaporation sources deliver larger C fragments. Therefore, we have to assume different incorporation kinetics for different C sources. Osten et al. [22] extended their model for the incorporation of substitutional C during epitaxial growth as a function of various parameters. Based on that model one should expect for a certain Si growth rate and C flux the behavior illustrated schematically in Figure 11.1. Generally, we can distinguish three different regions (labeled A to C in Figure 11.1). In the first region, we have complete substitutional incorporation, independent of the growth temperature. For the highest temperatures in this region, we might find a slight decrease in the substitutional carbon concentration (n_s) with increasing temperature. This behavior is similar to the one reported by Zerlauth et al. [20]. In the middle region (B), we observe a relatively strong decrease of n_s with increasing growth temperature, similar to the results reported in Refs. [18,19]. For even higher growth temperatures (region C), almost no substitutional carbon incorporation should be possible. If the substrate temperature is too high, a metastable regime no longer exists, and some deterioration in crystal quality (like a transition to 3D growth or a mixed alloy carbide phase) occurs. The observed behavior can be described using the growth rate r (limiting the time available for defect formation), the total carbon concentration

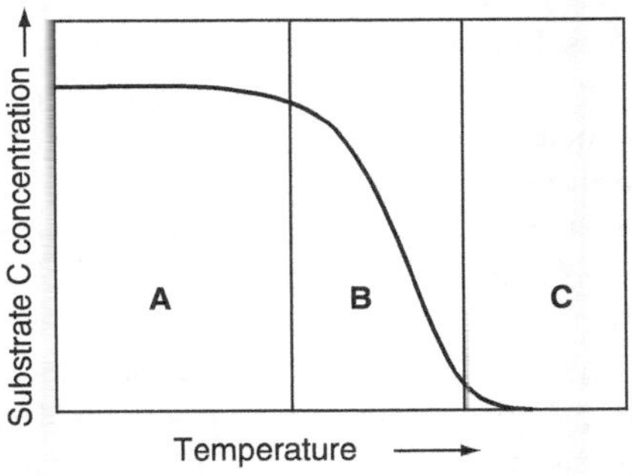

FIGURE 11.1 Schematic representation of substitutional carbon incorporation in $Si_{1-y}C_y$ layers as a function of growth temperature for given and constant Si and C fluxes. Three distinct regions are marked.

n_0, the growth temperature T, and an energy barrier E_c. Any incorporated C atom has to overcome this barrier to form an interstitial C-containing defect complex. The frequency of defect formation k is given by: $k = k_0 \exp(-E_c/kT)$. Based on all available data it seems reasonable to postulate that the energy barrier is lowered by the presence of surface strain, that is, E_c decreases with increasing n_0. This behavior can be modeled by a first-order exponential decay, leading to a value of E_c for n_0 approaching zero which is close to the barrier obtained for the diffusion of substitutional carbon in bulk silicon with very low carbon concentrations [7]. For the low carbon concentrations used to control dopant diffusion, the energy barrier is still large enough to suppress any defect formation within the temperature window for epitaxial growth. Thus, low carbon concentrations will be predominantly incorporated substitutionally. The preexponential frequency factor k_0 decreases over several orders of magnitude with increasing n_0. This factor mainly describes the ability of adatoms to migrate on the surface. Thus, the presence of carbon on the surface appears to reduce surface diffusion significantly. Extrapolating to $n_0 = 0$, $k_0(0)$ is in the same order of magnitude typically assumed for surface migration of adatoms on an unperturbed Si surface.

There is another interesting consequence arising from the decrease of E_c and k_0 with increasing C concentration, namely that the fraction of substitutionally incorporated carbon atoms decreases with increasing n_0. This leads to a saturation limit, i.e., any further increase in the carbon flux leads mainly to an increase in the number of interstitial defects, with the amount of substitutionally incorporated carbon remaining constant [22]. This behavior supports the often postulated existence of a limit for substitutional carbon incorporation [1]. This limit depends, of course, on the specific growth conditions and the specific carbon source.

So far, we discussed the incorporation of C into silicon. Liu and Osten [24] extended the investigations to substitutional C incorporation during MBE growth of $Si_{1-x-y}Ge_xC_y$ alloys. The strain introduced by the incorporation of C and Ge makes these processes even more complicated. They showed that the substitutional C incorporation can be dominated by the Ge or C concentration, depending on the total C concentration range. At low total C concentration, the Ge impact is insignificant; increasing total C concentration leads to less substitutional C incorporation efficiency. At high n_0, the impact of C concentration becomes marginal, and the presence of Ge inhibits the substitutional C incorporation. The very low substitutional C incorporation during $Ge_{1-y}C_y$ growth even at an extremely low growth temperature of 200°C (reported recently in Ref. [25]) is consistent with this prediction.

11.4 Control of Dopant Diffusion by Adding Carbon

The amount of carbon discussed in this chapter is in the order of dopant concentrations. To distinguish these materials from the $Si_{1-y}C_y$ or $Si_{1-x-y}Ge_xC_y$ systems discussed so far, we will use the same notation as for dopants, that is Si:C or SiGe:C. The carbon concentration is too low to affect significantly band alignment and strain of the epitaxially grown Si:C or SiGe:C layers. However, the used carbon concentration in the range of 10^{20} cm^{-3} is still orders of magnitude higher than the solid solubility limit, i.e., we have to consider metastable systems with carbon supersaturation.

A major challenge for the fabrication of ultrasmall devices for next silicon technology generations is the control of steep dopant profiles. In this context, it is tempting to apply supersaturated carbon as a diffusion-controlling agent in silicon. The fundamental question for applications of C-rich Si in devices is whether electronic device parameters are affected by possible C-related defects. Recently, Gossmann et al. [26] have announced the realization of MOSFET with reduced reverse short-channel effect due to suppressed transient diffusion in C-rich silicon. Little increase in leakage current was reported for devices with carbon. Successful applications of C-rich layers were reported for SiGe HBT [27,28]. The IHP group has demonstrated that transistors with excellent static and dynamic parameters can be fabricated with epitaxial SiGe:C layers [29,30]. The main result of this investigation was that carbon supersaturation can preserve steep boron doping profiles without degrading fundamental transistor parameters. The diffusion coefficient of B in Si is reduced by more than one order of magnitude when the concentration of substitutional C is elevated to about 10^{20} cm^{-3} [31]. Furthermore, transient-enhanced diffusion (TED) of B is strongly suppressed in C-rich Si [32].

In the following, we will describe the physical mechanism for suppressed boron diffusion in C-rich Si and SiGe (below 0.2% carbon) based on an undersaturation of Si self-interstitials [31,33,34]. Diffusion of dopants as well as Si self-diffusion, and diffusion of Ge and C in Si are mediated by point defects, i.e., vacancies and self-interstitials. It is generally accepted that both vacancy and interstitial mechanisms contribute to atomic transport in Si while the direct exchange mechanism of nearest-neighbor atoms plays a negligible role. The effect of C on B diffusion can be explained by coupled diffusion of B and point defects in Si. Diffusion of C out of C-rich regions causes an undersaturation of Si self-interstitials in the C-rich region, which in turn results in suppressed diffusion of boron.

Pronounced non-Fickian diffusion behavior was found experimentally for buried carbon with concentrations well above its solid solubility [34]. The basic mechanism for nonequilibrium point defect densities due to C outdiffusion is as follows [1,31]. Depending on growth conditions, carbon can be mainly substitutionally dissolved in Si. Diffusion of C in Si is via a substitutional–interstitial mechanism. Immobile substitutional carbon atoms (C_s) are transformed into mobile interstitial carbon (C_i) through the kick-out reaction with Si self-interstitials (I): $C_s + I \leftrightarrow C_i$. In addition, interstitial C can be formed in the dissociative Frank–Turnbull reaction, $C_s \leftrightarrow C_i + V$, where V is a vacancy. In order to conserve the total number of atoms the flux of interstitial C atoms out of the C-rich region has to be balanced by a flux of Si self-interstitials into this region or a flux of vacancies outwards. The individual fluxes are determined by the products of the diffusion coefficient D and the concentration C, which are all known. For C concentrations $C_C > 10^{18}\,\text{cm}^{-3}$, the transport coefficient of C may exceed the corresponding transport coefficients of Si self-interstitials and vacancies: $D_C C_C > D_I C_I^{eq}$ and $D_C C_C > D_V C_V^{eq}$, where C_I^{eq} and C_V^{eq} are the equilibrium concentrations of Si self-interstitials and vacancies, respectively. Consequently, outdiffusion of supersaturated C from C-rich regions becomes limited by the compensating fluxes of Si point defects, which leads to an undersaturation of self-interstitials and a supersaturation of vacancies in the C-rich region.

In Figure 11.2 measured C profiles in Si are compared with model calculation [34]. A 50-nm thick box-shaped C profile was grown using MBE and annealed in pure nitrogen atmosphere. Diffusion-limited rates were assumed for the two reactions described above and for Frenkel-pair generation. For the kick-out reaction, local equilibrium is established for typical point defect parameters. In contrast, the concentration of vacancies in the C-rich region is sensitive to the rate of the Frank–Turnbull reaction. Rücker et al. [34] fitted only the reaction rate to the measured diffusion profile of C. From this simulation, he found also a significant vacancy supersaturation. The point defect distribution after annealing resulting from the calculation is also shown in Figure 11.2. In the C-rich region, there is a significant reduction in the interstitial density accompanied by an increase in the vacancy density. This has consequences for all dopants diffusing via point defects. Boron diffusion in Si is via an interstitial mechanism. The effective diffusion coefficient of B is proportional to the normalized concentration of self-interstitials. Accordingly, interstitial undersaturation results in suppressed diffusion of B. Also the diffusion of phosphorous will be strongly reduced. On the contrary, the diffusion of dopants, which are diffusing by a vacancy mechanism (like As or Sb), will be enhanced due to carbon incorporation [34].

TED during the first annealing step after ion implantation accounts for a substantial fraction of dopant redistribution during device processing. The source of TED is an enhanced density of Si self-interstitials. In a subsequent publication, Rücker et al. [35] have extended this physical model to TED, i.e., nonequilibrium distribution of Si self-interstitials. The authors could show that the same mechanism holds also for TED. The effect of C on diffusion of donors and acceptors can correctly be described by a physically based model for coupled diffusion of C and Si point defects. In the simulation, they have used the "+1" model for implantation damage and coupled diffusion of C and Si point defects. Reduced diffusion of B in SiGe and the effect of the electric field due to hole confinement in SiGe were also included. The simulation reveals the strong suppression of TED in the C-containing structure. Again, the density of excess interstitials is reduced in the SiGe:C layer due to C outdiffusion.

The quality of epitaxial layers for electronic device application can also be evaluated by studying the carrier generation lifetime in MOS-like structures [36]. For LPCVD grown SiGe:C layer stacks similar to that used for HBT application lifetimes above 1 μs were measured, i.e., values that are sufficient for

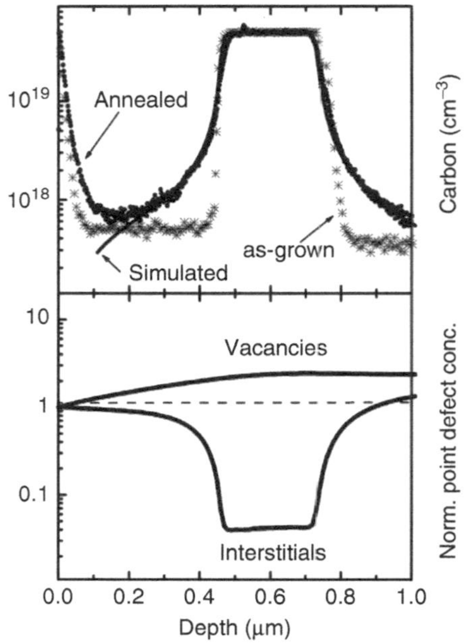

FIGURE 11.2 Diffusion profiles of buried carbon layers annealed in nitrogen atmosphere (900°C/45 min). SIMS profiles of as-grown (•) and annealed (*) samples are shown in the upper panel together with results of model calculation. The corresponding normalized point defect densities are shown in the lower panel.

successful application of those layers in high-performance devices with low noise. Recently, Heinemann et al. [37] could show that the electron mobility in the p-type base of npn bipolar transistors is not reduced by the incorporation of low C concentrations. They reported about a comparative analysis of vertical minority carrier transport in experimentally realized HBTs with Si, $Si_{1-x}Ge_x$ ($x = 7.5\%$), and $Si_{1-y}C_y$ ($y = 0.2\%$ or 1%) base layers. To overcome drawbacks of the simple transit time analysis, they used 2D device simulations to obtain an improved understanding of the measured high-frequency parameters. Taking into account the real doping profiles and device structures, and using a calibrated parameter set for strained SiGe, the simulation results for the Si, $Si_{1-x}Ge_x$, and $Si_{1-y}C_y$ ($y = 0.2\%$) base layer transistors reproduce very well the measured transit times (assuming the Si data for the electron mobility) in the heteroepiaxial base layers. In the case of higher carbon concentration ($y = 1\%$), the electron mobility is reduced by a factor of two.

11.5 Strain Manipulation in $Si_{1-x-y}Ge_xC_y$ Layers

Adding a constituent with a covalent radius much smaller than that of silicon to a layer containing the larger Ge ($r_{Si} = 1.17$ Å, $r_{Ge} = 1.22$ Å, but $r_C = 0.77$ Å) opens the possibility to manipulate the strain. It was shown that it is possible to form an "inversely distorted" SiGe cell, i.e., a cell that is tetragonally distorted due to tensile stress instead of the usual compressive stress by adding some carbon [16]. Using simple linear extrapolations between the different lattice constants and the appropriate relative concentration (Vegard's law for a ternary system), we can estimate the effective lattice constant in the $Si_{1-x-y}Ge_xC_y$ system as $a(x,y) = (1-x-y)a_{Si}+xa_{Ge}+ya_C$. The linear approximation (between Si, Ge, and diamond) for the average lattice constants results in a Ge:C ratio of 8.2 for complete strain compensation. *Ab initio* calculations predict values larger than 10 (see for example Ref. [38]). Diamond seems not to be the best choice for interpolating lattice parameter in $Si_{1-x-y}Ge_xC_y$. Osten et al. [16] proposed a linear interpolation between the lattice constants of Si, Ge, and β-SiC: $a_{SiGeC} = (1-x-2y)a_{Si}+xa_{Ge}+2ya_{SiC}$, which is illustrated in Figure 11.3. This approach yields that the

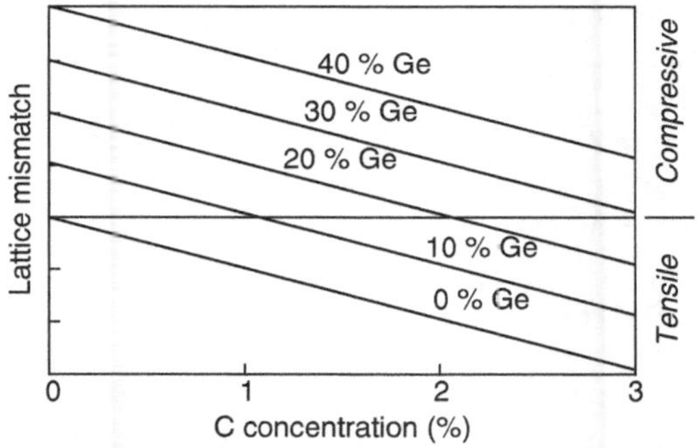

FIGURE 11.3 Estimated strain reduction in SiGe layers due to the presence of carbon (ternary Vegard's law).

compressing strain of 9.4 % Ge can be fully compensated by 1% C, i.e., strain compensation occurs for a Ge:C ratio of 9.4. Kelires [39] has calculated the lattice constant of $Si_{1-y}C_y$. He predicted some small bowing effects, which were also found experimentally by Berti et al. [40]. We define an "effective lattice mismatch" as $mf_{eff} = [a(x,y) - a_{Si}]/a_{Si}$. This quantity is directly accessible by x-ray diffraction. A positive mf_{eff} stands for a compressively strained material, $mf_{eff} <0$ indicates tensile strain, and $mf_{eff} = 0$ represents a fully strain-compensated $Si_{1-x-y}Ge_xC_y$ alloy. Such a layer should have on an average scale the cubic silicon structure.

11.6 Microscopic Structure of SiGeC Layers

The microscopic structure of a fully strain-compensated $Si_{1-x-y}Ge_xC_y$ layer has been investigated in detail in Ref. [41]. The Si—C bonds are stretched by about 7% with respect to the bond length in cubic silicon carbide, whereas the C atoms mainly strain the Si—Si bond in their neighborhood, with more distant Si—Si bonds less affected. Although the average lattice constant of the strain-compensated $Si_{1-x-y}Ge_xC_y$ layer is the same as that of pure silicon, the alloy contains large internal (microscopic) strain. Calculations support the picture of only partial relaxation of the Si—Si and Ge—Ge bonds towards a common bond length. The Si—C and Si—Ge bonds in strain-compensated $Si_{1-x-y}Ge_xC_y$ are strongly stretched. The calculated 7% stretching of the Si—C in strain-compensated $Si_{1-x-y}Ge_xC_y$ corresponds to 35% relaxation towards the average bond length of the alloy (predicted by Vegard's law). The observed dependence of Raman spectra on local strain and composition of $Si_{1-x-y}Ge_xC_y$ layers can be explained by the model calculations [41].

11.7 Strain-Stabilized Layers with High C Concentration

By combining theoretical and experimental results for local Si—C phonon modes in dilute $Si_{1-y}C_y$ alloys, information concerning the short-range order was obtained [42]. Calculations using an anharmonic Keating model predict satellite peaks near the vibrational frequency of an isolated C impurity, associated with second-, third-, etc., nearest-neighbor C—C pairs. By comparing theoretical spectra with those obtained by Raman and infrared absorption spectroscopies, it was concluded that the probability for a third-nearest-neighbor coordination of the C atoms is considerably above that for a purely random alloy. Maximizing the number of third-nearest neighbors leads to a $Si_{1-y}C_y$ system with $y \sim 20\%$. Recently, Rücker et al. [43] investigated the possible existence of such highly concentrated $Si_{1-y}C_y$ layers embedded in silicon. They demonstrated that $Si_{1-y}C_y$ layers with $y \sim 20\%$ can be grown pseudomorphically on a Si(0 0 1) substrate despite the large difference of the C and Si lattice constants. Such structures could only exist due to strain stabilization. It has been observed previously that semiconductor structures, which are not found in the bulk phase diagram, can be stabilized by the substrate-imposed strain in pseudomorphic epitaxial growth. The energy E_{epi} of an epitaxially confined structure can be decomposed as $E_{epi} = E_{equib} + E_{strain}$, where E_{equib} is the energy at the unconstrained equilibrium lattice constant and E_{strain} is the work needed to strain the material to match the Si substrate. An epitaxially stabilized structure X occurs when $E_{epi}(X)$ lies below that of the most stable bulk phase X_0 [here, $E_{epi}(\beta\text{-SiC})$]. Detailed structural investigations of such strain-stabilized $Si_{1-y}C_y$ layers with $y \sim 20$ % were presented by Ruvimov et al. [44].

11.8 Thermal Stability

$Si_{1-x}Ge_x$ alloys cannot be grown epitaxially on Si or Ge substrates without introducing large amounts of strain. Strain-relief occurs usually by plastic flow, i.e., by injection and propagation of misfit dislocations (see Chapter 9). $Si_{1-x}Ge_x$ alloys can be grown to much larger thickness than predicted by mechanical equilibrium theory. These layers become metastable. A later temperature treatment, as typical for further Si technological steps, causes also the strain-relieving formation of extended misfit dislocations. Therefore, the inherent strain limits the compatibility of heteroepitaxial SiGe systems to common Si technologies due to its lower thermal stability. In addition, the formation of extended misfit dislocations always changes the in-plane lattice constant (loss of pseudomorphy), enabling the construction of virtual substrates with a lattice constant differing from silicon.

The temperature stability of tensile-strained $Si_{1-y}C_y$ layers was investigated during and after post-growth annealing [11,12,45–47]. Despite the tensile strain in a 100-nm thick layer and the high carbon supersaturation, the samples were stable up to 850°C [12]. Beyond this temperature range, the substitutional carbon content started to decrease exponentially during isothermal annealing. From the temperature–time behavior of the measured changes in lattice spacing, Fischer et al. [12] extracted activation energy for that process of 3.2 eV. However, the layers remain always pseudomorphically to the Si substrate, i.e., they preserve the Si in-plane lattice constants. Although the number of possible nucleation centers for dislocation generation was quite high, no relaxation by the formation of extended misfit dislocation networks was detected [12,47]. The observed relaxation effect can be explained by precipitation of carbon or nucleation and diffusion-limited growth of SiC nanocrystals. Contrary to the mechanism of strain relief in $Si_{1-x}Ge_x$, in comparably strained $Si_{1-y}C_y$ epilayers, the main high-temperature process is precipitation, which does not alter the in-plane lattice constant. Thus, relaxed $Si_{1-y}C_y$ layers are not suitable as virtual substrates with a smaller in-plane lattice constant as silicon. Similar results were obtained for the thermal stability of compressive strained $Si_{1-x-y}Ge_xC_y$ alloys [48,49]. Carbon incorporation changes the thermal stability of SiGe layers significantly. As for pseudomorphic $Si_{1-y}C_y$ layers, plastic relaxation is suppressed by the dislocation pinning effect of C, leading to an increase in thermal stability of strained metastable $Si_{1-x-y}Ge_xC_y$ alloys compared to strain-equivalent $Si_{1-x}Ge_x$ layers.

Recently, the growth of a relaxed, only 1-μm thick, stepwise graded buffer based on a combination of $Si_{1-x}Ge_x$ and $Si_{1-x-y}Ge_xC_y$ with a low threading dislocation density was demonstrated [50]. For the nonoptimized buffer growth the authors found a threading dislocation density below $10^5\,cm^{-2}$. A stepped $Si_{1-x}Ge_x$ buffer with the identical thickness and strain profile grown with the same temperature ramp yield a threading dislocation density above $10^7\,cm^{-2}$. This buffer concept relies on the retardation of dislocation glide in $Si_{1-x-y}Ge_xC_y$ relative to strain-equivalent $Si_{1-x}Ge_x$ on silicon. Due to the very strong local strain fields around the individual carbon atoms, dislocation glide requires a higher energy in $Si_{1-x-y}Ge_xC_y$ than in undisturbed, strain-equivalent $Si_{1-x}Ge_x$ on silicon [51,52]. This indicates that the addition of carbon might be a promising way for new relaxed buffer concepts with low threading dislocation densities.

11.9 Impact of Carbon on Electrical Layer Properties

Assuming an average band structure for $Si_{1-x-y}Ge_xC_y$ alloys, Soref [53] has suggested an empirical interpolation between Si, Ge, and diamond for the bandgap. This technique results in an increase in the fundamental gap of $Si_{1-y}C_y$ layers with increasing y. Alternatively, Demkov and Sankey [54] have modeled $Si_{1-y}C_y$ alloys with supercells with random occupation of the lattice sites according to the alloy composition. They found that the fundamental gap is reduced when a few percent of carbon are added to the silicon lattice. Photoluminescence measurements on tensile-strained $Si_{1-y}C_y$ layers yield a reduction in the fundamental gap, with the main offset occurring in the conduction band [55]. From these experimental results, we can state that the virtual crystal approximation is not able to describe the changes in band structure for $Si_{1-y}C_y$ and $Si_{1-x-y}Ge_xC_y$ alloys correctly. Also the carbon effects cannot be reduced to changes in strain only as proposed earlier by Powell et al. [56]. To describe adequately the observed energy shifts for pseudomorphic C-containing layers, we have to consider at least strain-induced effects and effects due to alloying [57,58].

Extensive calculations including all contributions lead to the following conclusions for ternary alloys [58]: In tensile strained $Si_{1-y}C_y$, the fundamental bandgap is reduced, with the main offset occurring in the conduction band. This makes $Si_{1-y}C_y$ layers particularly suitable for electron confinement. The tensile strain lifts the sixfold degeneracy in the conduction band. Thus, the bandgap is determined by the separation between the light hole and the lowered twofold degenerate valley $\Delta(2)$. The energy splitting in the strained alloys transfers electrons from the upper to the lower valleys, which have a smaller effective mass in the plane. In addition, the energy splitting reduces intervalley scattering between twofold and fourfold degenerate valleys. Analogous to strained silicon, both effects should lead to an increase of in-plane electron mobility for low C contents in $Si_{1-y}C_y$. However, most available experimental data do not support these predictions, presumably due to scattering by interstitial C-containing defects. For samples with identical lattice distortion due to misfit strain and hence band alignment, Osten et al. [59] found differences in electron mobility of nearly a factor of two due to the different concentrations of interstitial carbon. Electron transport characteristics have been studied experimentally and theoretically using Monte Carlo (MC) simulations [60–62]. The authors found that the dependence of mobility on C content depends critically on the (unknown) value for the alloy scattering potential. For small alloy scattering potentials the authors found the following behavior for the electron mobility as a function of C concentration: With increasing carbon content the mobility first increases up to a maximum of approximately twice the value for silicon. Further increase in C content leads to a decrease in the mobility, which can be lower than the mobility in silicon, depending on the assumed scattering potential. Assuming a larger scattering potential, MC simulations yield a monotonic decrease in electron mobility with increasing carbon content due to the dominance of alloy scattering over the strain-induced mobility gain. Brunner et al. [63] reported that the mobility enhancement induced by strain is larger than the degradation caused by alloy scattering, leading to electron mobility values above the appropriate Si values, especially at low temperatures.

Recently, pMOSFETs using tensile-strained $Si_{1-y}C_y$ channel layers on Si substrates have been shown to provide mobility enhancement over Si epitaxial channel layers and Si bulk devices [64]. These layers did not exhibit the defects typically associated with a relaxed SiGe buffer structure. Furthermore, pMOSFET devices fabricated on these alloy layers demonstrate enhanced hole mobility over Si epi control and Si bulk. However, increased amounts of C may result in degraded device performance. Therefore, when fabricating such devices, a balance must be struck to minimize C-induced interface charges and alloy scattering rate.

The bandgap for the ternary alloys is always smaller than that of silicon, independent of strain state (Figure 11.4). The addition of C (Ge) into compressive strained $Si_{1-x}Ge_x$ (tensile-strained $Si_{1-y}C_y$) leads to a smaller reduction in bandgap narrowing than an equivalent strain reduction in the binary alloy (lower Ge or C content, respectively). The band gap of a fully strain-compensated $Si_{1-x-y}Ge_xC_y$ is a linear function of Ge (or C) concentration. Those layers are strain-free, i.e., there are no thickness limitations due to relaxation. This opens a new way for bandgap engineering of a material on Si(0 0 1) that is macroscopically strain-free.

11.10 Summary

We have briefly reviewed the basic growth issues and some mechanical and electrical material properties of $Si_{1-y}C_y$ and $Si_{1-x-y}Ge_xC_y$ layers grown pseudomorphically on Si(0 0 1). This new material promises

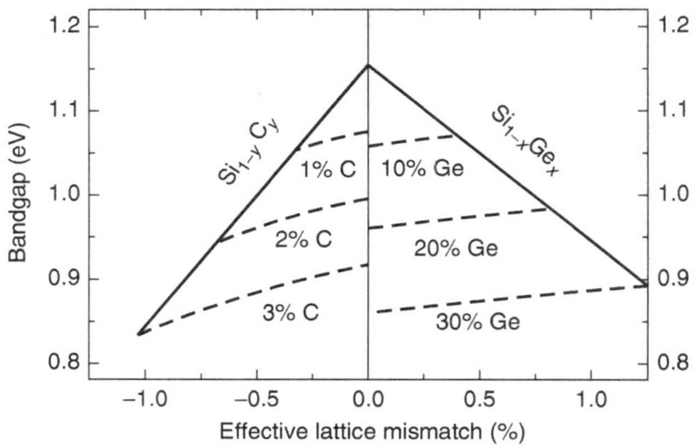

FIGURE 11.4 Bandgap narrowing for ternary $Si_{1-x-y}Ge_xC_y$ alloys, strained on Si(0 0 1) as a function of lattice mismatch. Both binary alloys (tensile strained $Si_{1-y}C_y$ and compressive strained $Si_{1-x}Ge_x$) are shown as references.

to alleviate some of the constraints of strained $Si_{1-x}Ge_x$, and open new possibilities for device applications of heteroepitaxial Si-based systems. The incorporation of carbon can be used (i) to improve SiGe layer properties, (ii) to obtain layers with new properties, and (iii) to control dopant diffusion in microelectronic devices.

Acknowlegdments

The author would like to thank D. Bolze, E. Bugiel, B. Dietrich, K.E. Ehwald, D. Endisch, G.G. Fischer, P. Gaworzewski, J. Griesche, B. Heinemann, M.C. Kim, W. Kissinger, D. Knoll, D. Krüger, G. Lippert, J.P. Liu, M. Methfessel, K. Pressel, H. Rücker, S. Scalese, R. Sorge, H.J. Thieme, B. Tillack, M. Weidner, and P. Zaumseil for their contributions to the research reviewed in this chapter.

References

1. ST Pantelides and S Zollner (Eds). *Silicon–Germanium Carbon Alloys, Growth Properties and Application.* New York: Taylor and Francis, 2002.
2. HJ Osten. *Carbon-Containing Layers on Silicon.* Zürich: Transtech Publications, 1999.
3. E Kasper and K Lyutovich (Eds). *Properties of Silicon Germanium and SiGe:Carbon.* London: EMIS Data reviews Series No 24, 2000.
4. KE Ehwald, D Knoll, B Heinemann, K Chang, J Kirchgessner, R Mauntel, Ik-Sung Lim, J Steele, B Tillack, A Wolff, K Blum, W Winkler, M Pierschel, F Herzel, U Jagdhold, P Schley, R Barth, and HJ Osten. Modular integration of high-performance SiGe:C-HBTs in a deep submicron, epi-free CMOS process. Technical Digest of the IEEE International Electron Device Meeting, Washington, 1999, pp. 561–564.
5. AR Powell, K Eberl, BA Ek, and SS Iyer. $Si_{1-x-y}Ge_xC_y$ growth and properties of ternary systems. *J. Crystal Growth* 127: 425–429, 1993.
6. HJ Osten. Supersaturated carbon in silicon and silicon germanium alloys. *Mater. Sci. Eng.* B 36: 268–274, 1996.
7. JL Regolini, F Gisbert, G Dolino, and P Boucaud. Growth and characterization of strain compensated $Si_{1-x-y}Ge_xC_y$ epitaxial layers. *Mater. Lett.* 18: 57–60, 1993.
8. S Im, J Washburn, R Gronsky, NW Cheung, KM Yu, and JW Ager. Optimization of Ge/C ratio for compensation of misfit strain in solid phase epitaxial growth of SiGe layers. *Appl. Phys. Lett.* 63: 2682–2684, 1993.
9. JW Strane, HJ Stein, SR Lee, BL Doyle, ST Picraux, and JW Mayer. Metastable SiGeC formation by solid phase epitaxy. *Appl. Phys. Lett.* 63: 2786–2788, 1993.

10. G Davis, RC Newman. Carbon in monocrystalline silicon, in TS Moss (ed.), *Handbook of Semiconductors*, Vol. 3, Amsterdam: Elsevier Science, 1994.
11. MS Goorsky, SS Iyer, K Eberl, FK LeGouech, J Angilello, and F Cardonne. Thermal stability of $Si_{1-x}C_x$/Si strained layer superlattices. *Appl. Phys. Lett.* 60: 2758–2761, 1992.
12. GG Fischer, P Zaumseil, E Bugiel, and HJ Osten. High temperature behaviour of strained $Si_{1-y}C_y$/Si heterostructures. *J. Appl. Phys.* 77: 1934–1937, 1995.
13. J Tersoff. Enhanced solubility of impurities and enhanced diffusion near crystal surfaces. *Phys. Rev. Lett.* 74: 5080–5083, 1995.
14. PC Kelires and J Tersoff. Equilibrium alloy properties by direct simulation: oscillatory segregation at the Si–Ge(100) 2x1 surface. *Phys. Rev. Lett.* 63: 1164–1167, 1989.
15. WJ Taylor, TY Tan, and U Goesele. Carbon precipitation in silicon: why is it so difficult? *Appl. Phys. Lett.* 62: 3336–3338, 1993.
16. HJ Osten, E Bugiel, and P Zaumseil. The growth of an inverse tetragonal distorted SiGe layer on Si(001) by adding small amounts of carbon. *Appl. Phys. Lett.* 64: 3440–3443, 1994.
17. K Pressel, B Dietrich, H Rücker, M Methfessel, and HJ Osten. Optical investigations of $Si_{1-y}C_y$ alloys. *Mater. Sci. Eng. B* 36: 157–170, 1996.
18. HJ Osten, M Kim, K Pressel, and P Zaumseil. Substitutional versus interstitial carbon incorporation during pseudomorphic growth of $S_{1-y}C_y$ on Si(001). *J. Appl. Phys.* 80: 6711–6715, 1996.
19. MW Dashiell, LV Kulik, D Hits, J Kolodzey, and G Watson. Carbon incorporation in $S_{1-y}C_y$ alloys grown by MBE using a single-graphite source. *Appl. Phys. Lett.* 72: 833–835, 1998.
20. S Zerlauth, H Seyringer, C Penn, and F. Schäffler. Growth conditions for complete substitutional carbon incorporation into layers grown by molecular beam epitaxy. *Appl. Phys. Lett.* 71: 3826–3828, 1997.
21. TO Mitchell, JL Hoyt, and JF Gibbons. Substitutional carbon incorporation in epitaxial $S_{1-y}C_y$ layers grown by chemical vapor deposition. *Appl. Phys. Lett.* 71: 1688–1691, 1997.
22. HJ Osten, J Griesche, and S Scalese. Substitutional carbon incorporation in epitaxial $Si_{1-y}C_y$ alloys on Si(001) grown by molecular beam epitaxy. *Appl. Phys. Lett.* 74: 836–838, 1999.
23. PC Kelires and E Kaxiras. Substitutional carbon impurities in thin silicon films: equilibrium structure and properties. *J. Vac. Sci. Technol.* B 16: 1687–1691, 1998.
24. JP Liu and HJ Osten. Substitutional carbon incorporation during $Si_{1-x-y}Ge_xC_y$ growth on Si(100) by molecular beam epitaxy: dependence on germanium and carbon concentration. *Appl. Phys. Lett.* 76: 3546–3548, 2000.
25. R Duschl, OG Schmidt, W Winter, K Eberl, MW Dashiell, J Kolodzey, NY Jin-Phillipp, and F Phillipp. Growth and thermal stability of pseudomorphic $Ge_{1-y}C_y$/Ge superlattices on Ge(0 0 1). *Appl. Phys. Lett.* 74: 1150–1152, 1999.
26. HJ Gossmann, C Rafferty, G Hobler, H Vuong, D Jacobson, and M Frei. Suppression of reverse short channel effect by a buried carbon layer. Technical Digest of the IEEE International Electron Device Meeting, Washington, 1998, pp. 725–728.
27. HJ Osten, G Lippert, D Knoll, R Barth, B Heinemann, H Rücker, and P Schley. The effect of carbon incorporation on SiGe HBT performance and process margins. Technical Digest of the IEEE International Electron Device Meeting, Washington, 1997, pp. 803–806.
28. LD Lanzerotti, JC Sturm, E Stach, R Hull, T Buyuklimanli, and C Magee. Suppression of boron transient enhanced diffusion in SiGe heterojunction bipolar transistors by carbon incorporation. *Appl. Phys. Lett.* 70: 3125–3127, 1997.
29. D Knoll, B Heinemann, HJ Osten, KE Ehwald, B Tillack, P Schley, R Barth, M Matthes, KS Park, Y Kim, and W Winkler. Si/SiGe:C heterojunction bipolar transistors in an epi-free well, single-polysilicon technology. Technical Digest of the IEEE International Electron Device Meeting, Washington, 1998, pp. 703–806.
30. H. Rücker, B Heinemann, R Barth, D Bolze, J Drews, U Haak, W Höppner, D Knoll, K Köpke, S Marschmeyer, HH Richter, P Schley, D Schmidt, R Scholz, B Tillack, W Winkler, HE Wulf, and Y Yamamoto. SiGe:C BiCMOS technology with 3.6 ps gate delay. Technical Digest of the IEEE International Electron Device Meeting, Washington, 2003, pp. 121–124.

31. H Rücker, B Heinemann, W Röpke, R Kurps, D Krüger, G Lippert, and HJ Osten. Suppressed Diffusion of boron and carbon in carbon-rich Silicon. *Appl. Phys. Lett.* 73: 1682–1684, 1998.
32. PA Stolk, DJ Eaglesham, HJ Gossmann, and JM Poate. Carbon incorporation in silicon for suppressing interstitial-enhanced boron diffusion. *Appl. Phys. Lett.* 66: 1370–1372, 1995.
33. R Scholz, U Gösele, JY Huh, and TY Tan. Carbon-induced undersaturation of silicon self-interstitials. *Appl. Phys. Lett.* 72: 200–202, 1998.
34. H Rücker, B Heinemann, D Bolze, D Knoll, D Krüger, R Kurps, HJ Osten, P Schley, B Tillack, and P Zaumseil. Dopant diffusion in C-doped Si and SiGe: physical model and experimental verification. Technical Digest of the IEEE International Electron Device Meeting, Washington, 1999, pp. 345–348.
35. H Rücker, B Heinemann, D Bolze, R Kurps, D Krüger, G Lippert, and HJ Osten. The impact of supersaturated carbon on transient enhanced diffusion. *Appl. Phys. Lett.* 74: 3377–3379, 1999.
36. HJ Osten, G Lippert, P Gaworzewski, and R Sorge. Impact of low carbon concentrations on the electrical properties of highly boron doped SiGe layers. *Appl. Phys. Lett.* 71: 1522–1524, 1997.
37. B Heinemann, D Knoll, GG Fischer, P Schley, and HJ Osten. Comparative analysis of minority carrier transport in npn bipolar transistors with Si, $Si_{1-x}Ge_x$, and $Si_{1-y}C_y$ base layers. *Thin Solid Films* 369: 347–351, 2000.
38. W Windl, OF Sankey, and J Menéndez. Theory of strain and electronic structure of $Si_{1-y}C_y$ and $Si_{1-x-y}Ge_xC_y$ alloys. *Phys. Rev. B* 57: 2431–2442, 1998.
39. PC Kelires. Monte Carlo studies of ternary semiconductor alloys: application to the $Si_{1-x-y}Ge_xC_y$ system. *Phys. Rev. Lett.* 75: 1114–1117, 1995.
40. M Berti, D De Salvator, AV Drigo, F Romanato, J Stangl, S Zerlauth, F Schäffler, and G Bauer. Lattice parameter in $Si_{1-y}C_y$ epilayers: deviation from Vegard's rule. *Appl. Phys. Lett.* 72: 1602–1604, 1998.
41. B Dietrich, HJ Osten, H Rücker, M Methfessel, and P Zaumseil. Lattice distortion in a strain compensated $Si_{1-x-y}Ge_xC_y$ layer on silicon. *Phys. Rev. B* 49: 17185–17190, 1994.
42. H Rücker, M Methfessel, B Dietrich, K Pressel, and HJ Osten. Phonons as a probe of short-range order in $Si_{1-y}C_y$ alloys. *Phys. Rev. B* 53: 1302–1309, 1996.
43. H Rücker, M Methfessel, E Bugiel, and HJ Osten. Strain-stabilized highly-concentrated pseudomorphic $Si_{1-y}C_y$ layers in Si. *Phys. Rev. Lett.* 72: 3578–3581, 1994.
44. S Ruvimov, E Bugiel, and HJ Osten. Structural characterization of Si_nC δ-layers embedded in a silicon matrix. *J. Appl. Phys.* 78: 2323–2327, 1995.
45. AR Powell, FK LeGoues, and SS Iyer. Formation of β-SiC nanocrystals by the relaxation of $Si_{1-y}C_y$ random alloy layers. *Appl. Phys. Lett.* 64: 324–326, 1994.
46. AR Powell, K Eberl, FK Legoues, and BA Ek. Stability of strained $Si_{1-y}C_y$ random alloy layers. *J. Vac. Sci Technol. B* 11: 1064–1068, 1993.
47. HJ Osten, D Endisch, E Bugiel, B Dietrich, GG Fischer, M Kim, D Krüger, and P Zaumseil. Strain relaxation in tensile strained $Si_{1-y}C_y$ layers on Si(0 0 1). *Semicond. Sci. Tech.* 11: 1678–1687, 1996.
48. P Warren, J Mi, F Overney, and M Dutoit. Thermal stability of $Si/Si_{1-x-y}Ge_xC_y/Si$ heterostructures grown by rapid thermal chemical vapor deposition. *J. Crystal Growth* 157: 414–419, 1995.
49. P Zaumseil, GG Fischer, K Brunner, and K Eberl. Comparison of the thermal stability of $Si_{0.603}Ge_{0.397}/Si$ and $Si_{0.597}Ge_{0.391}C_{0.012}/Si$ superlattice structures. *J. Appl. Phys.* 81: 6134–6140, 1997.
50. HJ Osten and E Bugiel. Relaxed $Si_{1-x}Ge_x/Si_{1-x-y}Ge_xC_y$ buffer structures with low threading dislocation density. *Appl. Phys. Lett.* 70: 2813–2815, 1997.
51. E Bugiel, S Ruvimov, and HJ Osten. TEM characterization of $Si_{1-x-y}Ge_xC_y$ and $Si_{1-y}C_y$ layers grown with molecular beam epitaxy on (0 0 1)Si substrates. *Int. Phys. Conf. Ser.* 146: 301–306, 1995.
52. HJ Osten and J Klatt. *In situ* monitoring of strain relaxation during antimony-mediated growth of Ge and $Ge_{1-y}C_y$ layers on Si(001) using refection high energy electron diffraction. *Appl. Phys. Lett.* 65: 630–632, 1994.
53. RA Soref. Optical band gap of the ternary semiconductor $Si_{1-x-y}Ge_xC_y$. *J. Appl. Phys.* 70: 2470–2472, 1991.

54. AA Demkov and OF Sankey. Theoretical investigation of random Si–C alloy. *Phys. Rev. B* 48: 2207–2214, 1993.
55. K Brunner, K Eberl, and W Winter. Near-band-edge photoluminescence from pseudomorphic $Si_{1-y}C_y$ quantum well structures. *Phys. Rev. Lett.* 76: 303–306, 1996.
56. AR Powell, K Eberl, BA Ek, and SS Iyer. $Si_{1-x-y}Ge_xC_y$ growth and properties of ternary systems. *J. Crystal Growth* 127: 425–429, 1993.
57. S Zollner. Theory of optical interband transitions in strained $Si_{1-y}C_y$ grown pseudomorphically on Si(001). *J. Appl. Phys.* 78: 5209–5211, 1995.
58. HJ Osten. Band gap changes and band offsets for ternary $Si_{1-x-y}Ge_xC_y$ alloys on Si(0 0 1). *J. Appl. Phys.* 84: 2716–2721, 1998.
59. HJ Osten, J Griesche, P Gaworzewski, and KD Bolze. Influence of interstitial carbon defects on electron transport in strained $Si_{1-y}C_y$ layers on Si(0 0 1). *Appl. Phys. Lett.* 76: 200–202, 2000.
60. HJ Osten and P Gaworszewski. Charge transport in strained $Si_{1-y}C_y$ and $Si_{1-x-y}Ge_xC_y$ alloys on silicon. *J. Appl. Phys.* 82: 4977–4981, 1997.
61. M Ershov and V Ryzhii. Monte Carlo study of electron transport in strained silicon–carbon alloy. *J. Appl. Phys.* 76: 1924–1926, 1994.
62. P Dollfus, S Galdin, P Hesto, and HJ Osten. Band offsets and electron transport calculation for strained $Si_{1-x-y}Ge_xC_y$/Si heterostructures. *J. Mater. Sci.: Mater. Electron.* 12: 245–248, 2001.
63. K Brunner, W Winter, K Eberl, NY Jin-Phillip, and F Phillip. Fabrication and band alignment of pseudomorphic $Si_{1-y}C_y$, $Si_{1-x-y}Ge_xC_y$ and coupled $Si_{1-y}C_y/Si_{1-x-y}Ge_xC_y$ quantum well structures on Si substrates. *J. Crystal Growth* 175/176: 451–458, 1997.
64. E Quinones, SK Ray, KC Liu, and S Banerjee. Enhanced mobility pMOSFETs using tensile-strained $Si_{1-y}C_y$ layers. *IEEE Electron. Device Lett.* 20: 338–340, 1999.

12

Contact Metallization on Silicon–Germanium

C.K. Maiti
Indian Institute of Technology Kharagpur

12.1 Introduction

During the last 40 years, Si-based complementary metal oxide semiconductor (CMOS) technology has emerged as the most predominant technology for the microelectronics industry. The success of the technology has largely been due to technological innovations that enabled continuous scaling of device dimensions, which have resulted in improvements in both device performance and packing density. However, this trend cannot continue indefinitely and we are approaching the fundamental limits of Si technology. To overcome the fundamental limits of Si, still maintaining compatibility with the conventional Si processing, several new materials such as strained-SiGe, strained-Si, and silicon–germanium–carbon (SiGeC) are currently being considered. Acceptance of a new material in CMOS technology requires a detailed understanding of the consequences, if incorporated.

Currently, self-aligned silicide technology is being used to form the source and drain regions to reduce the RC delay, increase driving current, and enhance the performance of deep submicron CMOS devices. In CMOS technologies, among many other metals, Ti, Co, Ni, Pt, and Pd have been studied in detail for silicidation. The reader may refer to an excellent recent review by Zhang and Ostling [1]. Properties of some commonly used metals and their silicides for salicide applications in Si CMOS technology are shown in Table 12.1 and Table 12.2 respectively.

According to the International Technology Roadmap for Semiconductors (ITRS) 2003 [2], the device scaling and increased functionality is expected to continue at least until 2016, when the minimum feature size is projected to be 22 nm. Currently one of the key challenges in device scaling is to develop new contact technologies for the MOSFETs as the source/drain (S/D) parasitic series resistance is becoming a significant fraction of the channel resistance. The ITRS roadmap predicts the requirements for source/drain contact regions as: (i) junction depth/sheet resistance (<100 ohm/sq. for 30 nm junctions), (ii) ultralow resistivity contacts (1.5×10^{-8} ohm-cm^2), (iii) excellent reverse leakage characteristics (less than 1% of the I_{off}), and (iv) integration compatibility with high-k gate dielectrics using the conventional CMOS process flow.

TABLE 12.1 Properties of Some Commonly Used Metals for Silicide Applications in Si CMOS Technology

Metal	Melting Point (°C)	Resistivity (μΩ-cm)	Schottky Barrier (eV) (on n-Si)	Reaction Temperature to Si (°C)	Thermal Stability on SiO_2 (°C)
W	3387	6–15	0.65–0.7	600–700	≥1000
TiN	2950	50–200	0.45–0.5	450–500	≥900
Ta	2996	15–20	0.55–0.6	450–500	~700
Mo	2610	6–15	0.6–0.7	400–700	≥900
Ti	1675	45–50	0.5–0.55	400–500	
Al	660	3–4	0.65–0.75	~250	~450

TABLE 12.2 Properties of Commonly Used Silicides for Salicide Applications

	$TiSi_2$	$CoSi_2$	NiSi	PtSi
Formation temperature (°C)	750–900	550–900	350–750	350–750
Silicide per nm of metal	2.3	2.6	1.8	1.3
Silicon consumption per nm of silicide	0.904	1.03	0.82	0.7
Resistivity (μΩ-cm)	13–20	14–20	14–20	28–35
Barrier height on n-Si (eV)	0.61	0.65	0.67	0.87

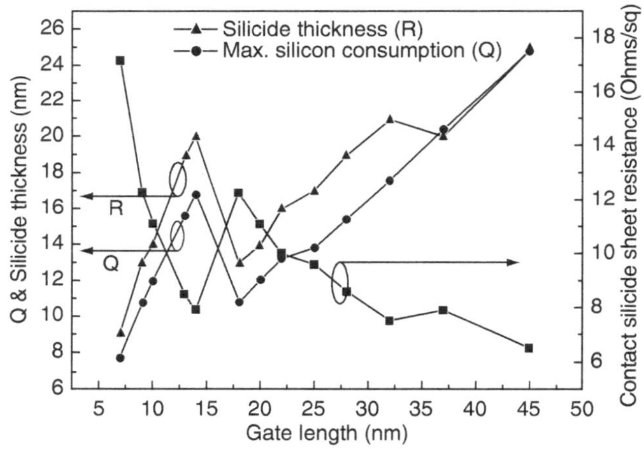

FIGURE 12.1 The predicted silicon consumption, silicide thickness, and the silicide sheet resistance values by ITRS-2003 are plotted as a function of gate length.

In addition to providing a low contact resistivity, future contact materials should satisfy several criteria, such as the formation of ultrathin silicide contacts, low silicide resistivity, and minimum temperature required to reach the low resistivity phase. Junction depth scaling requires that the thickness of the silicide be also scaled accordingly to reduce the silicon consumption and preserve the integrity of the junction. The trend can be seen in Figure 12.1, which shows the ITRS 2003 predictions for junction depth, silicide thickness, and silicide sheet resistance for future CMOS technology nodes.

12.2 Contact Metallization: Current Status

Recently, there have been concerns related to the increase of $TiSi_2$ film resistivity on submicron gate lines due to the retarded phase transition from the high-resistivity C49 phase to the low-resistivity C54 phase

[3,4]. Ni–silicide has been proposed as a potential candidate for advanced CMOS Si technology because Ni consumes less silicon (important for applications on ultrathin Si layers) [5], and has less risk of spiking in ultrashallow junctions. NiSi is used not only in CMOS but also in bipolar technology as the contact material because of its superior properties. NiSi does not show adverse line-width dependence of sheet resistance, which is often a big problem in $TiSi_2$ and sometimes in $CoSi_2$. Silicon consumption during the silicidation is the smallest among the Ti, Co, and Ni [6]. As such NiSi or nickel monosilicide are chosen as the gate, source, and drain contact material. The sheet resistance of the narrow line was found to even decrease because of the edge effect specific to NiSi [7], which is suitable for ultrashallow S/D junction formation for scaled CMOS. The mechanical stress of the NiSi film on silicon substrate is also the smallest. Contact resistance to p-type silicon, which was a problem for $TiSi_2$, is the smallest due to its lowest barrier height, and that to n-type silicon was found to be also small [8]. Bridging failure between the gate electrode and source/drain hardly occurs for the NiSi silicide due to its reaction mechanism. On the other hand, major concern for NiSi salicide is its low heat resistance. In fact, the maximum allowable process temperature after the salicidation is around 700–750°C. In addition, NiSi can be formed by one-step annealing at low temperature (400–600°C) without any agglomeration and shows a lower resistivity and junction leakage.

Aggressive scaling of the channel length and gate oxide thickness in a MOSFET leads to several problems, such as excessive tunneling current through the gate dielectric (100 A/cm^2), which limits the scaling of gate oxide thickness around 12–14 Å, polysilicon (poly-Si) gate depletion which increases the equivalent SiO_2 thickness by about 3 Å, high gate resistance, gate tunneling leakage current, and boron penetration into the channel region. Metal gate technology not only eliminates the gate depletion and boron penetration problems, but also greatly reduces the gate sheet resistance, particularly of interest in ac performance. Replacement of polysilicon gate electrodes in MOSFETs with metal thus promises many advantages. An approach analogous to the established dual poly-Si gate technology (dual metal) is desirable. The major challenge is to find two metals with suitable work functions and a way to integrate them in a CMOS process.

Alternative gate dielectrics such as HfO_2, ZrO_2, Ta_2O_5, and Si_3N_4 are being investigated for scaling gate dielectric thickness below 20 Å when direct tunneling current through SiO_2 may be too high for acceptable standby power consumption. Currently, the silicon nitride/oxynitride dielectric stack is in use due to its compatibility with CMOS processes. Therefore, a process that combines the advantages of dual metal gate and the high-k gate dielectric would be an attractive technology option for scaling the conventional MOSFET structures below 0.1 μm (beyond 65-nm technology node). For future technology nodes, junction formation at low temperatures is also a key necessity since high-k gate dielectrics of interest cannot withstand typical annealing temperatures used to remove implantation damage and activate the dopants.

12.3 Contact Metallization: Challenges

While the dual metal contact approach appears to be promising at a first glance, integration of two different self-aligned silicide processes is likely to be quite challenging. Furthermore, Fermi level pinning will most likely restrict the benefits of using materials with different work functions. Thus, there is an urgent need to develop new source/drain junction and contact technologies that limit the parasitic series resistance to a fraction of the MOSFET channel resistance (see Figure 12.2), and requires fundamentally new approaches. Of particular interest is the smaller bandgap of $Si_{1-x}Ge_x$, resulting in a smaller metal–semiconductor barrier height, which is a key advantage in reducing the contact resistivity of a metal–semiconductor contact.

Interest in $Si_{1-x}Ge_x$ alloys for performance enhancement of MOSFETs has been escalating in recent years as well [9]. Epitaxially grown $Si_{1-x}Ge_x$ layers on Si, under compression, display an enhanced effective hole mobility compared to bulk-Si. This hole mobility enhancement offers a superior channel properties compared to the conventional Si channel in a MOSFET. The SiGe alloys may be incorporated in a MOSFET to enhance the device performance as: (a) compressively strained channel, (b) relaxed

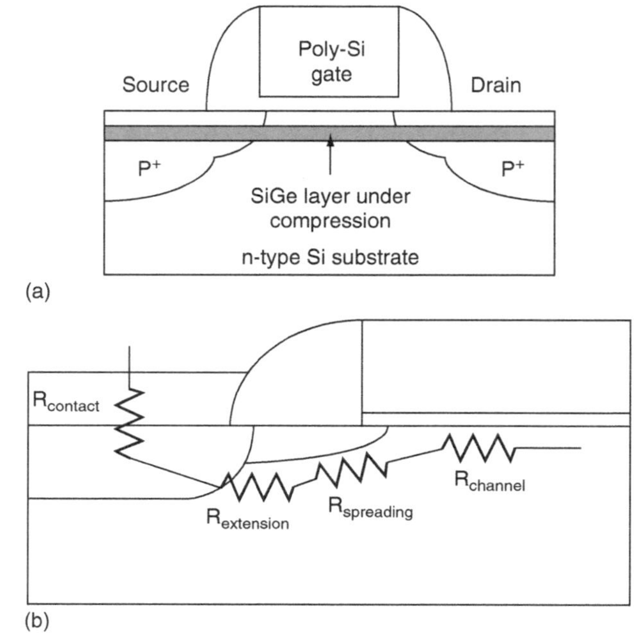

FIGURE 12.2 (a) Presence of $Si_{1-x}Ge_x$ alloy layer in different parts of a heterostructure MOSFET and (b) cross section of a MOS transistor illustrating the parasitic series resistance components.

virtual substrate for tensile strained-Si, (c) tunable work function in the gate, and (d) low-resistivity source/drain extension. Presence of $Si_{1-x}Ge_x$ alloys in different structural forms in the various parts of a MOSFET is shown in Figure 12.2. SiGe is advantageous because it can be selectively deposited as well as selectively etched compared to Si.

In order to realize the improvements in electron mobility and a usable conduction band offset, it is necessary for the film to be in biaxial tension. Instead of growing a thin $Si_{1-x}Ge_x$ layer on Si, a thin epitaxial layer of Si grown on a thick relaxed $Si_{1-x}Ge_x$ buffer layer can serve this requirement. Due to the lattice mismatch, the thin Si layer is in biaxial tension and exhibits an enhancement in both electron and hole mobilities, making it useful for both n- and p-MOSFETs. It also shows fairly large band offsets (on the order of 100 meV or more) in both valence and conduction bands relative to the relaxed $Si_{1-x}Ge_x$ on which it is grown. This allows the tailoring of the band structure to confine both holes and electrons. Strained-Si/SiGe heterolayers on Si wafers are currently being explored as a channel material for high performance heterostructure MOSFETs [10].

Developing robust technology for forming good metal(M)/SiGe source/drain contact regions and ultimately developing a metal silicidation process for application in future strained-Si and SiGe-channel hetero-FETs with high performance is of technological interest. Several new S/D technology using $Si_{1-x}Ge_x$ for future CMOS technology nodes have been proposed [11,12]. However, applying the conventional self-aligned silicide process to the strained-Si and SiGe MOSFETs is not straightforward due to the following reasons:

1. The strained-Si film is too thin to accommodate the conventional silicide. For example, if cobalt or nickel is used to make the silicide, then the Si film must be thicker than about 25 nm to accommodate the silicide.
2. The pseudomorphic strained-Si layer cannot be made thicker than 20 nm as the critical thickness depends upon the composition of the substrate, such as the mole fraction of Ge in the relaxed layer.
3. Since the strained-Si film is not thick enough to supply enough silicon to form the silicide, the silicide will consume some of the underlying SiGe buffer.

The conventional approach in the Si CMOS technology to the abovementioned problems is to make the silicon thicker in the source and drain regions prior to applying the salicide process. But the addition of Si must be selective, and should be limited only to the source and drain regions, which may be obtained by a selective deposition. Selective epitaxy is usually preferred to deposit Si only to the source, drain, and gate regions. Recently, a new self-aligned salicide process that does not require an increase in the thermal budget (high temperature) and is applicable to strained-Si/SiGe materials involving Ni, has been proposed [13].

12.4 Poly-SiGe Gate Technology

The fabrication of $Si_{1-x}Ge_x$ channel MOSFETs [14] always leads to the presence of $Si_{1-x}Ge_x$ in the source/drain regions (see Figure 12.2). Perhaps the most important advantage of using $Si_{1-x}Ge_x$ instead of Si as a junction material is that $Si_{1-x}Ge_x$ provides a promising solution to the contact resistance challenge by raising the possible active doping concentration and lowering the contact barrier height [11,15–19]. A poly-SiGe gate technology offers several advantages over conventional gate (poly-Si) conductors. First, poly-SiGe gate conductors require a relatively low-temperature rapid thermal anneal (RTA) to activate the dopants such as boron and phosphorous in the gate conductor. A low-temperature RTA facilitates the formation of ultrashallow source/drain junctions and the formation of ultratight pocket regions. Second, poly-SiGe gate conductors effectively suppress boron penetration into a gate conductor heavily doped with phosphorous. Third, poly-SiGe gate conductors can be utilized to adjust the threshold voltage of the transistor. The bandgap of $Si_{1-x}Ge_x$ lies between 0.67 eV (bandgap of Ge) and 1.12 eV (bandgap of Si) depending on the amount of Ge and the amount of strain in the lattice. The work function of a transistor is related to the concentration of Ge in the poly-SiGe gate conductor. This aspect is particularly advantageous in integrated circuits having transistors with several threshold voltage levels. Poly-$Si_{1-x}Ge_x$ gate technology offers a reduction in gate depletion compared to poly-Si as well.

Bang et al. [20] have reported sheet resistance, Hall mobility, and effective carrier concentration as a function of annealing parameters for boron and phosphorus ion implanted films of poly-Si, $Si_{0.75}Ge_{0.25}$ and $Si_{0.50}Ge_{0.50}$. The films were ion implanted with boron or phosphorus at dosages between 5×10^{14} and 4×10^{15} cm^2 and then thermally annealed between 550°C and 650°C from 0.25 to 120 min. The solid solubility of the most common dopants, such as B, P, and As is much higher in Si than in Ge. Boron-doped films showed decreasing minimum sheet resistance with increasing Ge mole fraction while phosphorus-doped films exhibited the reverse trend. However, the solubility of B in $Si_{1-x}Ge_x$ is found to drop when increasing the Ge fraction [21]. Supersaturation doping of $Si_{1-x}Ge_x$ with B is possible with the *in situ* doping technique using Si-, Ge-, and B-source gases simultaneously. For strained $Si_{1-x}Ge_x$, alloying with B up to 10 at.%, due to strain compensation effect [22,23], has been demonstrated by Gannavaram and Ozturk to yield extremely low specific resistivity and contact resistivity values [11] of 0.2-$\mu\Omega$-cm and 1.5×10^{-8} cm^2, respectively.

12.5 Silicidation of SiGe

In order to form contacts to source/drain junctions formed by the selective $Si_{1-x}Ge_x$ technology, a self-aligned contact process needs to be developed. Fortunately, $Si_{1-x}Ge_x$ alloys react with common silicide metals to form germanosilicides with properties much like those of silicides. To form these germanosilicide contacts, a good understanding of the metal/$Si_{1-x}Ge_x$ ternary solid-phase reactions is necessary. It is important to identify the most promising germanosilicides as self-aligned contacts to source/drain junctions formed by the selective $Si_{1-x}Ge_x$ technology. Towards this end, many research groups have made significant contributions towards the understanding of silicide formation of various metals in strained-SiGe and strained-Si layers. Results of studies of interaction between $Si_{1-x}Ge_x$ and other transition metals than Ti, Co, and Ni, such as Pd [24,25], Pt [26–29], Fe [30], Mo [31], W [32], and Zr [33–39] are available in the literature, but they will not be discussed in order to keep our focus on contact metallization using the SALICIDE technique in CMOS technology. Experimental SiGe germanosilicide results on Ti [40–45], Co [43,46–50], and Ni [51–56] have been reported. It is important to note that the majority of the published work on metal/$Si_{1-x}Ge_x$ solid-phase reactions involve undoped $Si_{1-x}Ge_x$ layers. For applications in strained-Si and SiGe, NiSi is chosen because Ni reacts with both strained-Si and SiGe and forms low sheet resistance silicide and germanosilicide in the source/drain area [54].

To study the solid-state interactions between Ti, Co, or Ni and $Si_{1-x}Ge_x$, the relevant phase diagrams are important. The available thermochemical data indicate that the germanides are less stable than their corresponding silicides with the same atomic composition but by replacing Ge with Si, as the heats of formation (enthalpy) or Gibbs free energies are lower (less negative) for the germanides. An approximate ternary phase diagram depicting the composition domains directly concerning the formation of $NiSi_{1-x}Ge_x$ as well as its interaction with $Si_{1-x}Ge_x$ is shown in Figure 12.3. The formation of germanides or germanosilicides tends to occur at a lower temperature than the formation of the silicides. The required experimental thermochemical data for the different phases involved in the pseudophase diagram were obtained from Ref. [55]. The pseudophase diagram shows that Ni, Si, Ge, NiSi, and NiGe surround the triangle. The existence and orientation of the tie lines provide the necessary information about the formation of Ni(Si,Ge) during low-temperature silicidation.

A comparative study on Ni and Co reacting with single-crystal $Si_{0.3}Ge_{0.7}$ also shows a clear advantage with Ni because a continuous $NiSi_{1-x}Ge_x$ layer was found, yielding the desired low-resistivity contact [57]. Figure 12.4 indicates that the processing window for $CoSi_2$ and $TiSi_2$ as 800–850°C. At low annealing temperatures, cobalt yielded a very high sheet resistance, presumably due to the formation of $Co(Si_{1-y}Ge_y)$ and retarded the formation of low resistive $CoSi_2$ phase. Because Ge atom hinders the transformation of $Co(Si_{1-y}Ge_y)$ structure to its lowest resistive Co-silicide phase $CoSi_2$ which forms at a high temperature, 900°C, and the germanosilicide layer shows severe agglomeration. But Ni is also found to be suitable in metal–semiconductor contact formation as its lowest resistive phase is attainable below 700°C as shown in Figure 12.4.

Figure 12.5a and b shows the SEM micrographs of the Ni-silicided films annealed at 500°C and 600°C for 60 s, respectively. The SEM micrographs show that the surface of the silicided film is still relatively smooth and the grain boundaries are not very prominent. Size of the grains gradually increases with increasing annealing temperature. It has been observed that agglomeration starts to take place above 700°C annealing and it results in a very abrupt increase in the sheet resistance [54,56].

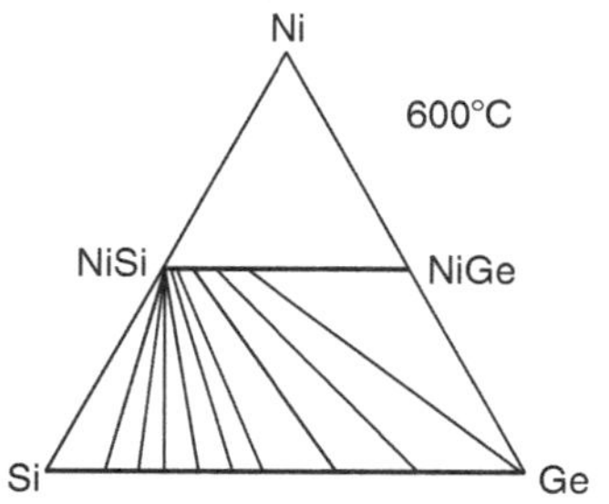

FIGURE 12.3 Pseudophase diagrams of the Ni–Si–Ge ternary system at 600°C. (After AR Saha, S Chattopadhyay, and CK Maiti. *Solid-State Electron* 48: 1391–1399, 2004.)

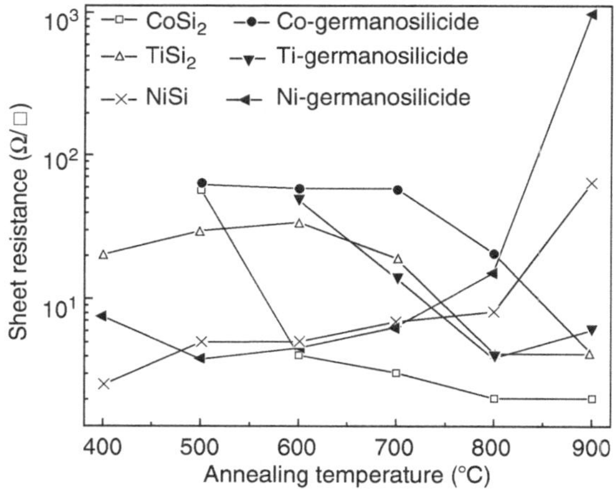

FIGURE 12.4 Variation of sheet resistance as a function of annealing temperature. (After AR Saha, S Chattopadhyay, and CK Maiti. *Solid-State Electron* 48:1391–1399, 2004.)

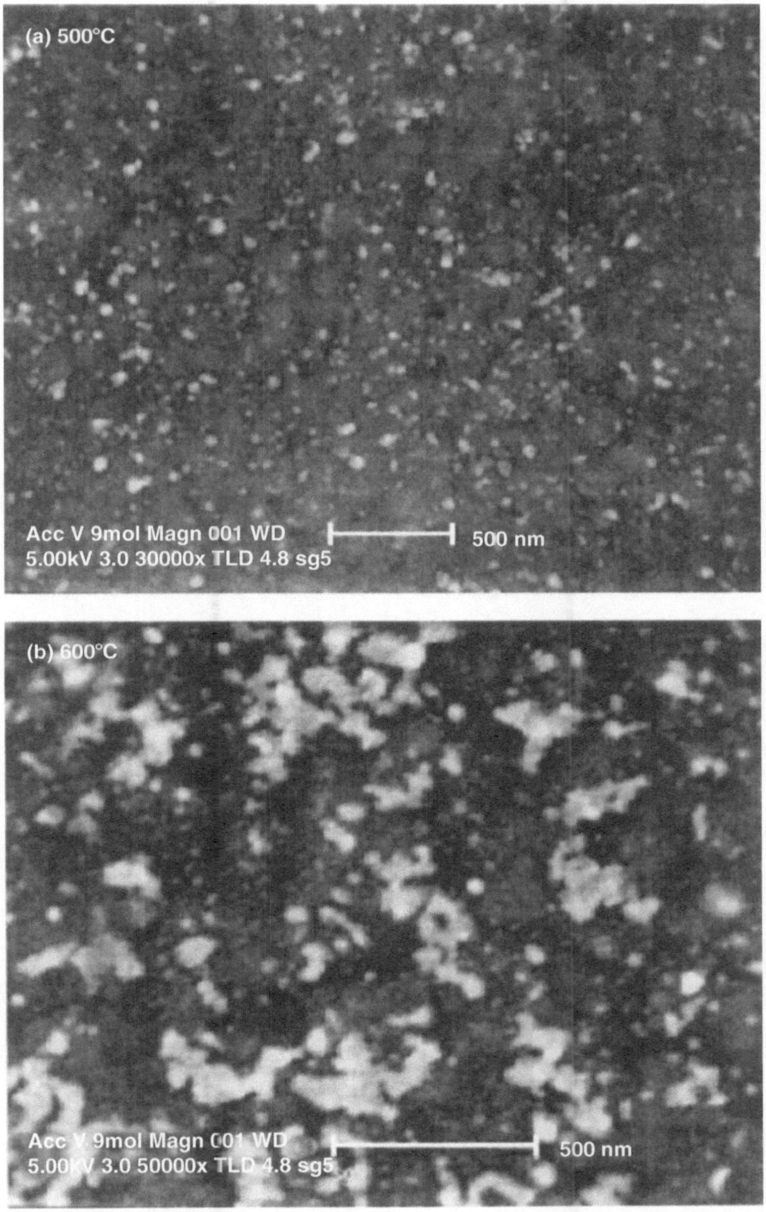

FIGURE 12.5 SEM micrographs of Ni-germanosilicided/$Si_{1-x}Ge_x$ film annealed at (a) 500°C and (b) 600°C. (After AR Saha, S Chattopadhyay, and CK Maiti. *Solid-State Electron* 48:1391–1399, 2004.)

The XTEM micrographs of Ni-silicided $Si_{1-x}Ge_x$ films annealed at 500°C and 600°C, for 60 s are shown in Figure 12.6a and b, respectively. The grain formation generally starts by grain boundary grooving and the atoms diffuse away from the highly curved grain boundaries. It is evident that Ge with higher diffusion coefficient in germanosilicide may out-diffuse into the grain boundaries and form a Ge-rich $Si_{1-x}Ge_x$ grain with an increase in the annealing temperature. At 500°C, a layer of relatively uniform $NiSi_{0.75}Ge_{0.25}$, with a thickness of about 80 ± 10 nm, was found. However, a very small amount of Ge-rich $Si_{1-x}Ge_x$ grains was detected, in between the $Ni(Si_{0.75}Ge_{0.25})$ grains. Similar results were also obtained for a

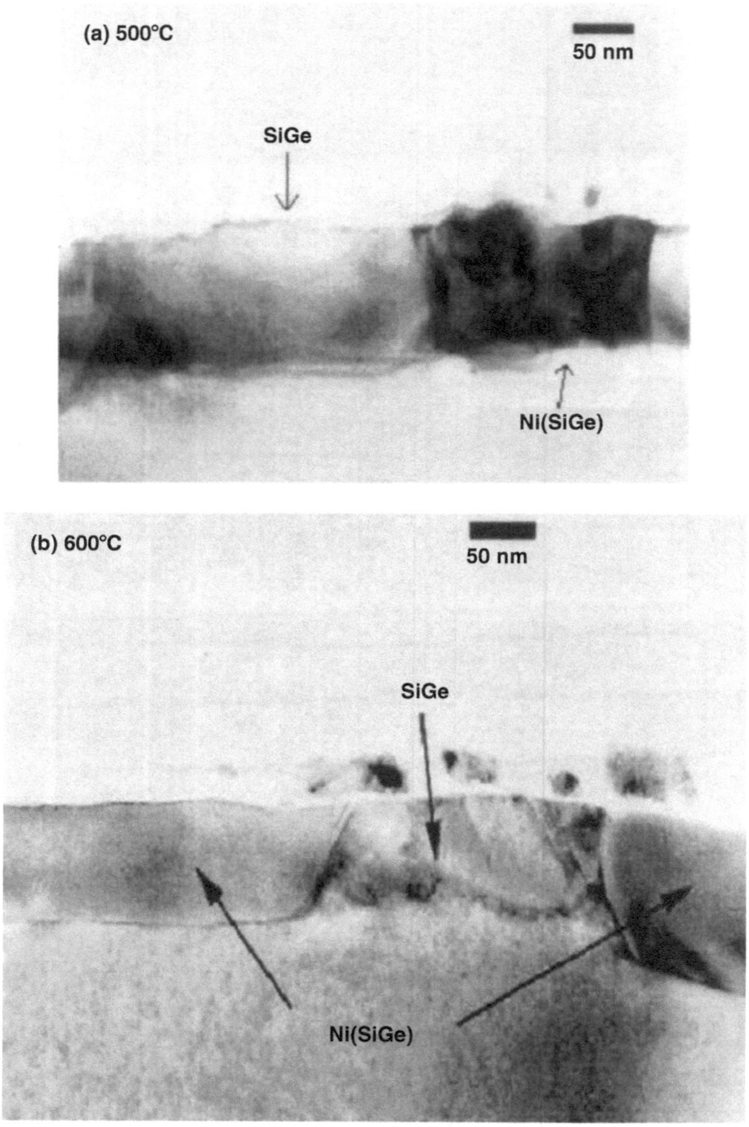

FIGURE 12.6 XTEM micrographs of the Ni-germanosilicided/$Si_{1-x}Ge_x$ films annealed at (a) 500°C and (b) 600°C. (After AR Saha, S Chattopadhyay, and CK Maiti. *Solid-State Electron* 48:1391–1399, 2004.)

lower anneal temperature (300°C) except that the $NiSi_{0.75}Ge_{0.25}$ was relatively uniform without the presence of any Ge-rich $Si_{1-x}Ge_x$ grains. At 600°C anneal temperature, Ge out-diffuses from the $Ni(Si_{0.75}Ge_{0.25})$ grain and penetrates into the grain boundaries and reacts with both Si and Ge. The distribution and thickness of such germanosilicide grains increase with the increase in annealing temperature. Due to thermal agglomeration, germanosilicide grains lose their columnar structure at the surface and at the interface of silicide–substrate. However, it is clear from the TEM images that the silicide/$Si_{1-x}Ge_x$ interface remains smooth within the annealing temperature range of 300–600°C [52,58,59].

Aldrich et al. [41,42] have studied the Ti/$Si_{1-x}Ge_x$ system, and found that C49 and C54 Ti/$Si_{1-y}Ge_y$ phases are formed with the Ge concentration of the $Si_{1-x}Ge_x$ substrate (when $x < 0.40$). At higher annealing temperatures, the Ge concentration in the germanosilicide decreases and a Ge-rich $Si_{1-z}Ge_z$

TABLE 12.3 Lattice Structure and Lattice Constants for Ni and Pt Silicide and Germanide

	Lattice Structure	a (Å)	b (Å)	c (Å)
NiSi	Orthorhombic	5.233	3.258	5.659
NiGe	Orthorhombic	5.381	3.428	5.811
PtSi	Orthorhombic	5.595	3.603	5.932
PtGe	Orthorhombic	5.719	3.697	6.084
PdSi	Orthorhombic	5.599	3.381	6.133

alloy precipitates along the grain boundaries, while germanium segregates out of the germanosilicide lattice, and is replaced by Si from the alloy. Lattice structure and lattice constants for Ni and Pt silicide and germanide are shown in Table 12.3.

Formation of Ni_2Si causes an intrinsic compressive stress [60] in agreement with the expectation of a volumetric expansion at the Ni_2Si–Si interface. Forces caused by the formation of NiSi are difficult to understand due to the presence of two reaction interfaces. The force changes during the silicide formation and was modeled by Zhang and d'Heurle [61]. However, when the Ni films are subjected to prolonged isothermal annealing, stresses that developed during silicide formation gradually relax.

Metal silicides and silicides/germanides have the drawback of less thermal stability and the consumption of silicon/germanium; while the latter is less of a problem for raised source/drains, scaling of vertical dimensions is never the less hampered. Studies on the effect of Ge concentration in relaxed $Si_{1-x}Ge_x$ on Ni silicidation process for different annealing temperatures have revealed that ternary compounds always form, and at the same time, Ge segregation has been observed when annealing temperature is above 400°C. Ge segregation with strain relaxation of the underlying $Si_{1-x}Ge_x$, which commonly appears in the metal/$Si_{1-x}Ge_x$ reactions using conventional annealing techniques, was effectively suppressed using pulsed laser annealing for Co [50] and Ni [51–53]. Due to rapid melt and solidification processes in laser annealing, the microstructure of the germanosilicide formed by pulsed laser annealing has been shown to be superior.

The segregated germanium atoms can react with Si atoms and result in a Ge-rich $Si_{1-x}Ge_x$ alloy adjacent to the germanosilicide. Therefore, the solid-phase reactions between metals and $Si_{1-x}Ge_x$ can, in general, be characterized as [42]

1. germanosilicide formation at lower temperatures; and
2. germanosilicide formation accompanied with germanium loss at higher temperatures. The two cases can be described with the following equations:

$$M + Si_{1-x}Ge_x \rightarrow M_y(Si_{1-x}Ge_x)_{1-y}$$

$$M_y(Si_{1-x}Ge_x)_{1-y} + Si_{1-x}Ge_x \rightarrow M_y(Si_{1-w}Ge_w)_{1-y} + (Si_{1-z}Ge_z)\ (w < x < z)$$

A schematic representation of this reaction sequence is shown in Figure 12.7. As shown, Ge loss from the germanosilicide occurs via Ge segregation at the grain boundaries, resulting in the formation of Ge-rich $Si_{1-z}Ge_z$ precipitates. The precipitates promote agglomeration and lead to higher sheet resistance due to high resistivity conduction barriers through the germanosilicide [42,62]. The structure, melting point, typical formation temperature, specific resistivity, and Schottky barrier heights to n-type Si for monosilicides of Ni, Pt, and Pd and monogermanides of Ni and Pt thin films are shown in Table 12.4.

As Ge segregation will degrade the contact performance, the control of the Ge segregation is crucial. Some reports have shown that the onset temperature for obtaining uniform low-resistivity Ni–SiGe is a strong function of the Ge concentration. In addition, it has further been shown that the RTA technology

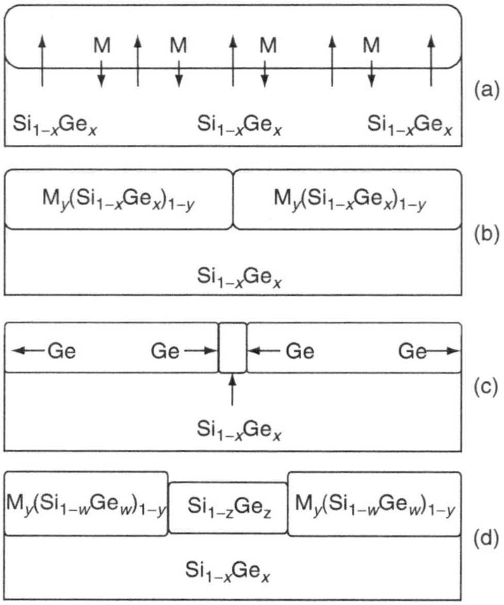

FIGURE 12.7 Model for the formation of $Si_{1-z}Ge_z$ grains between germanosilicide grains: (a) thermal reaction starts initially by the interdiffusion of metal (M), Si, and Ge, (b) formation of M-germanosilicide, (c) segregation of Ge from the germanosilicide grains, and (d) germanosilicide grains become discontinuous due to a rapid growth of Ge-rich grains.

TABLE 12.4 Structure, Melting Point, Typical Formation Temperature, Specific Resistivity, and Schottky Barrier Heights to n-type Si for Monosilicides of Ni, Pt, and Pd and Monogermanides of Ni and Pt Thin Films

Phase	Melting Point (°C)	Reaction Temperature (°C)	Resistivity (Ω-cm)	SBH (eV)
NiSi	980	250–400	14	0.65
NiGe	860	<400	—	
PtSi	1230	200–540	80	0.88
PtGe	1075	—	—	—
PdSi	972	850	30	—

has a very narrow process window for different Ge concentrations and may not be a viable manufacturable technology for fabricating uniform Ni–SiGe. In general, due to Ge segregation during germanosilicide formation, agglomeration and strain relaxation of the unreacted $Si_{1-x}Ge_x$ film take place.

At present, among all the germanosilicides studied, $NiSi_{1-x}Ge_x$ has offered the lowest resistivity and is capable of satisfying most of the needs of future CMOS technology. Since $NiSi_{1-x}Ge_x$ is a monogermanosilicide, it consumes less of the underlying $Si_{1-x}Ge_x$ compared to other digermanosilicides such as those of titanium or zirconium. However, the only concern for Ni germanosilicide appears to be its limited thermal stability. It has been observed that at temperatures equal to or higher than 500°C, Ge out-diffusion from the $NiSi_{1-x}Ge_x$ results in formation of Ge-rich $Si_{1-z}Ge_z$ precipitates, which could eventually transform into $Si_{1-z}Ge_z$ grains, which results in a rough top surface as well as a rough interface, high sheet resistance, and also high junction leakage.

A new approach has been proposed to improve the thermal stability of $NiSi_{1-x}Ge_x$. It has been reported that by inserting a thin layer of Pt between Ni and $Si_{1-x}Ge_x$ the thermal stability of the germanosilicide can be significantly improved through delaying the transition from NiSi to $NiSi_2$, which is responsible for the increased resistivity and the junction consumption [63–68]. It has also been observed that on heavily boron-doped $Si_{1-x}Ge_x$, the sheet resistance of the $NiSi_{1-x}Ge_x$ formed with a thin Pt interlayer is more stable than that of $NiSi_{1-x}Ge_x$ formed from pure Ni. However, the lowest sheet resistance value achieved using Ni with a thin Pt interlayer is slightly higher than the sheet resistance of the $NiSi_{1-x}Ge_x$ achieved using pure Ni.

For phosphorus-doped n^+-$Si_{1-x}Ge_x$, Pt incorporation might change the barrier height of the contact and preliminary results on n^+-$Si_{1-x}Ge_x$ obtained indicate that the Pt interlayer actually improves the contact resistance for Ni germanosilicide on n^+-$Si_{1-x}Ge_x$ while the contact resistance with pure Pt germanosilicide is higher. Incorporation of a small amount of Pt in Ni-$Si_{1-x}Ge_x$ via a thin Pt interlayer between Ni and $Si_{1-x}Ge_x$, in general, results in an improved thermal stability of Ni(Pt) germanosilicide.

12.6 Silicidation of SiGeC

The large mismatch between SiGe alloys and Si leads to severe restrictions in terms of critical thickness and thermal stability. The equilibrium critical thickness of a pseudomorphic SiGe layer depends on the total strain energy in the layer and restricts the applications where high Ge concentrations are needed. Addition of substitutional carbon to a SiGe alloy gives an extra degree of freedom to control the bandgap. Moreover, the strain introduced by the Ge atoms can be compensated by C atoms, which are smaller than Ge and Si atoms, resulting in a critical thickness that can be significantly increased.

Aubry-Fortuna et al. [35] have investigated the reactions between Zr and SiGeC alloys using RTA. The interactions of the metal films with the $Si_{1-x-y}Ge_xC_y$ alloys have been investigated by using various analytical tools. The authors concluded that Zr may be a good candidate for contacts on IV–IV alloys in terms of thermal stability.

The reaction of Co with epitaxial SiGeC/Si layers has been investigated and compared to the reaction of Co with SiGe/Si layers by Donaton et al. [69]. The sequence of phase formation is the same as the reaction of Co with monocrystalline Si, however, cobalt disilicide is formed at much higher temperatures. The presence of C delays the disilicide formation, as a result of C accumulation at the silicide–substrate interface during the reaction, which blocks the Co diffusion paths. XRD spectra of 17 nm Co on $Si_{0.811}Ge_{0.18}C_{0.009}$ after annealing for 30 s at different temperatures showed that after 500°C the predominant phase is Co_2Si in both cases, although the presence of the monosilicide (CoSi) has been observed. Further annealing at 600°C and 700°C results in a full transformation to CoSi. At 800°C both the mono- and the disilicide phases have been identified, with the $CoSi_2$ peaks being much more intense in the case of the reaction on SiGe than on SiGeC, indicating the presence of a higher fraction of this phase in the former film.

It has been observed that the presence of C retards the formation of the disilicide phase. During the reaction, C piles up at the silicide–substrate interface, which is believed to block the Co diffusion paths. The precise role of the C atoms in this retardation is not yet fully understood and is a subject that deserves further investigation. Also, in case of $Si_{1-x-y}Ge_xC_y$, the presence of substitutional C was found to be effective in reducing strain relaxation during solid-phase reactions to form Ti germanosilicide [44].

12.7 Summary

One of the key challenges for future CMOS technology is to form source/drain junctions and contacts with very small parasitic series resistance. Germanosilicides of different metals were reviewed with the objective of finding the best germanosilicide candidates for low-resistivity contacts to heavily doped $Si_{1-x}Ge_x$ alloys. Formation of germanosilicide of Ti, Co, and Ni contacts to $Si_{1-x}Ge_x$ alloys has been discussed in detail. Similar to other germanosilicides, Ni/$Si_{1-x}Ge_x$ suffers from Ge out-diffusion. It was found that Ge lost from the Ni/$Si_{1-x}Ge_x$ grains formed Ge-rich $Si_{1-z}Ge_z$ precipitates accumulating at the germanosilicide grain boundaries. Consequently, the formation temperature of Ni/$Si_{1-x}Ge_x$ needs to be kept below 450°C.

New junction formation technology based on selective deposition of $Si_{1-x}Ge_x$ alloys in source/drain regions is needed. Smaller bandgap of $Si_{1-x}Ge_x$ and poly-SiGe resulting in a smaller metal–semiconductor barrier height, which reduces the contact resistivity of a metal–semiconductor contact has been described. Ni and Pt germanosilicides are found to be attractive because they form both monogermanosilicides and consume much less $Si_{1-x}Ge_x$ than $Zr(Si_{1-x}Ge_x)_2$. A new approach to form ultrathin Ni/$Si_{1-x}Ge_x$ layers with enhanced thermal stability by incorporating a thin Pt interlayer between Ni and $Si_{1-x}Ge_x$ has also been discussed.

There is a growing interest to form strained-Si layers without a thick relaxed $Si_{1-x}Ge_x$ buffer. As the MOSFET dimensions are scaled down to tens of nanometers, the composition of the source/drain junctions will most likely have a strong impact on the strain in the channel region. In particular, the impact of $Si_{1-x}Ge_x$ junctions on the two sides of a MOSFET channel needs further investigation. However, the $Si_{1-x}Ge_x$ junction technology may provide a simple yet powerful means to form strained-Si channels without involving complex and costly fabrication procedures.

References

1. SL Zhang and M Ostling. Metal silicides in CMOS technology: past, present, and future trends. *Crit Rev Solid State Materials Sci* 28:1–129, 2003.
2. Semiconductor Industry Association, International Technology Roadmap for Semiconductors, San Jose, California, 2003.
3. J Lasy, J Nakos, O Chin, and P Geiss. Comparison of transformation to low-resistivity phase and agglomeration of $TiSi_2$ and $CoSi_2$. *IEEE Trans Electron Dev* 38:262–269, 1991.
4. H Jeon, CA Sukow, Honecutt, GA Rozgonyi, and RJ Nemanich. Morphology and phase stability of $TiSi_2$ on Si. *J Appl Phys* 71:4269–4276, 1992.
5. MC Poon, CH Ho, F Deng, SS Lau, and H Wong. Thermal stability of cobalt and nickel silicides. *Microelectron Reliab* 38:1495–1498, 1998.
6. T Morimoto, HS Momose, T Iinuma, I Kunishima, K Suguro, H Okano, I Katakabe, H Nakajima, M Tsuchiaki, M Ono, Y Katsumata, and H Iwai. A NiSi salicide technology for advanced logic devices. *IEEE IEDM Tech Dig* 653–656, 1991.
7. T Ohguro, S Nakamura, M Koike, T Morimoto, A Nishiyama, Y Ushiku, T Yoshitomi, M Ono, M Saito, and H Iwai. Analysis of resistance behavior in Ti- and Ni-salicided polysilicon films. *IEEE Trans Electron Dev* 41:2305–2317, 1994.
8. T Morimoto, T Ohguro, H S Momose, T Iinuma, I Kunishima, K Suguro, I Katakabe, H Nakajima, M Tsuchiaki, M Ono, Y Katsumata, and H Iwai. Self-aligned nickel-mono-silicide technology for high-speed deep submicrometer logic CMOS ULSI. *IEEE Trans Electron Dev* 42:915–922, 1995.
9. CK Maiti and GA Armstrong. *Applications of Silicon–Germanium Heterostructure Devices.* Bristol: Institute of Physics, 2001.
10. CK Maiti, NB Chakrabarti, and SK Ray. *Strained Silicon Heterostructures: Materials and Devices.* UK: IEE, 2001.
11. S Gannavaram, N Pesovic, and MC Ozturk. Low temperature ($< 800°C$) recessed junction selective silicon-germanium source/drain technology for sub-70 nm CMOS. *IEEE IEDM Tech Dig* 437–440, 2000.
12. MC Ozturk, J Liu, H Mo, and N Pesovic. Advanced $Si_{1-x}Ge_x$ source/drain and contact technologies for sub-70 nm CMOS. *IEEE IEDM Tech Dig* 375–378, 2002.
13. Q Xiang, JS Goo, J Pan, B Yu, S Ahmed, J Zhang, and MR Lin, Strained silicon NMOS with nickel–silicide metal gate. *IEEE VLSI Tech Symp Dig* 101–102, 2003.
14. SV Vandebroek, E Crabbe, B Meyerson, D Harame, P Restle, J Stork, and J Johnson. SiGe-channel heterojunction p-MOSFETs. *IEEE Trans Electron Dev* 41:94–101, 1994.
15. TJ King, JR Pfriester, JD Scott, JP McVittie, and KC Saraswat. A polycrystalline SiGe gate CMOS technology. *IEEE IEDM Tech Dig* 253–256, 1990.

16. TJ King, JR Pfiester, and KC Saraswat. A variable work function polycrystalline $Si_{1x}Ge_x$ gate material for submicrometer CMOS technologies. *IEEE Electron Dev Lett* 12:533–535, 1991.
17. YV Ponomarev, C Salm, J Schmitz, PH Woerlee, PA Stolk, and DJ Gravesteijn. Gate workfunction engineering using poly-(Si, Ge) for high-performance 0.18 m CMOS technology. *IEEE IEDM Tech Dig*, 1997, 829–832.
18. PE Hellberg, SL Zhang, HH Radamson, and W Kaplan. Threshold voltage control for pMOSFETs using an undoped epitaxial Si channel and a p^+-Si_xGe_{1-x} gate. *Solid-State Electron* 44:2085–2088, 2000.
19. WC Lee, B Watson, TJ King, and C Hu. Enhancement of PMOS device performance with poly-SiGe gate. *IEEE Electron Dev Lett* 20:232–234, 1999.
20. DS Bang, M Cao, A Wang, KC Saraswat, and TJ King. Resistivity of boron and phosphorous doped polycrystalline $Si_{1-x}Ge_x$ films. *Appl Phys Lett* 66:195–197, 1995.
21. PE Hellberg, A Gagnor, SL Zhang, and CS Petersson. Boron doped polycrystalline Si_xGe_{1-x} films: Dopant activation and solid solubility. *J Electrochem Soc* 144:3968, 1997.
22. B Tillack, P Zaumseil, G Morgenstern, D Kruger, B Dietrich, and G Ritter. Strain compensation in ternary $Si_{1-x\ y}Ge_xB_y$ films. *J Cryst Growth* 157:181–184, 1995.
23. S Gannavaram and MC Ozturk. Ultra-shallow p^+–n junctions for 35–70 nm CMOS using selectively deposited very heavily boron doped silicon-germanium films. *ECS Meeting Proc Abs* 2000, p. 73.
24. QZ Hong, JG Zhu, and JW Mayer. Solid phase epitaxy of a Ge–Si alloy on [111] Si through a Pd_2Si layer. *Appl Phys Lett* 55:747–749, 1989.
25. A Buxbaum, M Eizenberg, A Raizman, and F Schaffler. Interaction of Pd with strained layers $Si_{1-x}Ge_x$ epitaxially grown on Si(100). *Jpn J Appl Phys* 30:3590–3593, 1991.
26. QZ Hong and JW Mayer. Thermal reactions between Pt thin films and $Si_{1-x}Ge_x$ alloys. *J Appl Phys* 66:611–615, 1989.
27. H Kanaya, Y Cho, F Hasegawa, and E Yamaka. Preferential PtSi formation in thermal reaction between Pt and $Si_{0.80}Ge_{0.20}$ MBE layers. *Jpn J Appl Phys* 29:L850–L852, 1990.
28. P Zhong and Y Zheng. Study of Ge movement during thermal reactions between Pt and GeSi/Si heterostructures. *Appl Phys Lett* 62:3259–3261, 1993.
29. HK Liou, X Wu, U Gennser, VP Kesan, SS Iyer, KN Tu, and ES Yang. Interfacial reactions and Schottky barriers of Pt and Pd on epitaxial $Si_{1-x}Ge_x$ alloys. *Appl Phys Lett* 60:577–579, 1992.
30. H Chen, P Han, XD Huang, LQ Hu, Y Shi, and YD Zheng. Semiconducting Ge–Si–Fe alloy grown on Si(100)substrate by reactive deposition epitaxy. *Appl Phys Lett* 69:1912–1914, 1996.
31. JS Luo, WT Lin, CY Chang, PS Shih, and TC Chang. Pulsed KrF laser annealing of $Mo/Si_{0.76}Ge_{0.24}$. *Nuclear Inst Meth Phys Res B* 169:129–134, 2000.
32. V Aubry, F Meyer, R Laval, C Clerc, P Warren, and D Dutartre. Thermal stability of W on RTCVD $Si_{1-x}Ge_x$ films. *Appl Surf Sci* 73:285–289, 1993.
33. Y Yasuda, O Nakatsuka, and S Zaima. Interfacial reactions of Ti/ and $Zr/Si_{1-x}Ge_x/Si$ contacts with rapid thermal annealing. *Thin Solid Films* 373:73–78, 2000.
34. Z Wang, DB Aldrich, RJ Nemanich, and DE Sayers. Electrical and structural properties of zirconium germanosilicide formed by a bilayer solid state reaction of Zr with strained $Si_{1-x}Ge_x$ alloys. *J Appl Phys* 82:2342–2348, 1997.
35. V Aubry-Fortuna, M Barthula, JL Perrossier, F Meyer, V Demuth, HP Strunk, and O Chiax-Pluchery. Contacts on $Si_{1-x-y}Ge_xC_y$ alloys: electrical properties and thermal stability. *J Vac Sci Technol B* 16:1659–1662, 1998.
36. V Aubry-Fortuna, A Eyal, O Chaix-Pluchery, M Barthula, F Meyer, and M Eizenberg. Thermal stability and electrical properties of $Zr/Si_{1-x-y}Ge_xC_y$ contacts after rapid thermal annealing. *Appl Phys Lett* 73:1248–1250, 1993.
37. V Aubry-Fortuna, G Tremblay, F Meyer, Y Miron, Y Roichman, M Eizenberg, F Fortuna, U Hormann, and H Strunk. Phase formation and strain relaxation during thermal reaction of Zr and Ti with strained $Si_{1-x-y}Ge_xC_y$ epilayers. *J Appl Phys* 88:1418–1423, 2000.

38. V Aubry-Fortuna, O Chaix-Pluchery, F Fortuna, C Hermandez, Y Campidelli, and D Bensahel. Structural properties and stabilities of Zr and Ti germanosilicides formed by rapid thermal annealing. *J Appl Phys* 91:5468–5473, 2002.
39. V Aubry-Fortuna, M Barthula, F Meyer, A Eyal, C Cytermann, M Eizenberg, and O Chaix-Pluchery. Characterizations of Zr/$Si_{1-x-y}Ge_xC_y$ after rapid thermal annealing. *Proc Mater Res Soc* 1998, pp. 49–53, 1998.
40. JB Lai and LJ Chen. Effects of composition on the formation temperatures and electrical resistivities of C54 titanium germanosilicide in Ti–$Si_{1-x}Ge_x$ systems. *J Appl Phys* 86:1340–1345, 1999.
41. DB Aldrich, YL Chen, DE Sayers, RJ Nemanich, SP Ashburn, and MC Ozturk. Effect of composition on phase formation and morphology in Ti–$Si_{1-x}Ge_x$ solid phase reactions. *J Mater Res* 10:2849–2863, 1995.
42. DB Aldrich, YL Chen, DE Sayers, RJ Nemanich, SP Ashbun, and MC Ozturk. Stability of C54 titanium germanosilicide on a silicon–germanium alloy substrate. *J Appl Phys* 77:5107–5114, 1995.
43. Z Wang, DB Aldrich, YL Chen, DE Sayers, and RJ Nemanich. Silicide formation and stability of Ti/SiGe and Co/SiGe. *Thin Solid Films* 270:555–560, 1995.
44. R Eyal, R Brener, R Beserman, M Eizenberg, Z Atzmon, DJ Smith, and JW Mayer. The effect of carbon on strain relaxation and phase formation in the Ti/$Si_{1-xy}Ge_xC_y$/Si contact system. *Appl Phys Lett* 69:64–66, 1996.
45. JC Huang, JS Luo, WT Lin, CY Chang, and PS Shih. Effects of an interposed Mo layer on the interfacial reactions of Ti/$Si_{0.76}Ge_{0.24}$ by rapid thermal annealing and pulsed laser annealing. *Semicond Sci Technol* 15:941–945, 2000.
46. SP Ashburn, MC Ozturk, G Harris, and DM Maher. Phase transitions during solid-phase formation of cobalt germanide by rapid thermal annealing. *J Appl Phys* 74:4455–4460, 1993.
47. JS Luo, WT Lin, CY Chang, WC Tsai, and SJ Wang. Interfacial reactions of the Co/$Si_{1-x}Ge_x$ system. *Mater Chem Phys* 48:140–144, 1997.
48. O Nur, M Willander, HH Radamson, MR Sardela, Jr., GV Hansson, CS Patersson, and K Maex. Strain characterization of $CoSi_2$/n-$Si_{0.9}Ge_{0.1}$/p-Si heterostructures. *Appl Phys Lett* 64:440–442, 1994.
49. O Nur, M Willander, L Hultman, HH Radamson, GV Hansson, MR Sardela, and JE Greene. $CoSi_2$/$Si_{1-x}Ge_x$/Si(0 0 1) heterostructures formed through different reaction routes: silicidation-induced strain relaxation, defect formation and interlayer diffusion. *J Appl Phys* 78:7063–7069, 1995.
50. JS Luo, YL Hang, WT Lin, CY Chang, and PS Shih. Interfacial reactions of Co/$Si_{0.76}Ge_{0.24}$ and Co($Si_{0.76}Ge_{0.24}$)/$Si_{0.76}Ge_{0.24}$ by pulsed KrF laser annealing. *J Mater Res* 14:3433–3438, 1999.
51. JS Luo, WT Lin, CY Chang, PS Shih, and FM Pan. Annealing effects on the interfacial reactions of Ni on $Si_{0.76}Ge_{0.24}$ and $Si_{1-x-y}Ge_xC_y$. *J Vac Sci Technol A* 18:143–148, 2000.
52. JS Luo, WT Lin, CY Chang, and WC Tsai. Pulsed KrF laser annealing of Ni/$Si_{0.76}Ge_{0.24}$ films. *J Appl Phys* 82:3621–3623, 1997.
53. JS Luo, WT Lin, CY Change, and WC Tsai. Interfacial reactions of Ni on $Si_{0.76}Ge_{0.24}$ and Si by pulsed laser annealing. *Mater Chem Phys* 54:160–163, 1998.
54. HB Zhao, KL Pey, WK Choi, S Chattopadhyay, EA Fitzgerald, DA Antoniadis, and PS Lee. Interfacial reactions of Ni on $Si_{1-x}Ge_x$ ($x = 0.2, 0.3$) at low temperature by rapid thermal annealing. *J Appl Phys* 92:214–217, 2002.
55. T Jarmar, J Seger, F Ericson, D Mangelinck, U Smith, and SL Zhang. Morphological and phase stability of nickel-germanosilicide on $Si_{1-x}Ge_x$ under thermal stress. *J Appl Phys* 92:7193–7199, 2002.
56. KL Pey, WK Choi, S Chattopadhyay, HB Zhao, EA Fitzgerald, DA Antoniadis, and PS Lee. Thermal reaction of nickel and $Si_{0.75}Ge_{0.25}$ alloy. *J Vac Sci Technol A* 20:1903–1910, 2002.
57. CY Lin, WJ Chen, CH Lai, A Chin, and J Liu. Formation of Ni germeno-silicide on single crystalline $Si_{0.3}Ge_{0.7}$/Si. *IEEE Electron Dev Lett* 23:464–466, 2002.
58. S Zaima, O Nakatsuka, A Sakai, J Murota, and Y Yasuda. Interfacial reaction and electrical properties in Ni/Si and Ni/SiGe(C) contacts. *Appl Surf Sci* 224:215–221, 2004.

59. V Aubry, OC Pluchery, F Fortuna, C Hernandez, Y Campidelli, and D Bensahel. Structural properties and stability of Zr and Ti germanosilicides formed by rapid thermal annealing. *J Appl Phys* 91:5468–5473, 2002.
60. CJ Tsai and KH Yu. Stress evolution during isochronal annealing of Ni/Si system. *Thin Solid Films* 350:91–95, 1999.
61. SL Zhang and FM d'Heurle. Stresses from solid state reactions—a simple model silicides. *Thin Solid Films* 213:34–39, 1992.
62. KL Pey, S Chattopadhyay, WK Choi, Y Miron, EA Fitzgerald, DA Antoniadis, and T Osipowicz. Stability and composition of Ni-germanosilicided $Si_{1-x}Ge_x$ films. *J Vac Sci Technol B* 22:852–858, 2004.
63. D Manglelinck, JY Dia, JS Pan, and SK Lahiri. Enhancement of thermal stability of NiSi films on (100)Si and (111)Si by Pt addition. *Appl Phys Lett* 75:1736–1738, 1999.
64. LW Cheng, SL Cheng, LJ Chen, HC Chien, HL Lee, and FM Pan. Formation of Ni silicides on (001) Si with a thin interposing Pt layer. *J Vac Sci Technol A* 18:1176–1179, 2000.
65. JF Liu, HB Chen, JY Feng, and J Zhu. Improvement of the thermal stability of NiSi films by using a thin Pt interlayer. *Appl Phys Lett* 77:2177–2179, 2000.
66. DZ Chi, D Mangelinck, SK Lahiri, PS Lee, and KL Pey. Comparative study of current–voltage characteristics of Ni and Ni(Pt)-alloy silicided p^+/n diodes. *Appl Phys Lett* 78:3256–3258, 2001.
67. JF Liu, JY Feng, and J Zhu. Comparison of the thermal stability of NiSi films in Ni/Pt/(111)Si and Ni/Pt/(100)Si systems. *J Appl Phys* 90:745–749, 2001.
68. PS Lee, KL Pey, D Mangelinck, J Ding, DZ Chi, and L Chan. New silicidation technology with Ni(Pt) alloy for MOSFETs. *IEEE Electron Dev Lett* 22:568–570, 2001.
69. RA Donaton, K Maex, A Vantomme, G Langouche, Y Morciaux, AS Amour, and JC Sturm. Co silicide formation on SiGeC/Si and SiGe/Si layers. *Appl Phys Lett* 70:1266–1268, 1997.
70. AR Saha, S Chattopadhyay, and CK Maiti. Contact metallization on strained-Si. *Solid-State Electron* 48:1391–1399, 2004.

13

Selective Etching Techniques for SiGe/Si

S. Monfray and Thomas Skotnicki
ST Microelectronics

S. Borel
CEA LETI

13.1 Introduction

SiGe alloys are used extensively for their electrical properties (bandgap engineering by adjusting the Ge content, carriers mobility improvement). With the progress in deposition by epitaxial growth, a new application is emerging, based on the crystalline properties of the material. Indeed, from pure Si ($x=0$) to pure Ge ($x=1$), any $Si_{1-x}Ge_x$ alloy can be deposited on Si with the same crystal structure and the same lattice constant (in the growth plane) as the substrate, provided that it does not exceed the critical thickness for plastic relaxation (E_c). As a consequence, SiGe can be buried between mono-Si layers without modifying their conduction properties. This method is used in several advanced microelectronics architectures that require a thin single-crystal Si-film isolated from the substrate by a cavity: the SiGe layer is used as a sacrificial layer for lateral etching. In this approach, the selectivity is crucial because both similar materials are exposed to the complete etching process whereas only one has to be removed.

The objective of this chapter is to describe here the mainstream of the selective SiGe etching and its applications. In particular, we present the Silicon-on-Nothing (SON) process and its possible technological variants within the process of simplification, and improvement of versatility and robustness. This development effort converges towards an SON technological platform, allowing easy co-integration of SON and bulk transistors.

13.2 Process Development for Industrial Applications

Getting some selectivity between SiGe alloys and pure Si is challenging because of the similarity between the two materials. Indeed, they have the same crystal structure (diamond), their lattice constants are not so different (<1% for Ge contents <25%) and their components (Si and Ge) are very close in terms of reactivity to halogen-based chemistries (same external electronic configuration, same kind of volatile etching products).

Anyway, it is possible to discriminate between them in highly selective etching processes. In fact, there is no clear explanation about the mechanisms responsible for this selectivity. Neither the strain of SiGe deposited on Si, nor the relative strengths of Si—Si and Si—Ge bonds or electronic effects can explain the high etch rates (ERs) that can be obtained on SiGe as compared to that of pure Si.

This section gives a way of obtaining a tunnel by lateral etching of SiGe in a Si/SiGe/Si stack. It details a tool that can be used for this purpose, its functioning, but also a process with its performances and limitations (Figure 13.1).

Equipment

Shibaura CDE-80 is an industrial tool (Figure 13.2) dedicated to chemical dry etching by activated gases, i.e., by neutrals coming from a remote plasma.

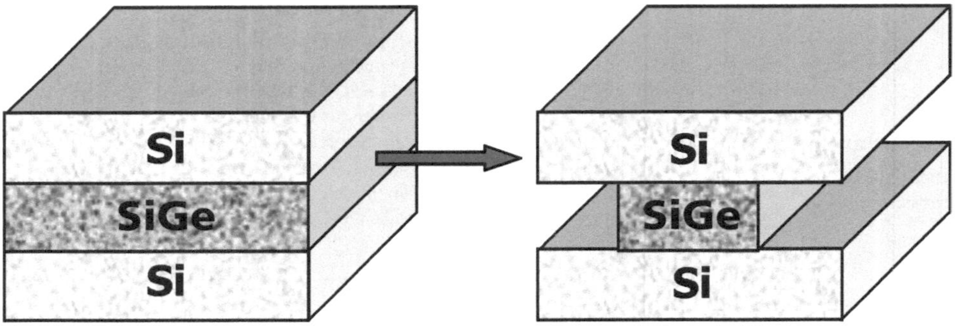

FIGURE 13.1 Selective lateral etching of SiGe in a Si/SiGe/Si stack (cross section).

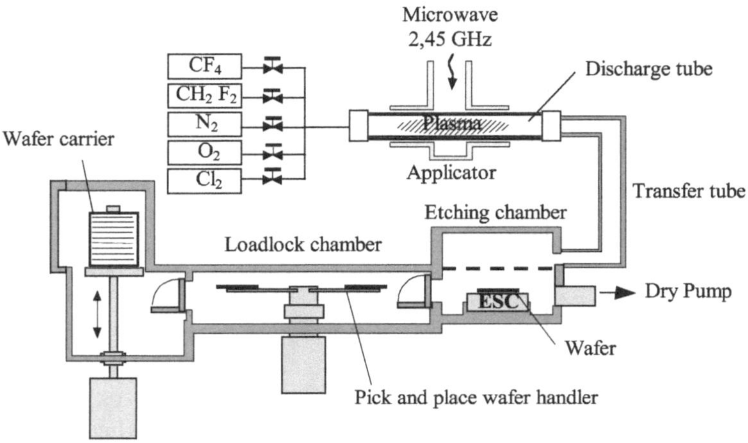

FIGURE 13.2 Schematic drawing of CDE-80 etching tool.

A gas mixture is injected in the etching chamber through a quartz tube (discharge tube) connected to a Teflon tube (transfer tube). The pressure of the system is regulated by an auto pressure controller connected to a dry pump. Then, a plasma is ignited in the discharge tube by using a 2.45 GHz microwave generator. The gas molecules are dissociated and ionized by energetic electrons, leading to several kinds of species. Some of these reactions are given below, as well as examples concerning CF_4:

- Production of *radicals* and *atoms*:

$$AB + e \rightarrow (AB)^* \qquad CF_4 + e \rightarrow (CF_4)^* \tag{13.1}$$

$$(AB)^* \rightarrow A + B \qquad (CF_4)^* \rightarrow CF_3 + \underline{F} \tag{13.2}$$

- Production of electrons and ions by dissociative ionization:

$$AB + e \rightarrow A^+ + B + 2e \qquad CF_4 + e \rightarrow {CF_3}^+ + F + 2e \tag{13.3}$$

$$A + e \rightarrow A^+ + 2e \qquad CF_3 + e \rightarrow {CF_3}^+ + 2e \tag{13.4}$$

As the plasma is cold and not highly ionized ($\alpha < 10^{-4}$), ions and electrons are much less numerous than molecules, radicals, and atoms. Moreover, their lifetime is much shorter, so only a few of them reach the Teflon tube:

- Loss of electrons and ions:

$$AB^+ + e \rightarrow A + B \qquad {CF_2}^+ + e \rightarrow CF + F \tag{13.5}$$

Neutral species reach the transfer tube and diffuse through it to the etching chamber. They also partially recombine:

- Loss of *radicals* and *atoms*:

$$A + B + M \rightarrow AB + M \qquad CF + \underline{F} + CF_4 \rightarrow CF_2 + CF_4 \tag{13.6}$$

As a consequence, the etching of the wafer consists of reactions between the material to etch and neutral species remaining from the discharge. It is purely chemical (no ion bombardment), that is why it is isotropic (no favored direction) and potentially selective (no sputtering). The etching is mainly driven by the composition of the gas mixture, but other parameters can be tuned in order to control it.

Influence of the Different Parameters

Gas Mixture

As mentioned above, this parameter is of primary importance because of the chemical nature of the process. The gases that are used must contain elements which have a spontaneous affinity for the material to etch, and they must form volatile products with it.

Silicon and germanium are commonly etched with halogen-based chemistries because of the easy formation of volatile SiX_4 and GeX_4 ($X = F$, Cl, Br) in reactive ion etching. In the case of chemical dry etching, there is no ion bombardment to assist the reaction, and all halogens are not of equal efficiency. Fluorine has a spontaneous action on Si or Ge, so fluorinated compounds are the basic gases for Si or SiGe isotropic etching.

Other gases may be added in order to modify the level of dissociation/recombination or the passivation layers at the surface of the wafer. For example, a small amount of oxygen contributes to the dissociation of CF_x by forming CO_2 and atomic fluorine, but an excess of O_2 leads to a passivating oxide layer that prevents the etching (Figure 13.3).

Microwave Power

The microwave power sets the level of dissociation in the plasma, i.e., the ratio between active species and inert molecules.

At low power, the discharge is unstable and the dissociation is poor. The small number of active species make the etch rate quite low. By increasing the power, the production of electrons is increased so the initiation of the discharge and its stability are easier, leading to a greater dissociation (Equations 13.1 to 13.4). The amount of active species is higher and the etch rate is increased (Figure 13.4).

Pressure

Pressure determines the rate of recombinations between the different species. An increase of the pressure corresponds to a decrease of the mean free path of the species (Figure 13.5). They undergo more collisions and may become inert (Equations 13.5 and 13.6). The result is a decrease of the etch rate.

Temperature

During the process, the quartz tube warms up because of energy dissipated by the discharge. For high-power processes, the tube temperature can exceed 150°C after a few minutes. The consequence is an

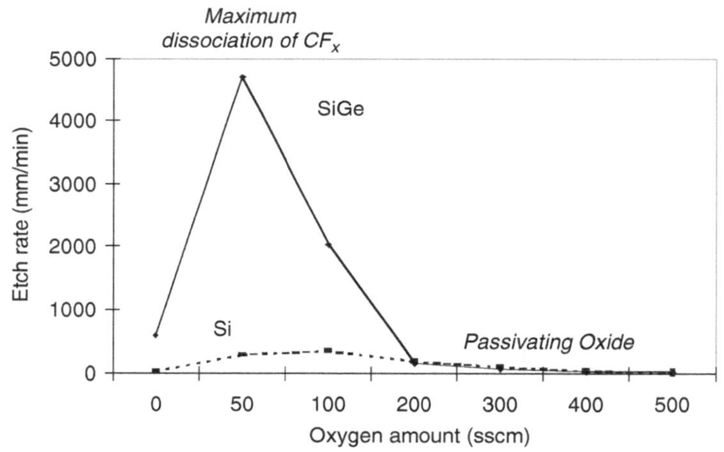

FIGURE 13.3 Effect of O_2 addition into pure CF_4.

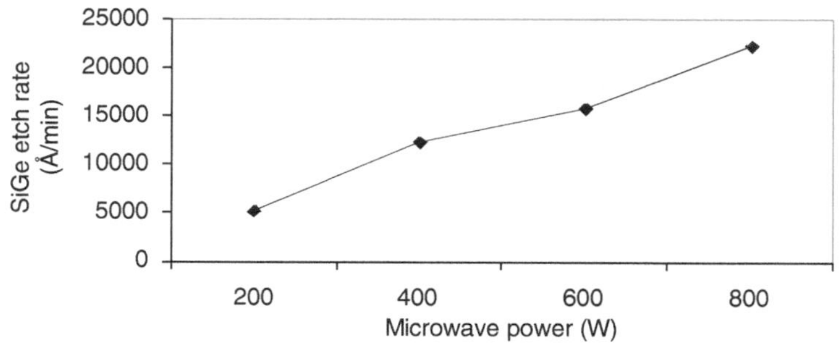

FIGURE 13.4 Effect of the microwave power on SiGe lateral etch rate.

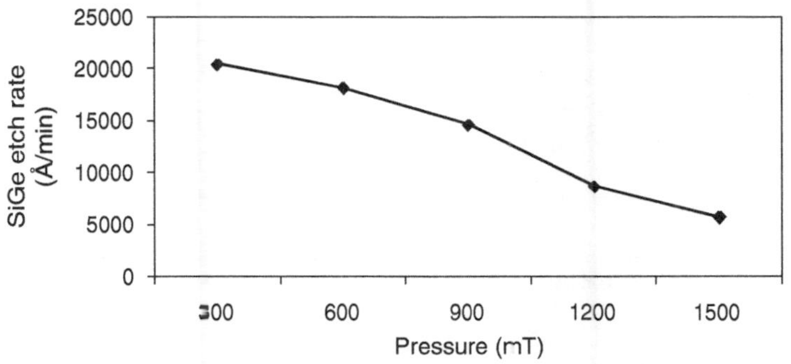

FIGURE 13.5 Effect of the pressure on SiGe lateral etch rate.

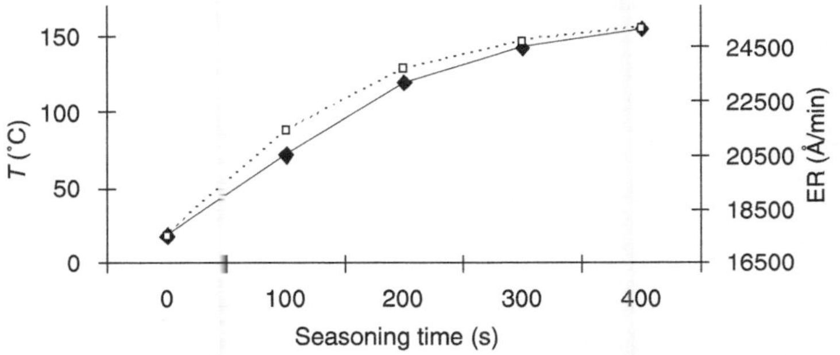

FIGURE 13.6 Effect of a seasoning on the quartz tube temperature and on SiGe lateral etch rate.

increase of the etch rate due to more dissociation of the gas mixture. But at low power, or for short etching processes, this phenomenon is negligible (Figure 13.6).

The temperature of the electrostatic chuck (ESC) that holds the wafer may also play an important role in the control of the reactions that take place on it. Indeed, some products are not volatile at low temperature but they can be produced, gasified, and pumped away at higher temperature (e.g., SiF_4 is volatile from $-86°C$ at 1 atm). In this process, the chuck temperature is water cooled by a chiller between 5°C and 25°C but its modifications are not easy, so it is kept constant at 10°C.

Process Development

Method

The process used for the realization of a tunnel in a Si/SiGe/Si stack has to etch SiGe with a high selectivity to pure Si. In other words, the ratio between the tunnel depth and the Si loss must be maximized. This ratio is considered as the most representative index for the selectivity (Figure 13.7):

$$\text{Selectivity} = \frac{\text{Tunnel depth}}{\text{Si loss}}$$

This complex structure is essential in order to evaluate the selectivity of a lateral etching process because any simplified method would lead to fallacious results (Figure 13.8). Actually, both materials must be present simultaneously on the wafer and the distance between them must be as small as possible. For example, the ratio between SiGe and Si etch rate measured on two independent wafers is lower than the tunnel depth/Si loss (Td/Sl) ratio. This suggests that the presence of SiGe has an influence on Si etch rate.

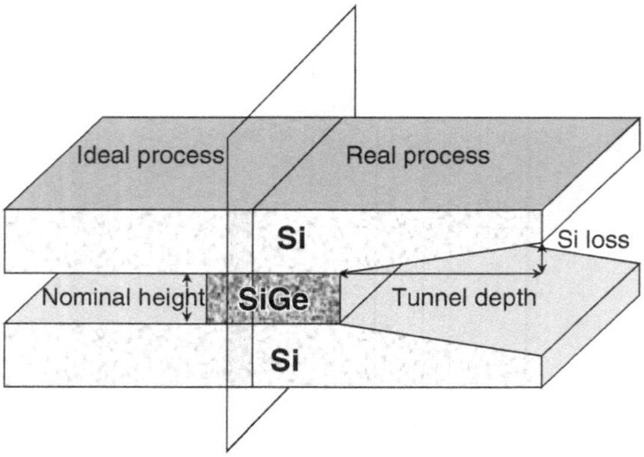

FIGURE 13.7 Selectivity characteristics of a tunnel etching process.

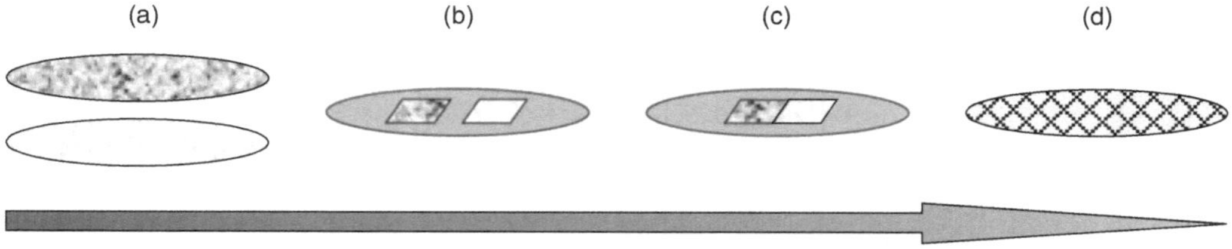

FIGURE 13.8 Relevance of the method used to evaluate the selectivity of a lateral etching process. (a) Two independent wafers covered with SiGe or Si; (b) SiGe and Si samples on an inert holder wafer; (c) the same SiGe and Si samples side by side; and (d) patterned wafer with Si/SiGe/Si stack.

Each tested recipe is performed in the same experimental conditions:

- FH wet treatment in order to remove the native oxide on both materials
- Etching of a patterned wafer with Si/SiGe/Si stack (20% Ge)
- Stripping in a O_2 plasma in order to remove the resist that protects the upper side of Si
- SEM cross-section observation in order to determine Td and Sl

It is then possible to compare the values of Td/Sl or the relative positions of each recipe on the graph in Figure 13.9.

Best Point

Many gas mixtures have been tested, using CF_4, CH_2F_2 or Cl_2 as the etching gas, adding N_2 or O_2 in different ratios. Finally, the chemistry that gives the best results is made of pure CF_4 (300 sccm).

The pressure is set to the maximum value in order to minimize the etch rate (see section on "Pressure") and better control the process. With 300 sccm of gas, the maximum pressure that can be reached is 1500 mT. A consequence of this high pressure is a selectivity improvement, compared with lower pressure recipes.

The same behavior is observed with power: at low power, the etch rate decreases (see section on "Microwave Power") and the selectivity between SiGe and Si is improved. A power of 200 W is applied because lower values lead to plasma instability (Table 13.1).

Performances

The graph of Figure 13.10, plotted after 60 sec of the process described above, shows a Td/Sl ratio around 40, which means that when a 100 nm-deep tunnel is etched in the SiGe layer, only 2.5 nm of Si are consumed beneath and underneath the entrance of the tunnel.

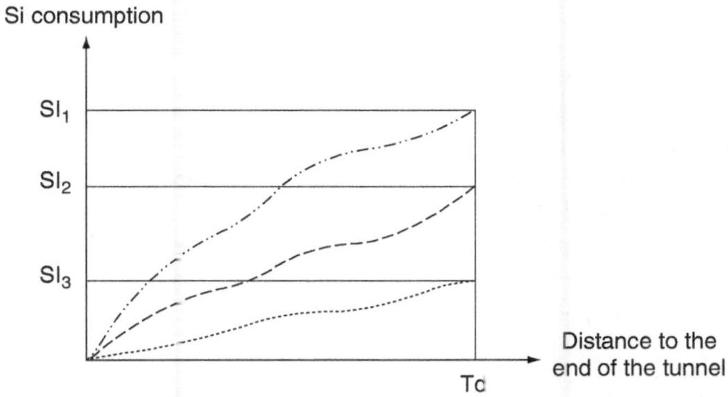

FIGURE 13.9 Comparison between different recipes, based on SEM cross-section images. Recipe 2 (dashed line) is more selective than recipe 1 (dashed–dotted line) but less than recipe 3 (dotted line).

TABLE 13.1 Process Parameters of the Best Point for Tunnel Etching

CF_4 Flow	Pressure	MW Power	ESC Temperature
300 sccm	1500 mT	200 W	10°C

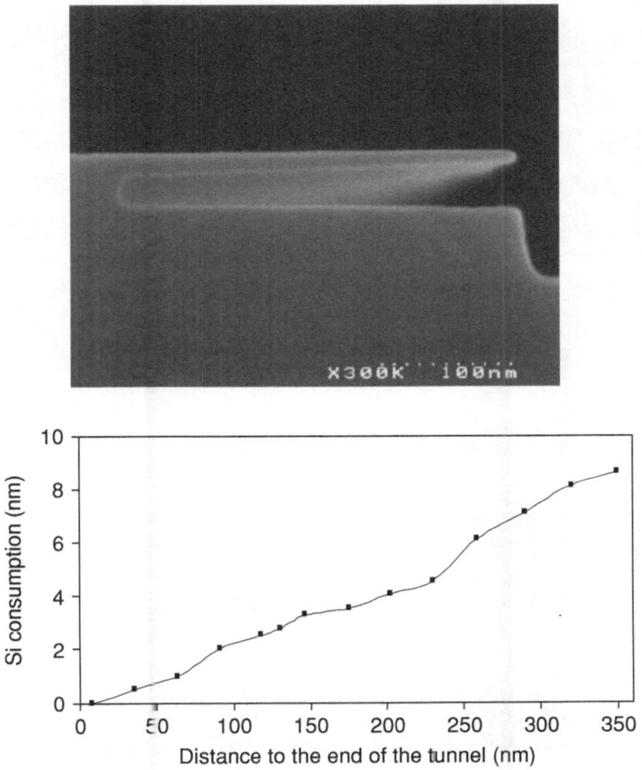

FIGURE 13.10 Characterization of the best process for tunnel etching. (a) SEM cross-section micrograph of the tunnel. (b) Si consumption all along the tunnel.

The following points are to be noted. The Si loss is higher if the material is doped with arsenic. On the contrary, if it is doped with boron, the Si is almost not impacted. As and B do not need to be activated to play their role, which suggests that this phenomenon is not electron-linked.

The process is sensitive to the germanium concentration in the alloy. The more Ge it contains, the faster it gets etched by fluorine atoms. A SiGe/Si multistack with Ge contents from 15% up to 30%, exposed to this process, shows the evolution of the etch rate as a function of [Ge] (Figure 13.11).

The real structure is represented below, after 10 sec of etching (Figure 13.12). The measurement of the different etch rates may show a nonlinear relationship between [Ge] and SiGe etch rate. But insofar as all tunnels have a different depth, some additional parameters may have to be taken into account, such as the accessibility of the etching species at the end of the tunnel.

Limitations

As mentioned in "Method," the presence of SiGe has an influence on Si etch rate since the selectivity obtained with two blanket wafers is very different from that obtained on a patterned wafer with stacked layers.

This is a limitation of the process because in almost all industrial applications, it is used for a complete removal of the SiGe sacrificial layer, which means that at the end of the process, there is only silicon on the wafer. At this moment, Si starts to be etched by fluorine species and the benefit of the selectivity may be rapidly lost.

The qualitative representations of Si and SiGe etch rate are given in Figure 13.13. They show that as long as some SiGe remains, the Si is preserved, but as soon as SiGe has been totally removed from the pattern, the Si etch rate increases.

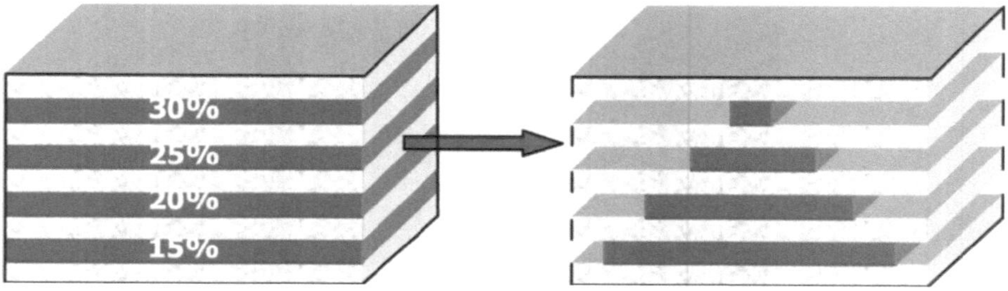

FIGURE 13.11 Structure used to show the impact of the Ge content on the SiGe ER.

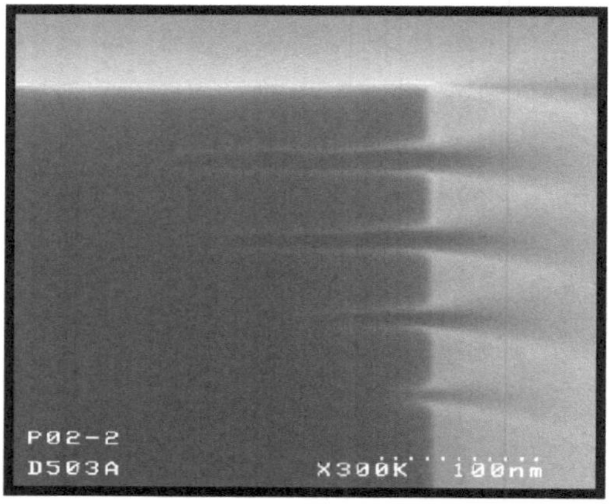

FIGURE 13.12 SEM cross-section micrograph of tunnels after etching of a SiGe/Si multistack with 15% $\leq$ [Ge] $\leq$ 30%.

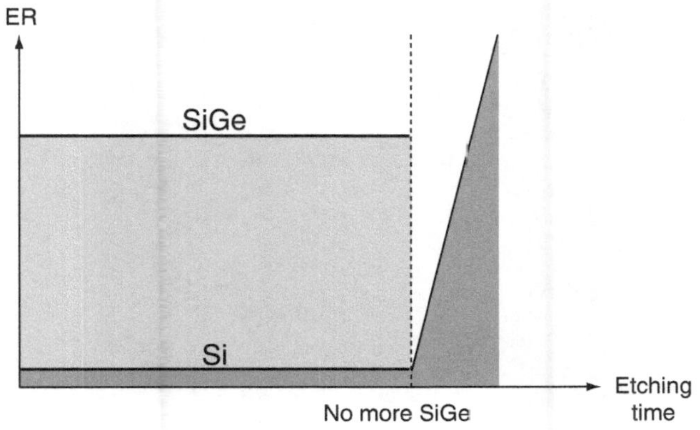

FIGURE 13.13 Qualitative evolution of Si and SiGe etch rate during the process.

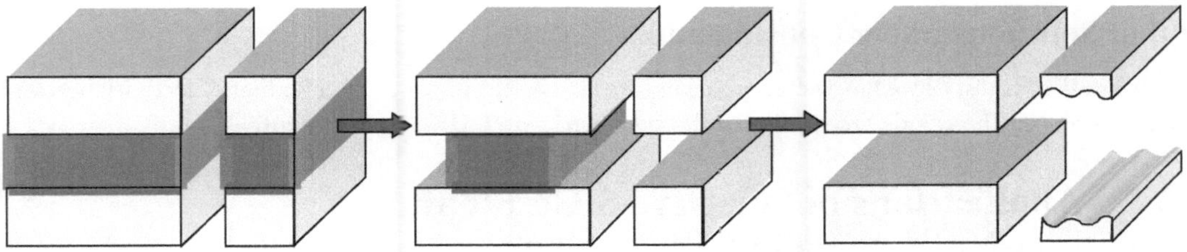

FIGURE 13.14 Si loss on small patterns during the etching of large patterns.

A consequence is that the process time must be adjusted in order to minimize the over-etch (necessary in case of bad uniformity of a tool).

Another consequence is that large patterns and small patterns cannot be achieved at the same time because large patterns require a longer process time, during which small patterns are over-etched (Figure 13.14).

Selective Passivation of Si

A possible way of skirting this issue could be the development of a process that passivates Si selectively to SiGe. It would offer a twofold advantage:

1. The entrance of the deep tunnels would be as high as their end (no bevel shape, good integrity of the transistor channel).
2. Small patterns could cohabit with large patterns (no boom in Si etch rate at SiGe disappearance).

As silicon oxide is chlorine-resistant whereas germanium oxide is not, a process using O_2 and Cl_2 may etch oxidized SiGe and stop on oxidized Si.

Selective Passivation of SiGe

A process using a mixture of $CF_4/CH_2F_2/N_2/O_2$ at high pressure (1500 mT) shows a Si etch rate about 8000 Å/min and a SiGe etch rate that cannot be measured (Figure 13.15). This almost infinite selectivity is due to a passivating layer on SiGe. This process enables the production of tunnels between two SiGe thin layers with no limitation of depth.

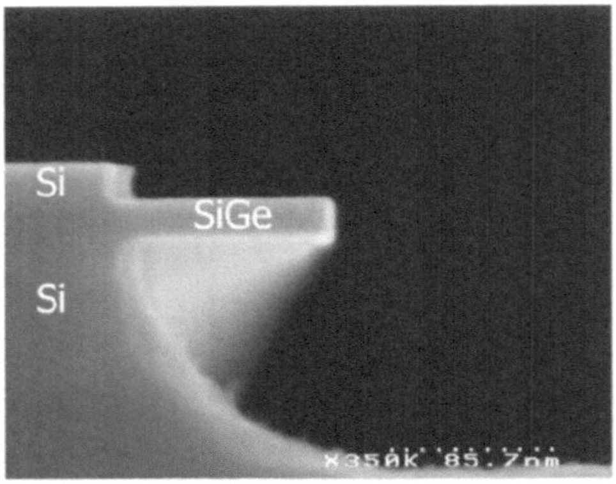

FIGURE 13.15 Use of a Si process (infinite selectivity to SiGe) on a Si/SiGe/Si stack.

Other Solutions under Consideration

Other kinds of tools, as well as wet approaches, are also tested in order to get a high selectivity between SiGe and Si, but the trends are similar to those obtained on CDE-80 equipment.

13.3 Applications of Selective SiGe Etching: Silicon-on-Nothing (SON) Technology

Introduction

Extremely thin (nanometric) films of monocrystalline silicon on insulator are widely recognized for their potential for end-of-roadmap CMOS transistors [1–7]. Various fabrication technologies (SIMOX, BESOI, Smart Cut, ELTRAN, ELO, etc.) have been developed permitting continuous downscaling of the Si film and the BOX, but the commercial availability of such thin films as 5–15 nm still faces problems of feasibility and dispersions. In contrast with any other SOI technology, in the SON process, the silicon film and buried insulator, both of nanometric scale, are defined by epitaxy on a bulk substrate.

Therefore, the SON process opens access to extremely thin films (the silicon channel as well as the BOX) at the same time offering the thickness control as fine as the resolution of the epi process (less than 1 nm). In addition, no more than one (bulk-type) wafer is needed to fabricate the SON layers, in contrast to two wafers in wafer-bonding-type techniques.

Successful and highly selective SiGe tunnel etching can be obtained by wet etching using a variety of solvents. Table 13.2 summarizes our own experiences with SON-tunnel wet etching. In spite of the good

TABLE 13.2 Efficiency of SiGe Wet-Etch Solvents Measured in SON Tunnel Lengths

Solution Composition	40 ml HNO_3			70 ml HNO_3		80 ml HNO_3
	20 ml H_2O_2 5 ml HF (5%)			40 ml acetic acide 20 ml HF (5%)		40 ml H_2O 10 ml HF (5%)
Etch time	1 min	3 min	10 min	3 min	10 min	3 min
SON SiGe 50%	21 nm	53 nm	—	—	—	—
SON SiGe 30%	—	<1 nm	10 nm	30 nm	70 nm	10 nm

selectivity and etching efficiency obtained with wet-etching, in our experiments a plasma-assisted isotropic dry etching process was used in order to ensure better process reproducibility.

As presented in the previous section, a well controlled and reproducible tunnel etching can also be obtained by dry etching. We have used for that a downstream plasma reactor (STRATA from GAZONICS) using NF_3, He, and N_2. The two important features of this reactor are that: (i) the reactive species are created by plasma and transported to the wafer where the reaction is purely chemical, (ii) the substrates are not biased that practically removes the danger of plasma damage to the wafer. The empty tunnel resulting from this selective etching is the key point for the realization of a thin single-crystal Si film isolated from the substrate by a cavity. Alternatively, we have also used the SHIBAURA reactor using a process very similar to that of STRATA except that CF4 replaces NF3.

The first application of the selective SiGe etching was dedicated to the realization of notched gates [8] in order to reduce the source and drain extensions overlap and also to improve the fringing capacitances. In this way, extremely aggressive gates were fabricated within the current state of the art of lithography process, Figure 13.16.

Implementation of the SON Process for MOSFET

The SON process is very suitable for fabricating thin-film SOI-like devices [9–15]. For this purpose, two selective epitaxial growth (SEG) steps need to be inserted into the conventional CMOS flow just after accomplishing the shallow trench isolation (STI) process. The first SEG step provides a SiGe layer, and the second a Si layer, both layers are grown selectively on active areas (Figure 13.17a). Next the process comes back to the conventional CMOS flow and runs the usual way up to the step of lowly doped drain (LDD) implants. After the first nitride spacer, a second (TEOS) sacrificial spacer is formed. Then, using the gate and spacers as masks, shallow trenches in active areas are etched to open access to the SiGe layer (Figure 13.17b) that is selectively etched out (Figure 13.17c). After this step, the gate stack (polysilicon, gate-oxide and the silicon channel) together with the adjacent spacers is hanging above an empty space—the tunnel remaining after the SiGe etch out.

As schematically shown in Figure 13.17d, the stability of the gate stack at this step is due to its bridging across the active area supported by the STI. The tunnel is next cleaned and its inner walls are passivated in an RTO step that is aimed at improving the interface quality. Optionally, this step can be preceded by a sacrificial oxidation aimed at smoothing and removing the inner-tunnel interface.

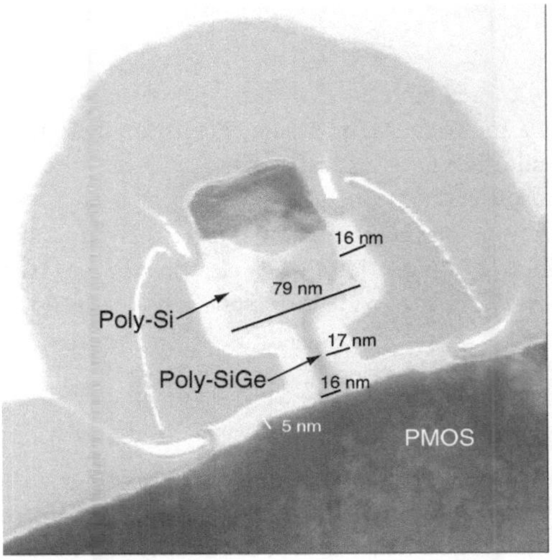

FIGURE 13.16 TEM cross section of a 16 nm gate length planar CMOS realized with a poly-SiGe selectively etched gate.

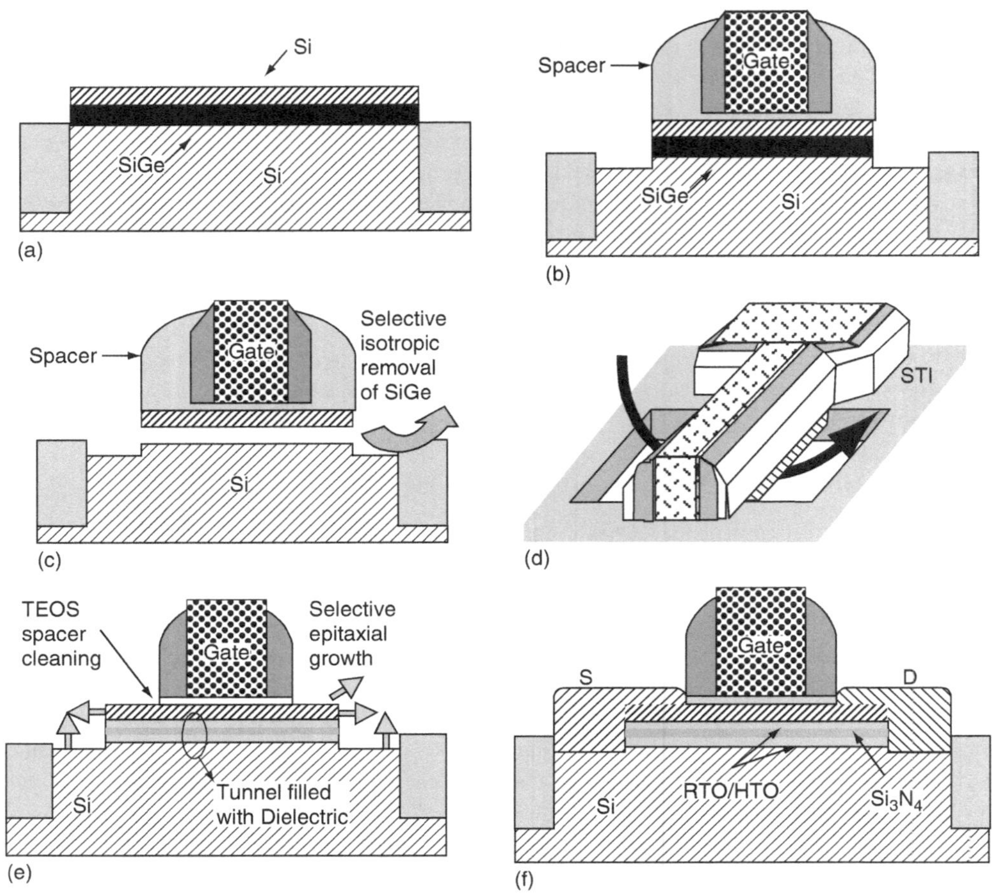

FIGURE 13.17 Key steps of the basic SON process in application to CMOS flow. (a) Selective SiGe/Si epitaxy; (b) standard gate-stack integration and SD etching; (c) SD-etching for the buried SiGe layer access and selective isotropic SiGe removal; (d) after SiGe removal, all the gate stack becomes sustained over the active area; (e) tunnel filling with dielectric and Si-cleaning before selective epi-growth; (f) source and drain area becomes contacted with the conduction channel.

Next, the entire structure is recovered with the first thin layer of deposited dielectric (HTO), sufficiently thin to leave a throughway via the tunnel. The second deposited dielectric (nitride) fulfills completely the tunnel but also recovers all other parts of the structure.

Thanks to the dual-dielectric structure, one can, however, etch-back the nitride (everywhere except inside the tunnel) in an anisotropic plasma etch using the HTO layer as an etch-stopper. The remaining HTO oxide is very thin and thus easy to get rid of when cleaning the S/D trenches preparing them for the SEG step. This cleaning step is selective to oxide thus preserving the nitride in the spacers and inside the tunnel. In contrast, the second TEOS spacer is etched out thus exposing the upper part of the extensions to the SEG. The more aggressive is the cleaning, the larger part of the extension is exposed to SEG (Figure 13.17e).

With this double-spacer and dual-dielectric process, wide range of cleaning conditions may be used without jeopardizing either the SEG quality, or the reunification between S/D and the channel. It should be realized that the additional surface of the channel exposed to SEG thanks to the removal of the second TEOS spacer is much larger than the vertical sidewall of the channel. Thanks to this developed exposure surface, the SEG starts not only at the bottom of the S/D trenches but also directly on the exposed parts of the extensions, thus guarantying a reliable reunification (Figure 13.17f).

In this way, we can see Figure 13.18 in which SON transistors show perfect morphology. On the color map of the entire structure are presented 80 nm transistors with silicon-conduction channel as thin as 20 nm with only 20 nm of buried dielectric. For very thin-film SOI devices the concern is about the bottom interface quality and the crystallinity of the film. The high-resolution TEM cross-section (Figure 13.19) made on a finished extremely thin 5 nm-thick SON transistor demonstrates that the bottom interface is almost as smooth as the upper one. This is certainly thanks to the very high SiGe etching selectivity but also due to the fact that the etching is a chemical reaction rather than aggressive ion etching process. This figure also illustrates the perfect monocrystalline structure of the channel (see the lattice regularity).

Another major advantage of SON devices is its compatibility with the silicidation process. In general, very thin-film SOI devices need SEG steps to allow S/D silicidation, leading to many issues: charging effects, temperature control on SOI substrates, start on amorphized areas after LDD implant, etc. In the case of SON process, thanks to the continuity between S/D area and the bulk, silicidation can occur without problems.

Figure 13.20 and Figure 13.21 show examples of electrical results obtained on 80 nm NMOS transistors with 30 Å of physical oxide thickness and on 38 nm PMOS transistors with 20 Å of physical oxide thickness. Concerning the output characteristics, an improvement of as much as 30% at high voltage ($V_g - V_{th} = 1.8$ V) (Figure 13.20) and as much as 130% at low voltage ($V_g - V_{th} = 0.6$ V),

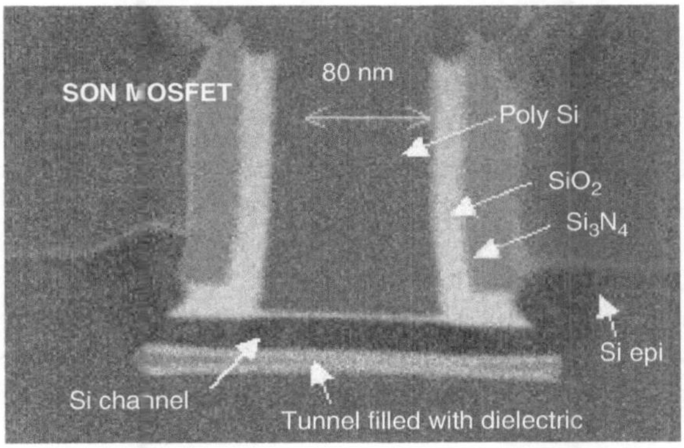

FIGURE 13.18 FIB/TEM color-map illustrating perfect SON morphology; note the uniform thickness of the buried dielectric and of the Si channel and the comfortable reunification between the channel extremities and the S/D domains.

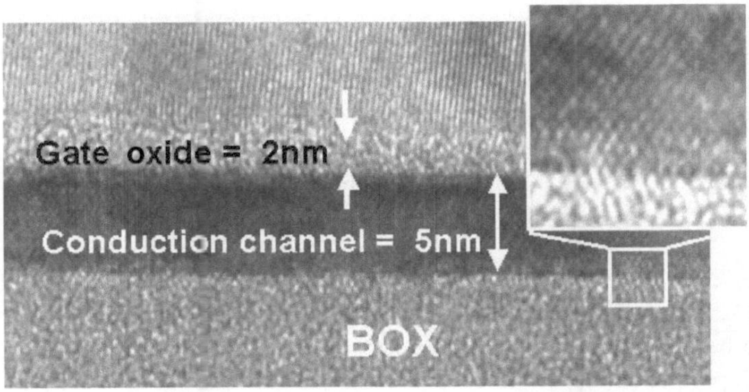

FIGURE 13.19 Magnification (TEM picture) of the buried-dielectric/channel/Gate_ox/poly-gate stack for a very thin Si-film ($T_{Si} = 5$ nm). Note good quality of bottom and upper tunnel interfaces and perfect crystallinity of the channel.

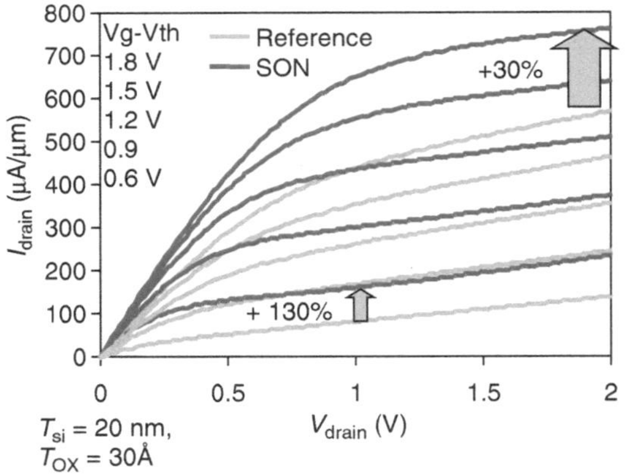

FIGURE 13.20 L_{in} – $I_d(V_d)$ of a 80-nm NMOS SON transistor. The drive current is improved by 30% compared to bulk reference. At low V_{dd}, the improvement reaches 130%.

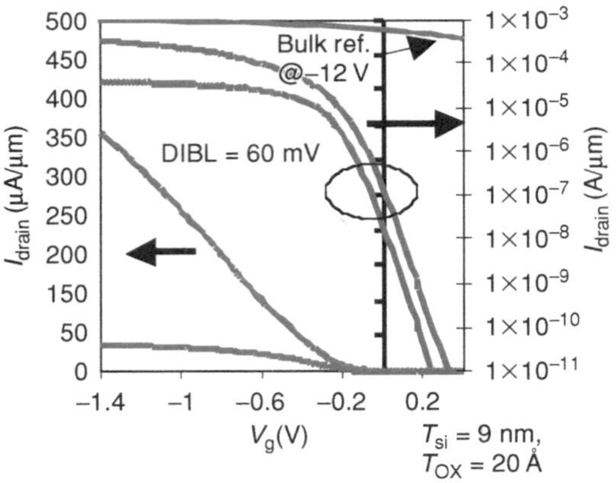

FIGURE 13.21 L_{in} – $I_d(V_g)$ and log-$I_d(V_g)$ of our 38 nm PMOS SON device. The architecture shows high-resistive SCE properties with only 60 mV of DIBL (pockets implanted). I_{off} current is only equal to 100 nA/μm.

has been measured, reflecting large improvement in transconductance (confirmed by direct measurements).

38 nm SON PMOS devices are presented with a 9-nm-thick Si-conduction channel. This architecture leads to a really SCE-resistive device, showing less than 60 mV of DIBL for such an aggressive gate length, (Figure 13.21). In absolute numbers, the drain current of those silicided SON pMOSFETs was measured as high as 340 μA/μm at $V_g = -1.4$ V (with a physical oxide thickness of 20 Å) with only 100 nA/μm I_{off}, which is consistent according to the ITRS roadmap. Those performances are really competitive compared to the state of the art for such aggressive gate length, and can be improved by reducing the gate oxide thickness. In fact, for such aggressive gate length, bulk references were showing so high SCE effects that $I_d(V_g)$ characteristics demonstrate quasi-short cuts between the extensions. This point shows that the SON architecture is definitely a major advantage for the aggressive ULSI area.

SON MOSFETs with Metal Gates

A major issue of very thin FD devices is that V_{th} decreases with the Si-film thickness (T_{Si}), leading to intolerably high off current.

The threshold voltage level of FD devices with thin films becomes consequently a serious issue that needs to be addressed. This is why metal gates may be mandatory for the end-of-the roadmap devices. The use of mid-gap gates allows V_{th} adjustment with lowly or undoped channels, and may be also useful for polydepletion and gate-resistance reduction. For SON application, the process of totally gates silicidation [16] can be easily applied:

The process of the total gate silicidation is the following: (a) the structure is covered with metal capping and (b) RTP is performed for silicide formation down to the gate oxide (Figure 13.22 and Figure 13.23).

SON architecture allows comfortable silicidation thanks to the metallurgical contact between S/D and the bulk. Figure 13.22 shows the implementation of the totally silicided gate process to SON devices, with reduced gate height in order to limit silicidation depth in the junctions. SON transistors were fabricated with this process and Figure 13.23 shows that the 800 Å height gates were totally transformed into $CoSi_2$.

Electrical performances are presented as illustration on 55 nm PMOS transistor with 5 nm of conduction-channel thickness on Figure 13.24, showing highly SCE-immune devices—only 60 mV

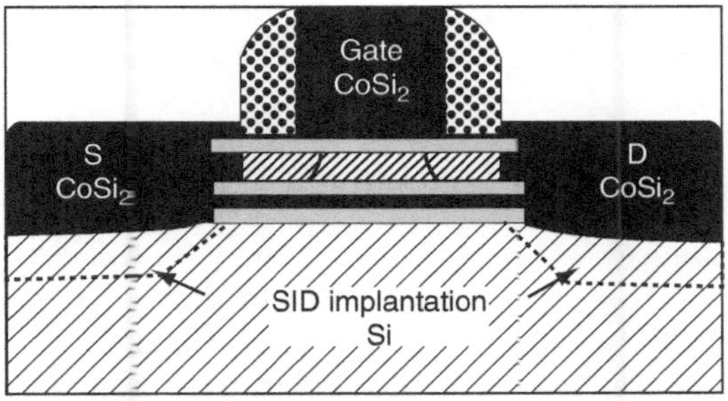

FIGURE 13.22 Implementation of the total gate silicidation process to SON devices. SON architecture allows total gate silicidation process without creating voids in the Si-film. This is thank to the metallurgical contact between S/D and bulk.

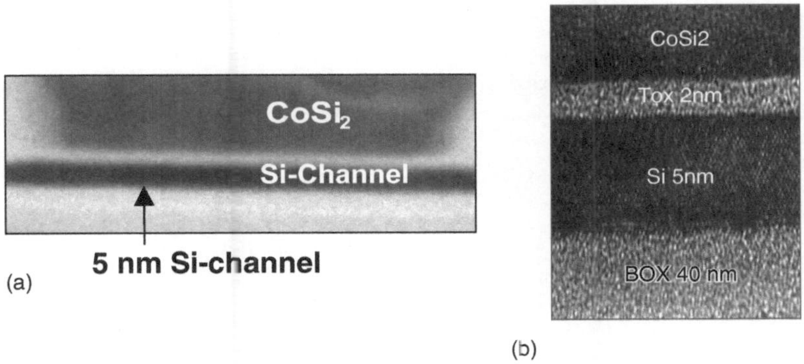

FIGURE 13.23 (a) Example of SON transistor with metal gate integration. (b) Plasmon and (c) HRTEM pictures showing the thickness of the Si-channel. Measured $T_{Si} = 5$ nm, with perfect top and bottom interfaces.

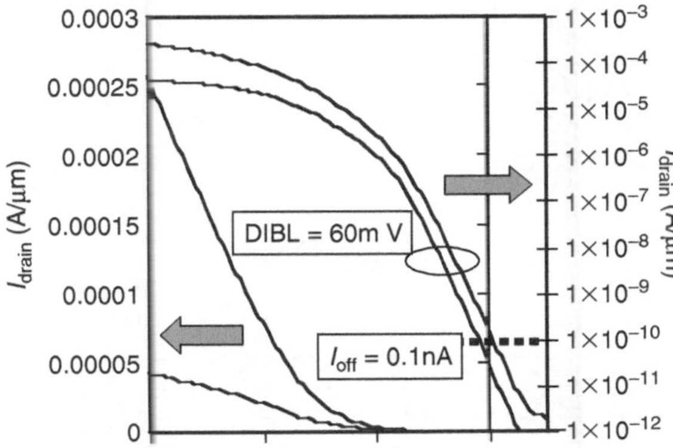

FIGURE 13.24 $-L_{in}-I_d(V_g)$ and $\log\text{-}I_d(V_g)$ of a 55-nm SON transistor with $CoSi_2$ gate. Data show high SCE-resistive devices (DIBL = 60 mV). Due to the V_{th} shift induced by the mid-gap gate, I_{off} is around 0.1 nA/μm. The drive current is over 250 μA/μm at $V_d = -1.2$ V with $T_{ox} = 20$ Å and over 350 μA/μm at $V_d = -1.4$ V.

of DIBL and a subthreshold slope of 73 mV/dec—in spite of deep S/D junctions. Such results are possible only on SON thanks to its intrinsic strength for SCE/DIBL suppression. Due to the V_{th} adjustment with mid-gap gate ($V_{th} = -0.47$ V at -0.1 V), I_{off} was reduced to 0.1 nA/μm for 350 μA/μm of drive current at -1.4 V.

SON without Channel Rupture: "Localized SOI"

This process removes the necessity of breaking the continuity between S/D and channel and thereby also that of reunification of these two regions by SEG of silicon. The indispensable access to the SiGe is obtained at the STI edge by means of a slight de-oxidation, or by a slight Si-etching. Thanks to the very high selectivity and sensitivity of the SiGe etching process, this access is sufficient for under-etching the entire active area (Figure 13.25). The remaining Si-cap layer is then supported by the gate stack that bridges the active area in the perpendicular direction, as in the basic SON process (Figure 13.17).

In this process, the tunnel is fulfilled in the same time as the spacer stack is deposited (oxide and Si_3N_4 deposition). During the standard spacers etching step, the nitride layer is kept in the previous tunnel. This very simple process allows us localized SOI-like structures, as we can see in Figure 13.26 and Figure 13.27, leading to a new way of co-integration of such devices with bulk transistors on the same chip. On the illustration, SOI-like transistors have a silicon-conduction channel thickness of 10 nm isolated from the bulk by a 30 nm buried dielectric layer. It is worth noting that this process is totally compatible with the standard bulk CMOS process and offers new perspective for localized architecture on the chip.

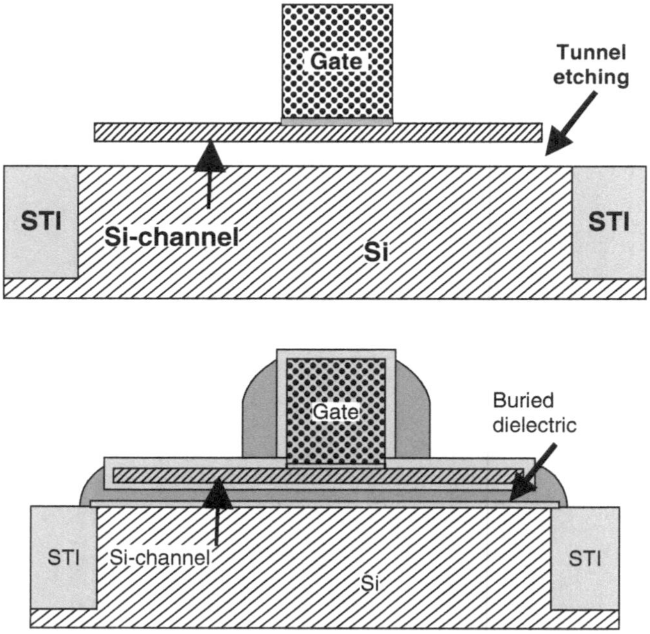

FIGURE 13.25 Concept of SON process without channel rupture: the buried SiGe layer is etched from STI edges and then the tunnel is fulfilled in the nitride spacers stack realization.

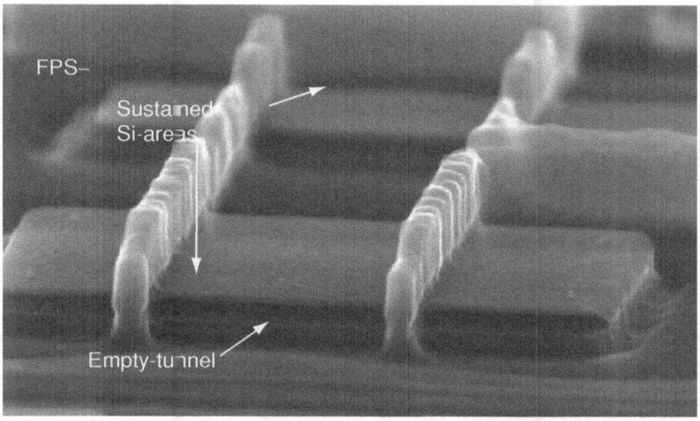

FIGURE 13.26 Morphological demonstration of the feasibility of the SON tunnel etching via the STI edges. Active areas are sustained thanks to the gates.

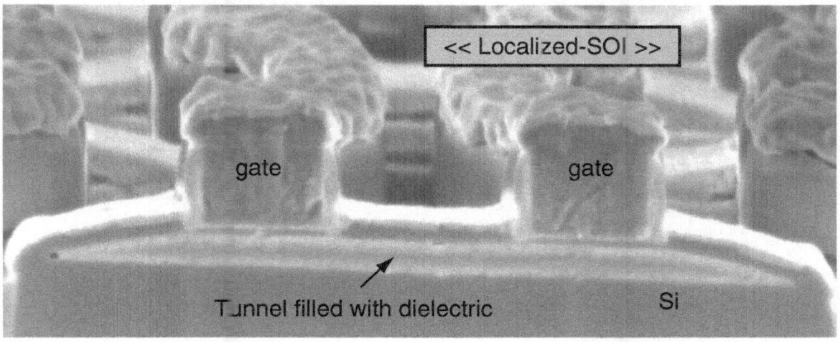

FIGURE 13.27 We have formed localized SOI-like areas. $T_{Si} = 10\,nm$ and T_{box} is around 30 nm.

Implementation of the SON Process to Gate All Around (GAA) MOSFETs

The SON process may also be easily adapted to fabrication of double gate or gate-all-around (GAA) devices. These kind of devices are reputed for their particularly strong immunity to short-channel effects as well as for their augmented transport efficiency. The main problems remaining to be solved are related to: (i) the thickness control of the Si channel, (ii) its surface roughness, and in some realization (wafer bonding) to (iii) a misalignment between the gates.

The GAA SON based process, shown in Figure 13.28, starts with a SEG of SiGe on STI isolated wafers. The SiGe/Si stack is next patterned by RIE into a strip expanding above the active area (future transistor channel) and overlapping on STI (future S/D). Then, the underlying SiGe is selectively etched out (Figure 13.29) as in the conventional SON process, and the entire bridge-like structure is recovered with dielectric (in our case thermal oxidation for ensuring gate-oxide quality). Next, the structure is recovered with poly-silicon that is patterned into a gate in a masked RIE step (Figure 13.30). The residuals of poly-silicon remaining in the corners under the bridge are neutral for a static operation of the device but contribute to overlap capacitance, thus necessitating as tight design rules as possible to reduce the gate-to-poly spacing. The structure is completed by ion-implantation of S/D junctions and extensions (optionally that of S/D after formation of spacers), silicidation and other conventional back-end steps.

Note that this process presents at least three important advantages: (i) the SiGe is not subject to any thermal steps (it is etched away just after deposition and channel patterning) thus limiting the danger of relaxation, (ii) the continuity between the channel and the S/D is never broken thus removing all the troubles we had with the reunification of these regions by SEG of Si, and (iii) the S/D junctions lay on

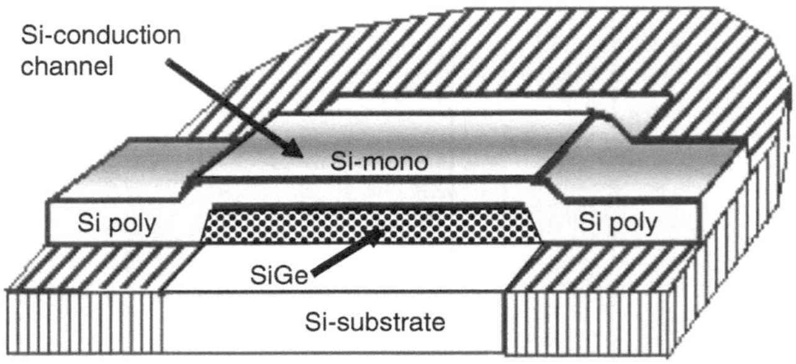

FIGURE 13.28 Epi stack: selective SiGe and nonselective Si. The channel of the GAA device is defined by photolithography and is followed by a strip patterning.

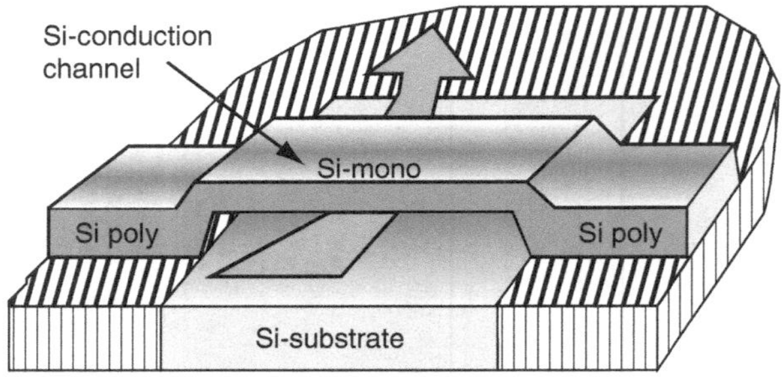

FIGURE 13.29 Selective removal of the SiGe layer.

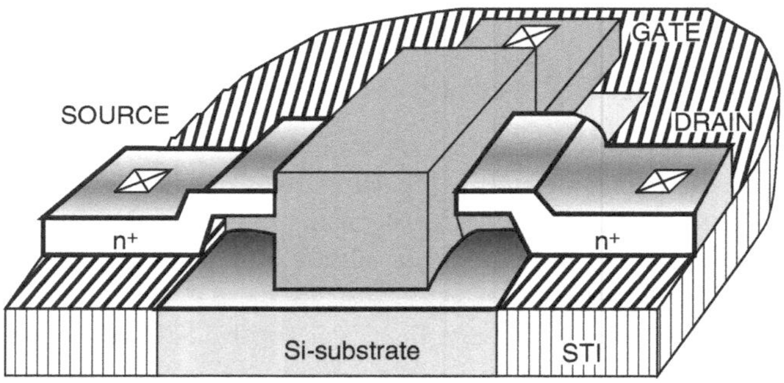

FIGURE 13.30 Gate oxidation, poly-Si deposition, and gate patterning.

field oxide (STI), thus promising reduced junction capacitance and leakage. The point still to improve is that the gates are not of the same size even if they are still self-aligned, thus leading to somewhat increased overlap capacitance relevant to the bottom gate. Nonetheless, SON technological steps can be modified in order to combine the bottom gate etching and S/D epitaxy to realize self-aligned top and bottom gate, see Figure 13.31.

Figure 13.32 demonstrates the feasibility of the GAA SON key steps as observed on our experimental samples. Worth noticing is that the nature of the silicon within the bridge-like structure changes from a monocrystal above the active area to polycrystal above the STI. This is not expected to be a problem as

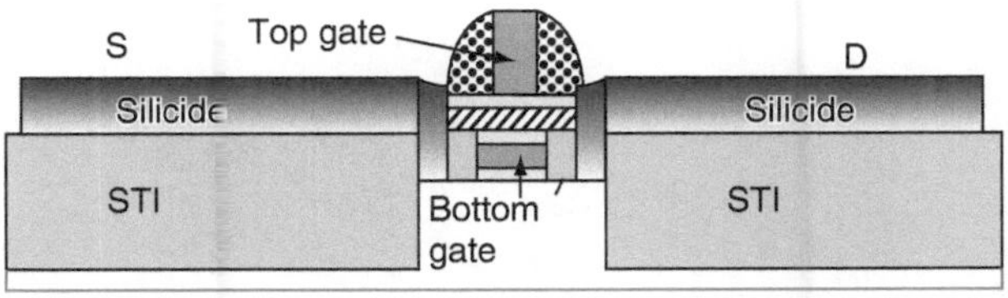

FIGURE 13.31 GAA transistor with top and bottom self-aligned gates.

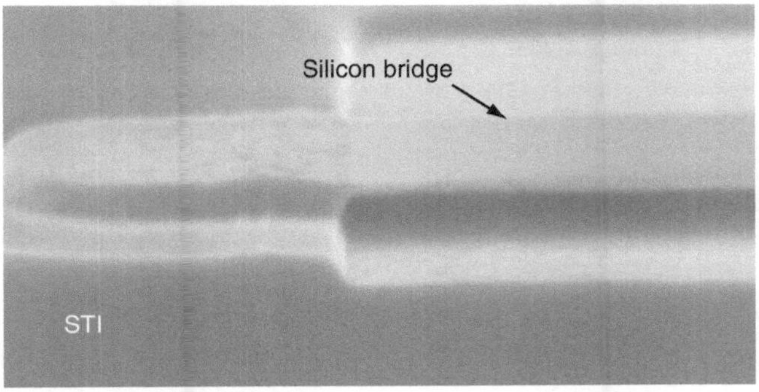

FIGURE 13.32 Sustained silicon-conduction channel.

long as the effective channel region is taken in the central, monocrystalline part of the bridge. Note that the bridge-like silicon strip never communicates with the initial Si-substrate. Thanks to that, it can be doped with very high doses and comfortably annealed for efficient dopant activation, without the danger of junction diffusion as in the case of bulk devices. The last photographs (TEM) (Figure 13.33 and Figure 13.34) show a perpendicular cross-section in the central part of the bridge after reoxidation, polydeposition and gate patterning. The oxide film is pretty conformal all-around the Si strip, and the poly fills the tunnel homogeneously without voids.

Finally, the devices obtained by this process have shown exceptional performances (Figure 13.35), with drive current reaching 2000 $\mu A/\mu m$ for 283 $nA/\mu m$ I_{off}.

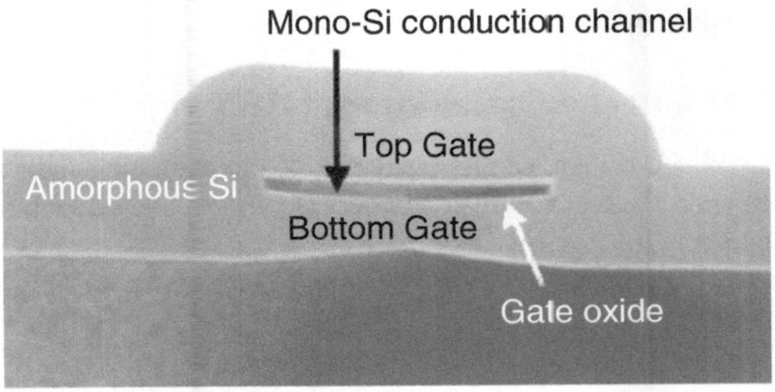

FIGURE 13.33 *In situ* doped amorphous Si surrounding the Si-channel.

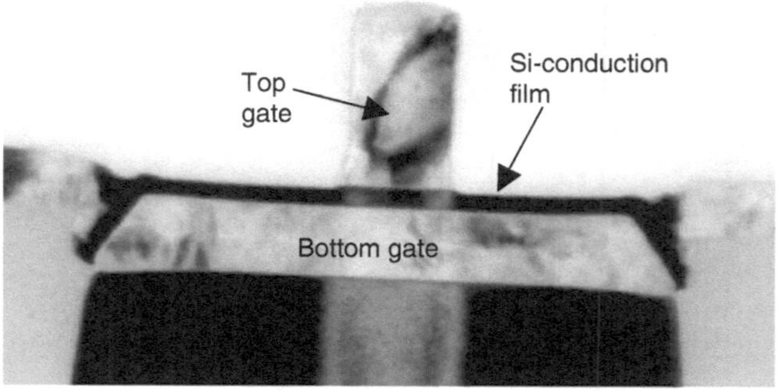

FIGURE 13.34 Gate all around transistor after gate patterning. The measured Si-film thickness is 16 nm.

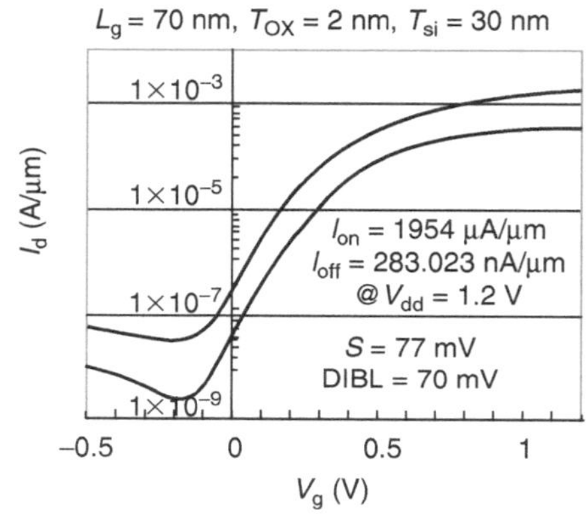

FIGURE 13.35 Example of $I_d(V_g)$ for a 70 nm gate length SON GAA transistor. The measured drive current is the highest ever measured on such devices.

Implementation to Multigate Devices

The principle of the SON technology, which consists in a selective etching of SiGe layer, can be extended to the etching of multilayer. In this way, we can imagine structures where several empty tunnels can be performed and filled with gate oxide and poly-Si or metal gates (Figure 13.36 and Figure 13.37). This opens access to new devices with multiconduction channels, which can considerably boost the drive current for high-performance applications.

SON and Microelectromechanical Systems (MEMS)

The fabrication of mono-silicon bridges can be easily adapted to electromechanical micro switch, if the flexure beam can be obtained by selective SiGe etching. This opens new possibilities for such applications, and in particular for the integration of front-end MEMS.

Considering the state-of-the-art, the maximal vibration frequencies that can be obtained with flexure beam are in the range of 10 to 100 MHz. With the introduction of monocrystalline beams and very small gap between the detection electrode and the flexure beam, frequencies in the range of several GHz can be reached. The possibilities that are proposed with the SON technology in this domain are also of great interest (see Figure 13.38 and Figure 13.39).

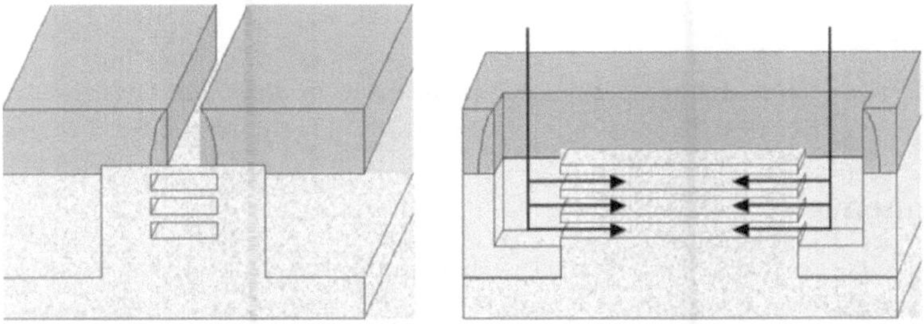

FIGURE 13.36 Application of SON-GAA technology to multigate devices. The access to the buried SiGe multilayer is performed from the STI area.

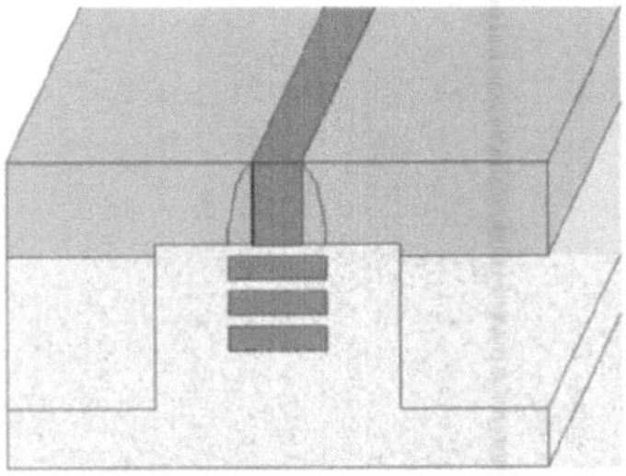

FIGURE 13.37 The empty tunnels are filled by poly-Si (or metal) to realize the multigate structure.

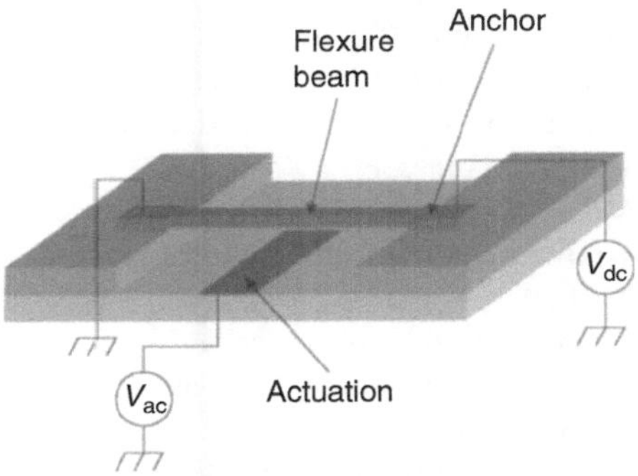

FIGURE 13.38 Electromechanical micro-switch principle: the flexure beam is realized with the SON process, allowing a thin monocrystalline beam.

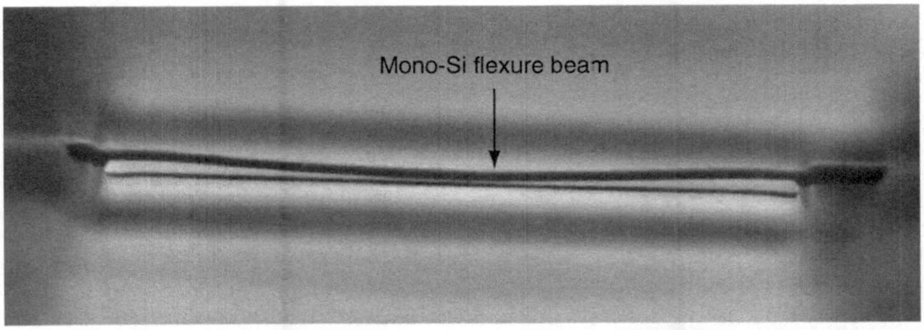

FIGURE 13.39 Example of morphological realization of a thin monocrystalline flexure beam with the SON technology.

13.4 Summary

The selective Si/SiGe etching has been presented here and extensively detailed in chemical dry etching. This method can be used in several advanced microelectronics architectures that require a thin single-crystal Si film isolated from the substrate by a cavity: the SiGe layer is used as a sacrificial layer for lateral etching. This approach is demonstrated as a new innovative technique, which is particularly suitable for the realization of localized SOI on bulk using the SON technology. Furthermore, the capability to perform sustained mono-Si areas over an empty cavity opens a wide range of applications for this technique, in particular for the realization of double gate devices, MEMs in nanoscale dimensions, new type of nanodetectors and sensors.

In conclusion, the selective etching techniques for SiGe can be considered as a fundamental way for new and innovative approaches, demonstrating in this way a large potential for future applications.

Acknowledgments

The author would like to thank P. Coronel; D. Chanemougame; D. Dutarte; A. Talbot; S. Harrison; F. Leverd for their contributions.

References

1. Wong, H.-S.P.; Frank, D.J.; Solomon, P.M. Device design considerations for double-gate, ground-plane, and single-gated ultra-thin SOI MOSFET's at the 25 nm channel length generation, *IEDM Tech. Dig.*, pp. 407–410, 1998.
2. Schulz, T.; Rosner, W.; Risch, L.; Langmann, U. 50nm vertical sidewall transistors with high channel doping concentrations, *IEDM Tech. Dig.*, pp. 61–64, 2000.
3. Ieong, M.; Jones, E.C.; Kanarsky, T.; Ren, Z.; Dokumaci, O.; Roy, R.A.; Shi, L.; Furukawa, T.; Taur, Y.; Miller, R.J.; Wong, H.-S.P. Experimental evaluation of carrier transport and device design for planar symmetric/asymmetric Double-Gate/Ground plane CMOSFETs, *IEDM Tech. Dig.*, pp. 441–444, 2001.
4. Chau, R.; Kavalieros, J.; Doyle, B.; Murthy, A.; Paulsen, N.; Lionberger, D.; Barlage, D.; Arghavani, R.; Roberds, B.; Doczy, M. A 50nm Depleted-Substrate CMOS transistor (DST), *IEDM Tech. Dig.*, pp. 621–624, 2001.
5. Esseni, D.; Mastrapasqua, M.; Fiegna, C.; Celler, G.K.; Selmi, L.; Sangiorgi, E. An experimental study of low field electron mobility in double-gate, ultra-thin SOI MOSFETs, *IEDM Tech. Dig.*, pp. 445–448, 2001.
6. Majima, H.; Saito, Y.; Hiramoto, T. Impact of quantum mechanical effects on design of nano-scale narrow channel n- and p-type MOSFETs, *IEDM Tech. Dig.*, pp. 733–736, 2001.
7. Balestra, F.; Benachir, M.; Brini, J.; Ghibaudo, G. Analytical models for subthreshold swing and threshold voltage for thin and ultra thin film SOI MOSFETs, *IEEE Trans. Electron Devices*, Vol. 37, no. 11, pp. 2303–2311, Nov. 1990.
8. Boeuf, F.; Skotnicki, T.; Monfray, S.; Julien, C.; Dutartre, D.; Martins, J.; Mazoyer, P.; Palla, R.; Tavel, B.; Ribot, P.; Sondergard, E.; Sanquer, A. 16 nm planar NMOSFET manufacturable within state-of-the-art CMOS process thanks to specific design and optimization, *IEDM Tech. Dig. Proc.*, pp. 637–640, 2001.
9. Jurczak, M.; Skotnicki, T.; Paoli, M.; Tormen, B.; Regolini, J.-L.; Morin, C.; Schiltz, A.; Martins, J.; Pantel, R.; Galvier, J. SON (Silicon On Nothing)—a new device architecture for the ULSI Era, *Symp. VLSI Tech. Dig.*, pp. 29–30, 1999.
10. Monfray, S.; Skotnicki, T.; Morand, Y.; Descombes, S.; Talbot, A.; Dutartre, D.; Leverd, F.; Le Friec, Y.; Palla, R.; Pantel, R.; Haond, M.; Nier, M.-E.; Vizioz, C.; Louis, D. Highly-performant 38nm SON (Silicon-On-Nothing) P-MOSFETs with 9nm-thick channels, *IEEE SOI Conference Proc.*, pp. 22–25, 2002.
11. Monfray, S.; Skotnicki, T.; Tavel, B.; Morand, Y.; Descombes, S.; Talbot, A.; Dutartre, D.; Jenny, C.; Mazoyer, P.; Palla, R.; Leverd, F.; Le Friec, Y.; Pantel, R.; Haond, M.; Charbuillet, C.; Vizioz, C.; Louis, D.; Buffet, N. SON (Silicon-On-Nothing) P-MOSFETs with totally silicided ($CoSi_2$) Polysilicon on 5nm-thick Si-films: the simplest way to integration of Metal Gates on thin FD channels, *IEDM Tech. Dig. Proc.*, pp. 263–266, 2002.
12. Skotnicki, T.; Jurczak, M.; Gwoziecki, R.; Lenoble, D.; Ribot, P.; Paoli, M.; Martins, J.; Tormen, B.; Grouillet, A.; Pantel, R.; Galvier, J. Heavily doped and extremely shallow junctions on insulator—by SONCTION (SilicON Cut-off juncTION) process *IEDM Tech. Dig.*, pp. 513–516, 1999.

13. Monfray, S.; Skotnicki, T.; Morand, Y.; Descombes, S.; Paoli, M.; Ribot, P.; Talbot, A.; Dutartre, D.; Leverd, F.; Lefriec, Y.; Pantel, R.; Haond, M.; Renaud, D.; Nier, M.-E.; Vizioz, C.; Louis, D. First 80nm SON (Silicon-On-Nothing) MOSFETs with perfect morphology and high electrical performance, *IEDM Tech. Digest Proc.* pp. 645–648, 2001.
14. Jurczak, M.; Skotnicki, T.; Paoli, M.; Tormen, B.; Martins, J.; Regolini, J.L.; Dutartre, D.; Ribot, P.; Lenoble, D.; Pantel, R.; Monfray, S. SON (Silicon On Nothing)—an Innovative Process for Advanced CMOS, *IEEE Trans. Electron Dev.* Vol. 47, Nov. 2000, p. 2179, 2000.
15. T. Skotnicki Silicon On Nothing (SON)—a robust technological platform for ULSI area, *ECS Proc.*, 2001–2003, p. 391.
16. Tavel, B.; Skotnicki, T.; Pares, G.; Carriere, N.; Rivoire, M.; Leverd, F.; Julien, C.; Torres, J.; Pantel, R. Totally silicided (CoSi2) polysilicon: a novel approach to very low-resistive gate (~2Ohms/square) without metal CMP nor etching, *IEDM Tech. Dig.*, pp. 825–828, 2001.

A.1

Properties of Silicon and Germanium

John D. Cressler
Georgia Institute of Technology

The energy band structures of Si and Ge are depicted in Figure A.1.1, together with (1) their carrier effective mass parameters (Table A.1.1) and (2) their bulk structural, mechanical, optical, and electrical properties (Table A.1.2) [1–3].

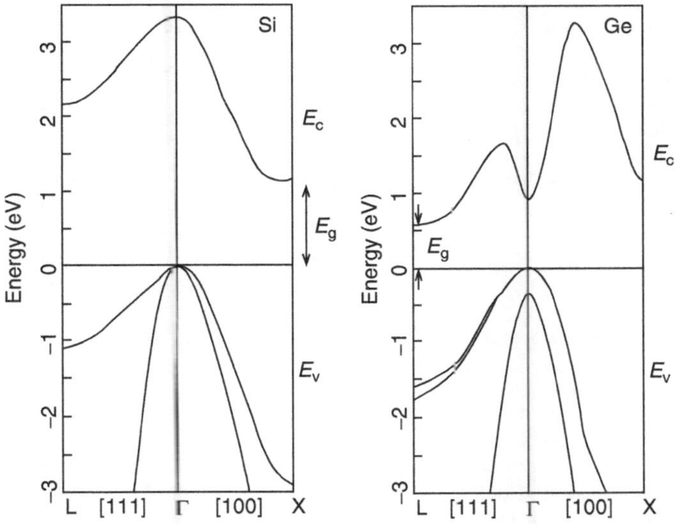

FIGURE A.1.1 Energy band structure, showing the principal conduction and valence bands of Si and Ge as a function of *k*-space direction. (From M Shur. *Physics of Semiconductor Devices.* Englewood Cliffs, NJ: Prentice-Hall, 1990. With permission.)

TABLE A.1.1 Carrier Effective Mass Parameters for Si and Ge

Parameter	Units	Silicon	Germanium
Effective electron mass (m_n^*)	($\times m_o$)		
Longitudinal (4.2 K)		0.9163	1.58
Transverse (4.2 K)		0.1905	0.082
Density-of-states (4.2 K)		1.062	—
Density-of-states (300 K)		1.090	—
Effective hole mass (m_p^*)	($\times m_o$)		
Heavy hole (4.2 K)		0.537	0.28
Light hole (4.2 K)		0.153	0.044
Density-of-states (4.2 K)		0.59	—
Density-of-states (300 K)		1.15	—

TABLE A.1.2 Properties of Bulk Si and Ge

Parameter	Units	Silicon	Germanium
Atomic number	—	14	32
Atomic density	(atoms/cm^3)	5.02×10^{22}	4.42×10^{22}
Atomic weight	(g/mole)	28.09	72.6
Density	(g/cm^3)	2.329	5.323
Electronic orbital configuration	—	(Ne) $3s^2 3p^2$	(Ar) $3d^{10} 4s^2 4p^2$
Crystal structure	—	Diamond	Diamond
Lattice constant (298 K)	(Å)	5.43107	5.65791
Dielectric constant	—	11.7	16.2
Breakdown strength	(V/cm)	3×10^5	1×10^5
Electron affinity	(V)	4.05	4.00
Specific heat	(J/g-°C)	0.7	0.31
Melting point	(°C)	1412	1240
Intrinsic Debye length (300 K)	(μm)	24	0.68
Index of refraction	—	3.42	3.98
Transparency region	(μm)	1.1–6.5	1.8–15
Thermal conductivity (300 K)	(W/cm-°C)	1.31	0.60
Thermal expansion coefficient (300 K)	(°C^{-1})	2.6×10^{-6}	5.9×10^{-6}
Young's modulus	(dyne/cm^2)	1.9×10^{12}	—
Energy bandgap (low doping)	(eV)	1.12 (300 K) 1.17 (77 K)	0.664 (291 K) 0.741 (4.2 K)
Equivalent conduction band minima	—	6	8
Effective electron mass (300 K)	($\times m_o$)	1.18	—
Effective hole mass (300 K)	($\times m_o$)	0.81	—
Intrinsic carrier density (300 K)	(cm^{-3})	1.02×10^{10}	2.33×10^{13}
Effective conduction band DoS (300 K)	(cm^{-3})	2.8×10^{19}	1.04×10^{19}
Effective valence band DoS (300 K)	(cm^{-3})	1.04×10^{19}	6.00×10^{18}
Electron mobility (300 K)	(cm^2/V-sec)	1450	3900
Hole mobility (300 K)	(cm^2/V-sec)	500	1900
Electron diffusivity (300 K)	(cm^2/sec)	37.5	100
Hole diffusivity (300 K)	(cm^2/sec)	13	49
Optical phonon energy	(meV)	63	37
Phonon mean free path length	Å	76	105
Intrinsic resistivity (300 K)	(Ω cm)	3.16×10^5	47.62

References

1. JD Cressler and G Niu. *Silicon–Germanium Heterojunction Bipolar Transistors.* Boston, MA: Artech House, 2003.
2. M Shur. *Physics of Semiconductor Devices.* Englewood Cliffs, NJ: Prentice-Hall, 1990.
3. R. Hull, editor. *Properties of Crystalline Silicon.* London: EMIS Datareviews Series, Number 20, INPSEC, 1999.

A.2

The Generalized Moll–Ross Relations

John D. Cressler
Georgia Institute of Technology

The classical solution for the collector current density in a Si BJT, derived by Shockley, necessarily assumes a constant base doping profile. In this case, for low-injection conditions, the drift component of the minority carrier transport equation can be neglected, and the minority carrier diffusion equation solved under the Shockley boundary conditions. The resultant equation, under the assumptions of negligible neutral base recombination and forward-active bias, is the well-known expression

$$J_C = \frac{qD_{nb}}{N_{ab}^- W_b} n_{io}^2 e^{\Delta E_{gb}^{app}/kT} \left(e^{qV_{BE}/kT} - 1 \right). \tag{A.2.1}$$

Here, D_{nb} is the minority electron diffusivity, N_{ab}^- is the ionized base doping level, n_{io}^2 is the low-doping intrinsic carrier density, given by,

$$n_{io}^2 = N_C N_V e^{-E_{go}/kT}, \tag{A.2.2}$$

and $\Delta E_{gb}{}^{app}$ is the heavy-doping induced bandgap narrowing.

The path to the generalization of this result to the "real-world" case of a nonconstant base doping profile (Figure A.2.1) is nonobvious, and even a cursory glance at the problem is enough to convince one that it cannot follow the original path in Shockley's approach. The complexity of this problem results from the addition of the field-driven transport, which is now no longer negligible due to the doping-gradient-induced field. The clever solution to this problem was first presented in the classic paper by Moll and Ross in 1956, the so-called "Moll–Ross relation" [1]. Unfortunately, that solution made two undesirable assumptions: (1) that the minority electron mobility (hence, diffusivity) is constant across the quasineutral base and (2) that the intrinsic carrier density is constant across the quasineutral base. The latter assumption, in particular, fails in the presence of a heavily doped base (i.e., real life), since the apparent bandgap narrowing is inherently position dependent across the base, and hence the effective bandgap in the base is also position dependent.* In essence, then, the problem becomes one of solving for the collector current density in the presence of *both* nonconstant base doping and nonconstant base bandgap, and is particularly relevant to the graded-base SiGe HBT. This problem remained unsolved for

*In fairness, Moll and Ross cannot be blamed for the second assumption since doping-induced bandgap narrowing had not yet been discovered.

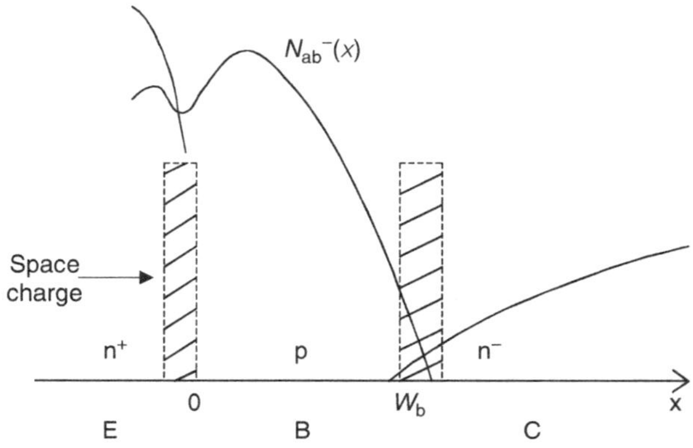

FIGURE A.2.1 Schematic nonconstant base doping profile used in the derivations.

almost 30 years until the seminal paper by Kroemer in 1985 [2]. Since Kroemer's "generalized Moll–Ross relations" are the starting point for both the dc and ac analysis of the graded-base SiGe HBT (Chapter 4, see *Silicon Heterostructure Devices*), we present that elegant derivation here (showing all of the mathematical steps that Kroemer neglected to include in his paper).

The assumptions in Kroemer's solution include: (1) 1-D transport, (2) transport by both drift and diffusion, (3) low-injection conditions (i.e., $n_b(x) \ll N_{ab}^-(x)$ for all x across the base), (4) negligible neutral base recombination, and (5) forward-active bias. Importantly, however, there are no assumptions on the position dependence of the base doping profile or the base bandgap.*

We begin from the generalized drift–diffusion minority electron transport equation, as expressed in terms of the minority electron quasi-Fermi potential

$$J_n = q\mu_n n \nabla \phi_n, \tag{A.2.3}$$

which for our 1-D Si BJT problem reduces to

$$J_C = q\mu_{nb}(x) n_b(x) \frac{d\phi_n(x)}{dx}. \tag{A.2.4}$$

In the quasineutral base, the majority carrier (hole) quasi-Fermi potential (ϕ_p) in low-injection is constant, such that

$$J_C = q\mu_{nb}(x) n_b(x) \frac{d}{dx}\left(\phi_n(x) - \phi_p\right), \tag{A.2.5}$$

and from the generalized Shockley boundary condition

$$n_b(x) p_b(x) = n_{ib}^2(x) e^{q(\phi_n(x) - \phi_p)/kT}, \tag{A.2.6}$$

which can be rewritten as

$$\frac{kT}{q} \ln\left\{\frac{n_b(x) p_b(x)}{n_{ib}^2(x)}\right\} = \phi_n(x) - \phi_p. \tag{A.2.7}$$

*Interestingly, additional generalizations to Kroemer's result have been recently offered [3]. Let it never be said that the final word in device physics is ever in.

Taking the derivative of both sides we have

$$\frac{kT}{q}\left\{\frac{n_{ib}^2(x)}{n_b(x)p_b(x)}\right\}\frac{d}{dx}\left\{\frac{n_b(x)p_b(x)}{n_{ib}^2(x)}\right\} = \frac{d}{dx}\left(\phi_n(x) - \phi_p\right). \tag{A.2.8}$$

Substituting this result back into Equation A.2.5, we obtain

$$J_C = q\mu_{nb}(x)n_b(x)\frac{kT}{q}\frac{n_{ib}^2(x)}{n_b(x)p_b(x)}\frac{d}{dx}\left\{\frac{n_b(x)p_b(x)}{n_{ib}^2(x)}\right\} \tag{A.2.9}$$

We now integrate this expression from some arbitrary point in the base profile to the neutral base boundary (W_b) to obtain

$$\int_x^{W_b}\frac{J_C}{qD_{nb}(x')}\frac{p_b(x')}{n_{ib}^2(x')}dx' = \left.\frac{n_b(x')p_b(x')}{n_{ib}^2(x')}\right|_x^{W_b} \tag{A.2.10}$$

Under the assumptions of negligible neutral base recombination (i.e., J_C is a constant to the integration), and using the fact that in forward-active bias,

$$p_b(W_b)n_b(W_b) \simeq n_{ib}^2(W_b) \tag{A.2.11}$$

we find

$$\frac{J_C}{q}\int_x^{W_b}\frac{p_b(x')dx'}{D_{nb}(x')n_{ib}^2(x')} = 1 - \frac{n_b(x)p_b(x)}{n_{ib}^2(x)} \tag{A.2.12}$$

At the emitter–base boundary ($x = 0$), we know from the generalized Shockley boundary condition that

$$n_b(0)p_b(0) = n_{ib}^2(0)e^{q(\phi_n(0)-\phi_p(0))/kT}, \tag{A.2.13}$$

and

$$\phi_n(0) - \phi_p(0) = V_{BE}, \tag{A.2.14}$$

so that we obtain

$$\frac{J_C}{q}\int_0^{W_b}\frac{p_b(x)dx}{D_{nb}(x)n_{ib}^2(x)} = 1 - e^{qV_{BE}/kT}, \tag{A.2.15}$$

and thus finally,

$$J_C = -\frac{q(e^{qV_{BE}/kT} - 1)}{\int_0^{W_b}\frac{p_b(x)dx}{D_{nb}(x)n_{ib}^2(x)}}, \tag{A.2.16}$$

This is the "generalized Moll–Ross relation"* for the collector current density in a bipolar transistor with nonconstant base doping and arbitrary position-dependence of the base bandgap. Observe that if

*I personally would have no problem calling this elegant result the "Kroemer relation."

we allow $p_b(x) = N_{ab}^-(x) = N_{ab}^- = \text{constant}$, then we obtain Equation A.2.1, as expected (the extra negative sign simply accounts for the fact that the electron flow is in the opposite direction of the positive current flow).

As detailed in Chapter 4 (see *Silicon Heterostructure Devices*), this fundamental result is the starting point of the derivations for collector current density, the current gain, and the output conductance in a graded-base SiGe HBT. In this case, in addition to the bandgap-narrowing-induced position dependence in the base bandgap, we have an additional contribution from the Ge-strained layer (Figure A.2.2). This Ge contribution easily enters the generalized Moll–Ross relation via n_{ib}^2 in Equation A.2.16. For more detail on the resultant derivations and the assumptions and approximations involved, the reader is referred to Ref. [4].

An additional desirable feature of Kroemer's approach is that we can also easily obtain an analytical expression for the base transit time in a device with nonconstant base doping and bandgap. Under a quasistatic assumption we can generally define the base transit time as

$$\tau_b = \frac{-q}{J_C} \int_0^{W_b} n_b(x)\mathrm{d}x. \tag{A.2.17}$$

From Equation A.2.12 and neglecting the unity factor, we can solve for $n_b(x)$ as

$$n_b(x) = \frac{-J_C}{q} \frac{n_{ib}^2(x)}{p_b(x)} \int_x^{W_b} \frac{p_b(x')\mathrm{d}x'}{D_{nb}(x')n_{ib}^2(x')} \tag{A.2.18}$$

Substituting this result into Equation A.2.17, we finally obtain

$$\tau_b = \int_0^{W_b} \frac{n_{ib}^2(x)}{p_b(x)} \left\{ \int_x^{W_b} \frac{p_b(y)\mathrm{d}y}{D_{nb}(y)n_{ib}^2(y)} \right\} \mathrm{d}x \tag{A.2.19}$$

Again, observe that if we allow $p_b(x) = N_{ab}^-(x) = N_{ab}^- = \text{constant}$, we obtain the classical result for a BJT with constant base doping

$$\tau_b = \frac{W_b^2}{2D_{nb}}, \tag{A.2.20}$$

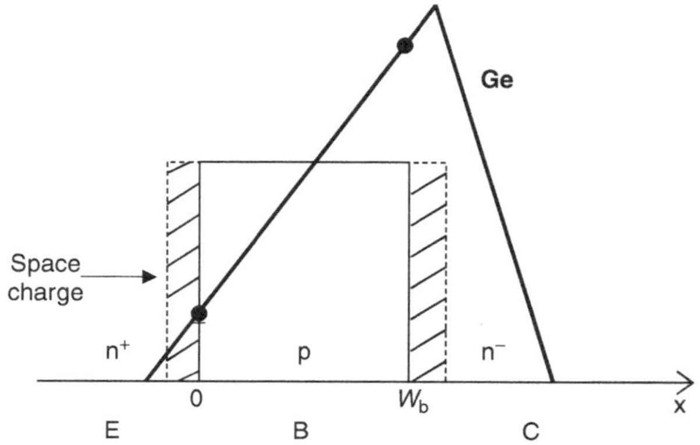

FIGURE A.2.2 Schematic drawing of a position-dependent, graded Ge base profile in a SiGe HBT.

as expected. This second generalized Moll–Ross relation is the starting point for the derivation of base transit in a graded-base SiGe HBT (Figure A.2.2), as detailed in Chapter 4 (see *Silicon Heterostructure Devices*) and Ref. [4].

References

1. J. Moll and I. Ross. The dependence of transistor parameters on the distribution of base layer resistivity. *Proceedings of the Institute of Radio Engineers* 44:72, 1956.
2. H Kroemer. Two integral relations pertaining to electron transport through a bipolar transistor with a nonuniform energy gap in the base region. *Solid-State Electronics* 28:1101–1103, 1985.
3. SN Mohammad. Generalization of the Moll–Ross relations for heterojunction bipolar transistors. *Solid-State Electronics* 46:589–591, 2002.
4. JD Cressler and G Niu. *Silicon–Germanium Heterojunction Bipolar Transistors.* Boston, MA: Artech House, 2003.

A.3

Integral Charge-Control Relations

Michael Schröter
University of California at San Diego

A.3.1 Introduction

One of the most important requirements for compact models is an accurate description of the devices' main current. In an npn bipolar transistor, this is the (generally time dependent) collector current i_C which is given by the transport of electrons from emitter to collector. For time (and frequency) dependent *quasistatic* (q.s.) operation, which is the case in the vast majority of practical applications, the current flowing at the collector terminal can be partitioned into a q.s. transfer current i_T and a charging current supplying all charge storage elements connected to the collector terminal. Since the latter elements are described and represented separately in a model, a formulation of i_T, which equals I_C under d.c. conditions, is of major interest. Before a general relation for i_T is derived, a brief historical perspective of the respective theory is given below.

The obvious starting point for such an equation is the transport and continuity equation. Since the time derivative term in the latter is taken into account separately by the charge storage elements, and the impact of recombination on i_T in modern bipolar transistors is negligibly small, a spatially constant electron current density j_{nx} results (and is observed in device simulation) throughout the structure of a one-dimensional (1D) transistor. This fact can be used to solve the transport equation first for the carrier density and associated charge at a *given transfer current*, and then reformulate the result to obtain the transfer current at a *given charge*. The first solution of this kind was published by Moll and Ross in 1956 [1, Equation (13)],

$$i_T = q\mu_{nB} n_{iB}^2 V_T \frac{\exp\left(V_{B'E'}/V_T\right)}{\int_{x_l}^{x_u} h(x)p(x)\mathrm{d}x} \tag{A.3.1}$$

with $p(x) = N_B(x)$, $h(x) = 1$, $V_{B'E'}$ as internal base–emitter voltage, and $[x_l, x_u]$ as base region. The relation provided a great deal of insight into the dependence of i_T (and also the transit time) on doping profile and material parameters in the base region. An improved and more complete expression was later used for one of the first TCAD papers by Ghosh et al. in 1967 [2]. Here, the material dependent function $h(x) = (\mu_{nB} n_{iB}^2) / (\mu_n n_i^2)$ was included for the first time, but the integral was evaluated numerically, thus not providing a formulation suitable for compact modeling. Only shortly thereafter, in 1970 Gummel published his well-known paper about the integral charge-control relation (ICCR) [3]. For the first time, a compact formulation for i_T was derived, that contained in a consistent relation between i_T and stored charge and with all bias-dependent nonideal effects that were relevant for transistors at that time. In one way or the other, the ICCR has been used until today in any reasonably physics-based compact model. Note that the Scharfetter–Gummel discretization, which was first published and applied to a bipolar transistor, also assumes a constant current density between discrete points, which is the main reason for its numerical stability and wide use for device simulation.

An extension of the Moll–Ross relation with special emphasis on a *spatially variable bandgap* in the base layer was derived by Kroemer in 1985 [4]. Here also the doping density was replaced by the bias-dependent hole density as in Equation A.3.1, but no equation suitable for compact modeling was given. In 1993, for both the latter equation and the ICCR a generalization was presented in Ref. [5], with application to compact modeling. The respective generalized ICCR (GICCR) will be discussed in more detail in subsequent sections. During its derivation, the various results mentioned above will be obtained and be referred to again.

Since the theories above all apply only to a 1D transistor structure with reasonably smooth material and bandgap changes, and to quasistatic operation, extensions for practical cases are of interest, which will be briefly discussed in A.3.5. The considerations in this chapter apply to Si and SiGe technologies. In the latter, two fundamentally different types of doping profiles can be distinguished which are presently being manufactured and illustrated schematically in Figure A.3.1. The conventional emitter doping (CED) type has a similar profile shape as BJTs, but contains in addition a Ge distribution that increases from the BE junction to the BC junction. In contrast, the low-emitter concentration (LEC) type contains a box Ge profile (or at least a sufficiently high step at the BE junction) that allows to significantly increase the base doping and to lower the emitter doping.

A.3.2 Derivation of a General Relationship

Although the GICCR can be derived for the 3D case, the considerations here will be restricted to the 1D case in order to demonstrate the concept. The assumptions required for the derivation are summarized below:

(a) A one-dimensional transistor structure (cf. Figure A.3.1) is considered, with the emitter contact ($x = 0$) at the mono-silicon surface and the collector contact ($x = x_C$) at the transition from the lightly-doped collector region to the buried layer. This 1D structure does not include the emitter and *external* collector series resistance.

(b) Volume recombination within the above 1D transistor region is negligible.

(c) The time derivative is negligible, which corresponds to quasistatic operation.

(d) Effects such as thermionic emission and tunneling across the junctions are neglected. They can be accounted for though by separate terms and can be combined with the GICCR solution.

Assumptions (a), (b), and (c) together with the electron continuity equation lead to a spatially independent electron current density,

$$J_n = \text{const}(x) = -J_T = \frac{-I_T}{A_E}, \tag{A.3.2}$$

which can be expressed by the transfer current and area (emitter) A_E of the 1D-transistor. The derivation starts with the electron transport equation,

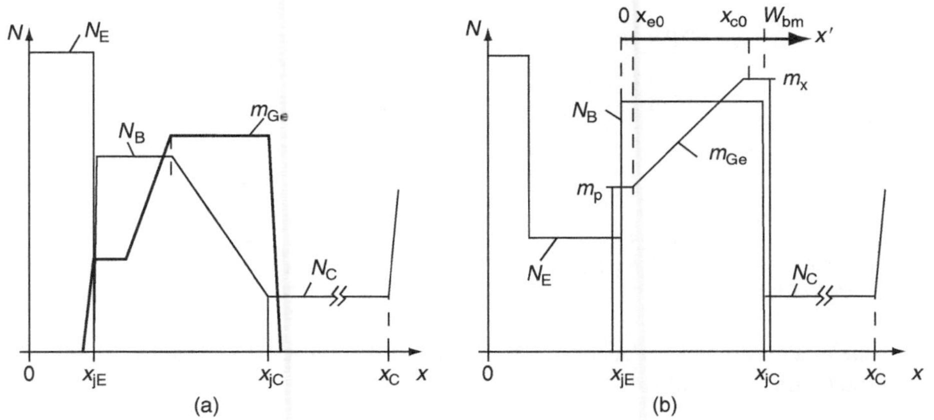

FIGURE A.3.1 Sketch of doping and Ge profiles for (a) CED transistors and (b) LEC transistors.

$$J_n = -q\mu_n n \frac{\mathrm{d}\varphi_n}{\mathrm{d}x} \tag{A.3.3}$$

where μ_n, n and φ_n are the electron mobility, density, and quasifermi potential. Inserting $n = n_i \exp[(\psi - \varphi_n) / V_T]$ into (A.3.3) gives

$$J_n = -q\mu_n n_i \exp\left(\frac{\psi}{V_T}\right) \exp\left(-\frac{\varphi_n}{V_T}\right) \frac{\mathrm{d}\phi_n}{\mathrm{d}x}. \tag{A.3.4}$$

Note that the effective intrinsic carrier density n_i accounts for bandgap differences caused by high-doping effects and bandgap-engineering. This topic will be discussed later. Using the transformation

$$\frac{\mathrm{d}\exp\left(-\varphi_n/V_T\right)}{\mathrm{d}x} = -\frac{1}{V_T} \exp\left(-\frac{\varphi_n}{V_T}\right) \frac{\mathrm{d}\varphi_n}{\mathrm{d}x} \tag{A.3.5}$$

leads to

$$J_n = qV_T\mu_n n_i \exp\left(\frac{\psi}{V_T}\right) \frac{\mathrm{d}\exp\left(-\varphi_n/V_T\right)}{\mathrm{d}x}. \tag{A.3.6}$$

Separation of the differential variables and rearranging terms gives

$$\frac{J_n}{qV_T\mu_n n_i} \exp\left(-\frac{\psi}{V_T}\right) \mathrm{d}x = \mathrm{d}\left(\exp\left(-\frac{\varphi_n}{V_T}\right)\right). \tag{A.3.7}$$

Extension of the l.h.s. by the product $n_i \exp(\varphi_p/V_T)$ allows to replace the inconvenient term $\exp(-\psi/V_T)$ by the hole density, yielding

$$\frac{J_n}{qV_T} \frac{p}{\mu_n n_i^2 \exp\left(\varphi_p/V_T\right)} \mathrm{d}x = d\left(\exp\left(-\frac{\varphi_n}{V_T}\right)\right). \tag{A.3.8}$$

Integration of the above equation over the general spatial interval $[x_l, x_u]$ gives

$$\int_{x_l}^{x_u} \frac{J_n}{qV_T} \frac{\exp(-\varphi_p/V_T)}{\mu_n n_i^2} p \, dx = \int_{\exp(\varphi_n(x_l)/V_T)}^{\exp(\varphi_n(x_u)/V_T)} d\left(\exp\left(-\frac{\varphi_n}{V_T}\right)\right). \tag{A.3.9}$$

The result for the r.h.s. is

$$\int_{\exp(\varphi_n(x_l)/V_T)}^{\exp(\varphi_n(x_u)/V_T)} d\left(\exp\left(-\frac{\varphi_n}{V_T}\right)\right) = \left[\exp\left(-\frac{\varphi_n(x_u)}{V_T}\right) - \exp\left(-\frac{\varphi_n(x_l)}{V_T}\right)\right]. \tag{A.3.10}$$

The exact value of φ_n on the r.h.s. depends on the choice of the integration limits and will be discussed later. First though, the left-hand-side of (A.3.9) is integrated,

$$\int_{x_l}^{x_u} \frac{J_n}{qV_T} \frac{\exp(-\varphi_p/V_T)}{\mu_n n_i^2} p \, dx = \frac{-J_T \exp(-V_{B'E'}/V_T)}{qV_T \mu_{nr} n_{ir}^2} \int_{x_l}^{x_u} h(x)p(x)dx, \tag{A.3.11}$$

with $h = h_g h_i h_\varphi$ as the *normalized* weighting function of the hole density, and its components

$$h_g = \frac{\mu_{nr} n_{ir}^2}{\mu_n(x) n_i^2(x)}, \quad h_i = \frac{-J_n(x)}{J_T}, \quad h_\varphi = \exp\left(\frac{V_{B'E'} - \varphi_p(x)}{V_T}\right). \tag{A.3.12}$$

μ_{nr} and n_{ir} are the mobility and intrinsic carrier density, respectively, of a reference material.

Equating (A.3.10) with (A.3.11), and solving for the desired transfer current yields the "master" equation

$$J_T = qV_T^2 \mu_{nr} n_{ir}^2 \frac{\exp\left(\frac{V_{B'E'}}{V_T}\right)\left[\exp\left(-\frac{\varphi_n(x_l)}{V_T}\right) - \exp\left(-\frac{\varphi_n(x_u)}{V_T}\right)\right]}{\int_{x_l}^{x_u} h_g h_i h_\varphi \, p dx}, \tag{A.3.13}$$

from which different analytical formulations can be derived. The choice of the reference material (or transistor region) and its associated values for $\mu_{nr} n_{ir}^2$ is arbitrary and will be discussed later.

The integration interval is undefined yet, and so is the impact of the various weighting functions on the integral. An attractive choice for the integration limits is $[0, x_C]$, which corresponds to the entire 1D transistor region. As a result of this choice, the electron quasifermi potentials in the numerator assume their known internal terminal values (e.g., for common-emitter configuration),

$$\varphi_n(x_l) = \varphi_n(0) = 0 \quad \text{and} \quad \varphi_n(x_u) = \varphi_n(x_C) = V_{C'E'}, \tag{A.3.14}$$

with $V_{C'E'}$ as (internal) collector–emitter voltage.

The denominator integral is more difficult to oversee. Thus, Figure A.3.2 shows the spatial dependence of the weighting functions with the hole density superimposed. Since the hole quasifermi potential equals $V_{B'E'}$ over the entire base and adjacent space–charge regions, h_φ equals 1 where $p(x)$ is significant. Similarly, the electron current density is constant and equals J_T even in a larger interval, except in a small region close to the emitter contact, where it increases slightly due to the back injection of holes and the corresponding recombination; the maximum deviation can be $1/B_f$. Therefore, without introducing a significant error, it is possible for both BJT and HBT to assume an average value for the functions h_i and h_φ that is very close to 1. According to Figure A.3.2, the main spatial dependence of the weighting function h is caused by h_g via the bandgap variation of n_i. A smaller contribution to the spatial dependence comes from the doping and field dependence of the mobility.

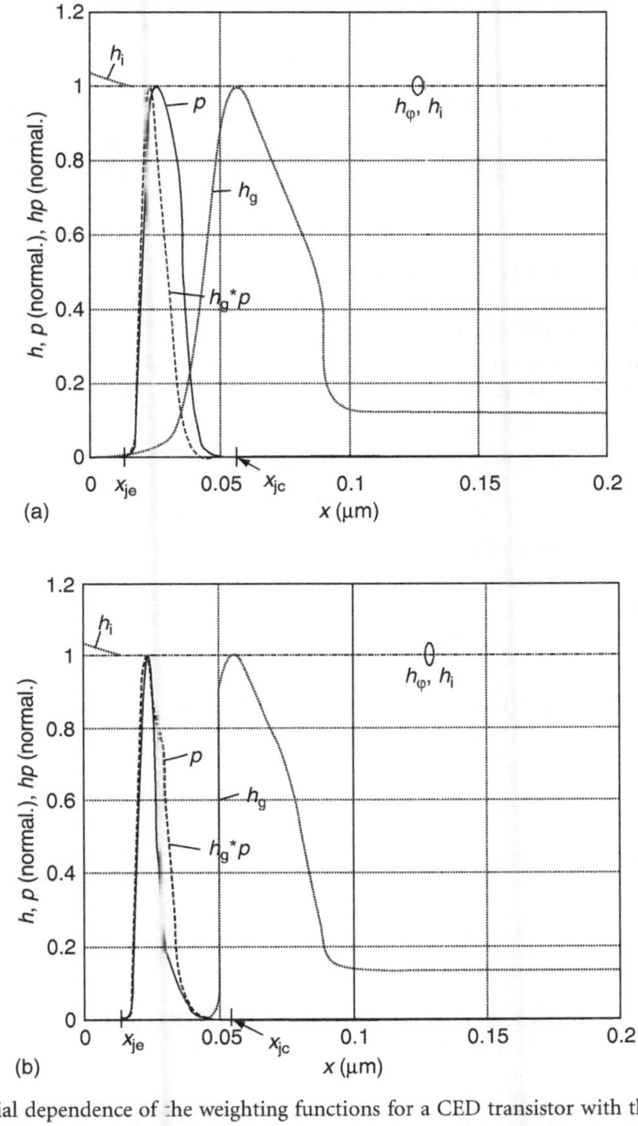

FIGURE A.3.2 Spatial dependence of the weighting functions for a CED transistor with the same doping profile: (a) 70 GHz BJT without Ge; (b) 120 GHz HBT with Ge. Bias point $(J_C = J_T, V_{B'C'}) = (0.1\,\text{mA}/\mu\text{m}^2, 0\,\text{V})$. The hole density p, weighting function h_g, and their product $h_g p$ have been normalized to the respective maximum value. x_{je} and x_{jc} denote the emitter and base junction depth.

Inserting (A.3.14) and h_g from (A.3.12) into (A.3.13), extension by q as well as using average values $\overline{h}_i$ and $\overline{h}_\varphi$ yields the basic formulation

$$J_T = q^2 V_T \frac{\mu_{nr} n_{ir}^2}{\overline{h}_i \overline{h}_\varphi} \frac{\exp(V_{B'E'}/V_T) - \exp(V_{B'C'}/V_T)}{q \int_0^{x_C} h_g p \mathrm{d}x}. \qquad \text{(A.3.15)}$$

from which practically relevant equations can be derived as shown later. The denominator is not yet suitable for (compact) modeling, but at this point can be prepared for further evaluation by partitioning it into a bias-independent and a bias-dependent portion

$$q\int_0^{x_C} h_g p \mathrm{d}x = q\int_0^{x_C} h_g p_0 \mathrm{d}x + q\int_0^{x_C} h_g \Delta p \mathrm{d}x, \tag{A.3.16}$$

where p_0 is the hole density in equilibrium and Δp is the hole density *change* in the transistor with respect to equilibrium. Δp consists of depletion and minority components. While the latter density is distributed over the whole base region, the former densities are located around the junctions and related to the (ionized) base doping.

A.3.3 Homojunction Transistors

In homojunction transistors, the variation of h_g is only caused by high-doping effects. Thus, h_g is fairly constant in the region around the peak hole density (cf. Figure A.3.2a), which contributes most to the integral. Defining an average value

$$\bar{h}_g = \frac{\mu_{nr} n_{ir}^2}{\mu_n n_i^2}, \tag{A.3.17}$$

the denominator of (A.3.15) becomes

$$q\int_{x_E}^{x_C} \frac{\mu_{nr} n_{ir}^2}{\mu_n n_i^2} p(x)\mathrm{d}x = \bar{h}_g \bar{Q}_p \tag{A.3.18}$$

with the hole charge *per area* stored in the 1D transistor structure,

$$\bar{Q}_p = q\int_0^{x_C} p(x)\mathrm{d}x. \tag{A.3.19}$$

Inserting (A.3.18) and (A.3.19) into (A.3.15) yields

$$J_T = c\frac{\exp(V_{B'E'}/V_T) - \exp(V_{B'C'}/V_T)}{\bar{Q}_p} \tag{A.3.20}$$

which has the same form as Gummel's ICCR in Ref. [3], but a different definition of the integration region and, thus, of the charge and controlling voltages. The factor

$$c = q^2 V_T \frac{\mu_{nr} n_{ir}^2}{\bar{h}_i \bar{h}_\varphi \bar{h}_g} \cong q^2 V_T \overline{\mu_n n_i^2} \tag{A.3.21}$$

is assumed to be constant over bias. Usually, the most right term is used only, which is obtained by setting $\bar{h}_i = \bar{h}_\varphi = 1$ and is an excellent assumption for the 1D case. Since the hole density is concentrated mostly in the base, the value for $\bar{h}_g$ and $\overline{\mu_n n_i^2}$ is close to that for the base material.

The charge in (A.3.19) can be divided into a bias-independent and a bias-dependent component, $\overline{Q}_p = \overline{Q}_{p0} + \Delta\overline{Q}_p$. The bias-independent charge $\overline{Q}_{p0}$ is defined at $V_{B'E'} = V_{B'C'} = 0$ and consists of holes stored mostly in the neutral base region $0 \le x' \le w_{B0}$ at equilibrium:

$$\bar{Q}_{p0} = q\int_0^{x_C} p_0(x)\mathrm{d}x \cong q\int_0^{w_{B0}} p_0(x')\mathrm{d}x' \approx q\int_0^{w_{B0}} N_B(x')\mathrm{d}x'. \tag{A.3.22}$$

(Note the use of a different coordinate x' in the neutral base region, cf. Figure A.3.1b.) The bias-dependent charge component represents all holes, Δp, that for a nonequilibrium bias condition enter the base region (i) to charge the depletion regions, (ii) to compensate the electrons injected into the (neutral) base, and (iii) for the injection into emitter and collector,

$$\Delta\bar{Q}_p = q\int_0^{x_C} \Delta p(x)dx = \bar{Q}_{pE} + \bar{Q}_{jE} + \bar{Q}_{nB} + \bar{Q}_{jC} + \bar{Q}_{pC} \qquad (A.3.23)$$

Here, $\overline{Q}_{jE}$ and $\overline{Q}_{jC}$ are the BE and BC depletion charge (per unit area), respectively, and the other components are the minority charges (per unit area) stored in E, B, and C. As a consequence of the choice of the chosen integration limits, $\Delta\overline{Q}_p$ can actually be measured via the terminals. From a physics-based point of view, the biggest advantage of the extended ICCR (A.3.20) is its firm link between the quasistatic transfer current and the *total* hole charge, which resembles the charge-controlled operation principle of a bipolar transistor in a very compact form.

Examples for the accuracy of various solutions are shown in Figure A.3.3 for a typical Gummel plot. In all cases, the denominator integral (e.g. $\overline{Q}_p$ in (A.3.20)) has been taken directly from device simulation. As expected, the basic Equation A.3.15 follows exactly the simulated behavior of J_T (crosses). For the older generation profile ("25 GHz transistor" in Figure A.3.3a), the practically interesting approximation (A.3.20) is very accurate at low and medium current densities, and shows only slight deviations at high current densities, which are tolerable though and typical for homo-junction bipolar processes (e.g. [6,7]). The reason for these small deviations is caused by the assumption of a bias-independent value for $\overline{h}_g$. However, if the minority charge contributions of emitter and collector, $\overline{Q}_{pE}$ and $\overline{Q}_{pC}$, are neglected in (A.3.20), the model current becomes far too large. Thus, including only the base charge is not a feasible approach.

For the profile of a modern generation Si(Ge) transistor ("70 GHz BJT" in Figure A.3.3b), Equation A.3.20 starts to deviate already at medium current densities, affecting the accuracy of the transconductance modeling. This is caused by the increased difference in doping and associated variation of h_g across the base region, which requires a different weighting factor for Q_{p0} and Q_{jE} along the same lines as discussed in the next section on HBTs, applying the corresponding Equation A.3.28 gives again a much higher accuracy.

The assumption of a bias-independent value for $\overline{h}_g$ deserves some consideration. High-doping effects cause both n_i and μ_n (which is also field dependent) to depend on x within the base region, with the change of n_i partially being compensated by the change of μ_n. Obviously though, as long as the spatial distributions of p and $\overline{\mu_n n_i^2}$ do not change with bias, $\overline{h}_g$ remains bias independent. This holds quite well for low and medium current densities where p is concentrated in the base region, and the mobility depends only very slightly on bias. Toward high current densities, however, p spreads out into the collector and partially also into the emitter region. In the collector, for instance, high-doping effects can be neglected but the field (and thus bias) dependence of μ_n is significant, leading to a different weighting of the hole density in this region. As a consequence, the average value $\overline{h}_g$ defined at low current densities is not quite correct anymore at high current densities and causes the observed small deviation from the exact current. Note though that the field dependence of the mobility around the BC junction decreases at high current densities.

At this point, the choice of the integration limits shall be briefly discussed. Fundamentally, there is an infinite number of possible integration intervals. Besides the one selected above (i.e., entire 1D transistor), the other most important choice in literature (e.g., Refs. [3,4]) is the base region, i.e., $x \in [x_{jE}, x_{jC}]$. In this case, $\overline{h}_g$ is only a very weak function of bias, but there are serious disadvantages of this choice: (i) the unknown values for the electron quasi-fermi potentials at the base region boundaries, i.e., $\varphi_n(x_{jE})$ and $\varphi_n(x_{jC})$ in the numerator of (A.3.13), are not easily accessible by measurements; (ii) a separate determination (i.e., measurement) of the *base* hole charge, in particular the base minority charge, is difficult at medium to high current densities.

The original and most famous expression of the ICCR as given by Gummel in Ref. [3], which has the same form as (A.3.20), is in fact based on considering only the base region for the integration and assuming contributions from outside that region to be negligible. Although this was a reasonable assumption at that time for the charge, it was not for the voltage drop toward the contacts.

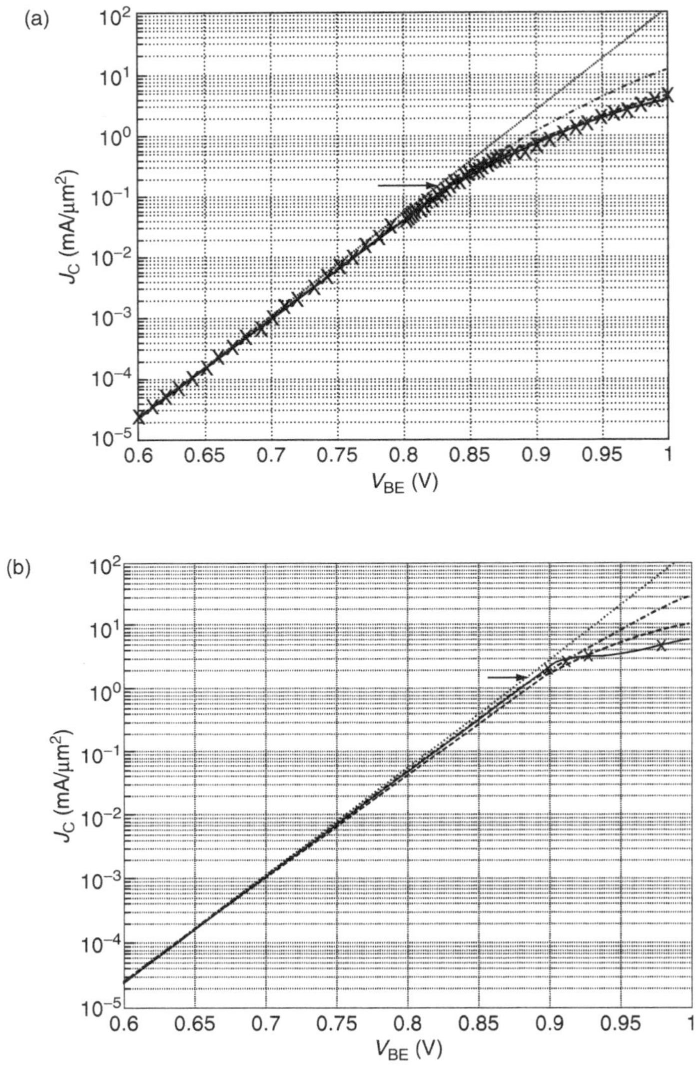

FIGURE A.3.3 Transfer current density for (a) 25 GHz BJT and (b) 70 GHz BJT: comparison between device simulation (symbols) and analytical calculations from Equation A.3.15 (solid line), Equation A.3.24 (dotted line), Equation A.3.20 (dashed line), as well as Equation A.3.20 without $Q_{pE} + Q_{pC}$ as dash-dotted line in (a) and Equation A.3.28 as dash-dotted line in (b). The arrow indicates the current density at which the f_T peak occurs.

An often found further simplification is to replace the hole density in (A.3.19) by the base doping density N_B. As a consequence, the hole charge equals the expression on the most r.h.s. of (A.3.22) and does not contain any bias-dependent "nonideal" effects anymore. Neglecting the voltage drops toward the contacts, Moll and Ross [1] first derived the corresponding expression for the transfer current

$$J_T = qV_T \, \overline{\mu_{nB} n_{iB}^2} \, \frac{\exp(V_{B'E}/V_T) - \exp(V_{B'C'}/V_T)}{\int_0^{W_B} N_B dx'} \tag{A.3.24}$$

This is a more general relation compared to the classical transistor theory along the lines of Ebers and Moll [9], and aided the understanding that the transfer current is linked to the *integral* of the base doping and not to some *spot* value. As shown in Figure A.3.2, Equation A.3.24 produces an almost ideal characteristic due to the missing bias-dependent charge components.

A.3.4 Heterojunction Transistors

A comparison for the weighting function h_g between a SiGe and a Si transistor with the same doping profile was already shown in Figure A.3.2. While the spatial dependence of the hole QFP and electron current density and, thus, of h_φ and h_i is still similar to BJTs, the intentional bandgap differences in HBTs lead to additional variations of n_i^2 within the integration interval $[0, x_C]$. As a consequence, it is impossible to define a single average value $\bar{h}_g$ over this whole interval. Thus, the denominator integral in (A.3.15) now has to be formulated in terms of charge components in *separate transistor regions that are multiplied with an average value of the weighting function* h_g taken over the respective region. In HBTs the transition from one bandgap value to another usually takes place close to the junctions. Therefore, partitioning of the respective integral $q\int_0^{x_c} h_g p dx$ into neutral and space–charge regions is *one* reasonable choice in order to obtain separate expressions, in which the weighting function is sufficiently independent on location and bias. This leads generally to

$$q\int_0^{x_C} h_g p dx = \bar{h}_{g,B0}\bar{Q}_{p0} + \bar{h}_{g,E}\bar{Q}_{pE} + \bar{h}_{g,jE}\bar{Q}_{jE} + \bar{h}_{g,B}\bar{Q}_{nB} + \bar{h}_{g,jC}\bar{Q}_{jC} + \bar{h}_{g,C}\bar{Q}_{pC} \tag{A.3.25}$$

$$\text{with} \quad \bar{h}_{g,v} = \frac{\mu_{n0r} n_{ir}^2}{\mu_{nv} n_{iv}^2} \quad \text{and} \quad v = \{B0, E, jE, B, jC, C\}. \tag{A.3.26}$$

Choosing the base region as reference and assuming $\bar{h}_{g,B} = \bar{h}_{g,B0}$, all weighting factors can be divided by $\bar{h}_{g,B}$ giving for the normalized remaining weighting factors

$$h_v = \frac{\bar{h}_{g,v}}{\bar{h}_{g,B}} = \frac{\overline{\mu_{nB} n_{iB}^2}}{\mu_{nv} n_{iv}^2} \quad \text{with} \quad v = \{E, jE, jC, C\}, \tag{A.3.27}$$

which are considered to be model parameters. Inserting the above expression and charge components into (A.3.15), and making again the valid assumption $\bar{h}_i = \bar{h}_\varphi = 1$ yields the GICCR as the final expression for 1D-HBTs [5]

$$J_T = q^2 V_T \overline{\mu_{nB} n_{iB}^2} \frac{\exp(V_{B'E'}/V_T) - \exp(V_{B'C'}/V_T)}{\bar{Q}_{p,T}} \tag{A.3.28}$$

with the modified ("transfer current related") charge density

$$\bar{Q}_{p,T} = \bar{Q}_{p0} + \Delta\bar{Q}_{p,T} = \bar{Q}_{p0} + h_E\bar{Q}_{pE} + h_{jE}\bar{Q}_{jE} + \bar{Q}_{nB} + h_{jC}\bar{Q}_{jC} + h_C\bar{Q}_{pC}, \tag{A.3.29}$$

in which—as for BJTs—still a bias-independent hole charge, Q_{p0}, and a bias-dependent portion can be distinguished. According to Refs. [5,8], and also as Figure A.3.4 shows, (A.3.28) with (A.3.29) using bias-independent values for h_v leads to a significant improvement over the conventional ICCR (A.3.20). Thus, the GICCR results in an accurate description of the transfer current characteristics and the respective derivatives over the entire bias range of interest (up to high current densities). Again, as in Figure A.3.3, the charges and weighting factors have been calculated directly from device simulation results to avoid any errors introduced by analytical approximations.

For "ideal" doping profiles like in Figure A.3.1, the weighting factors h_v in (A.3.2) can be calculated analytically using (A.3.15). This aids the physical understanding of the impact of the Ge profile on device characteristics and shall be demonstrated below for two Ge profile examples. However, the calculation can be extended to other profiles but just becomes more elaborate.

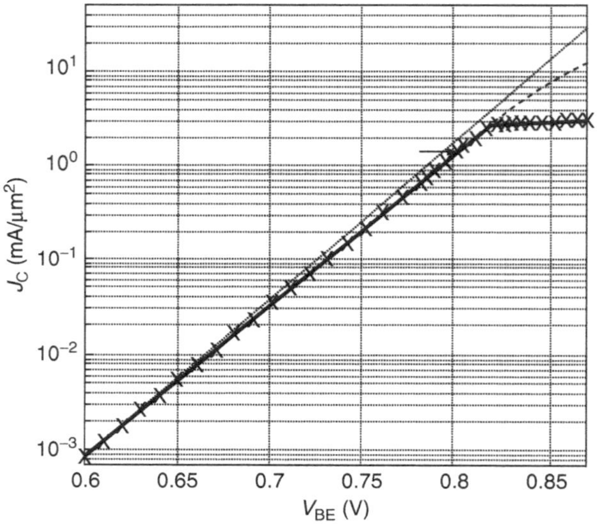

FIGURE A.3.4 Transfer current density for the 120 GHz HBT: comparison between device simulation (symbols) and analytical calculations from Equation A.3.15 (solid line), Equation A.3.28 (dashed line), Equation A.3.24 (dotted line), and Equation A.3.28 without $Q_{pE} + Q_{pC}$ (dash-dotted line).

Box Ge Profile in the (Metallurgical) Base Region

In this case, $\overline{\mu_n n_i^2}$ is constant over the base region and the portion of the space–charge regions that are associated with the base. For the sake of simplicity, low injection is assumed first. Setting $\overline{\mu_{nr} n_{ir}^2} = \mu_{nB} n_{iB}^2$ yields $h_g = 1$ already in the integral over the base region and, thus, $h_{jE} = h_{jC} = 1$. In the emitter and collector region, $h_E = \mu_{nB} n_{iB}^2 / \mu_{nE} n_{iE}^2 \gg 1$ and $h_C = \mu_{nB} n_{iB}^2 / \mu_{nC} n_{iC}^2 \gg 1$, respectively, due to the much larger bandgap in those regions. As a consequence of this bandgap, the hole charges $\overline{Q}_{pE}$ and $\overline{Q}_{pC}$ are very small and do not significantly impact the dynamic transistor behavior, regardless of the respective doping profile. However, according to (A.3.28) these charges can have a significant impact on the transfer current due to the large weighting factors h_E and h_C. The respective terms $h_E \overline{Q}_{pE}$ and $h_C \overline{Q}_{pC}$ in (A.3.28) actually cause the "saturation" of the $I_C(V_{BE})$ characteristics at high injection observed in Figure A.3.2b. In HBTs with a box Ge profile, Equation A.3.28 can be simplified to [5]

$$J_T = q^2 V_T \overline{\mu_{nB} n_{iB}^2} \frac{\exp(V_{B'E'}/V_T) - \exp(V_{B'C'}/V_T)}{\bar{Q}_p + (h_E - 1)\bar{Q}_{pE} + (h_C - 1)\bar{Q}_{pC}}, \quad (A.3.30)$$

which contains the total hole charge as a lumped variable and correction factors in the denominator. In Ref. [8], the product $(h_C - 1)\bar{Q}_{pC}$ is described directly by a compact expression rather than separately by the weighting factor and charge component.

Trapezoidal Profile in the (Metallurgical) Base Region

Consider the Ge profile in Figure A.3.1b. For the sake of simplicity it is assumed that the Ge mole fraction m_{Ge} and the respective bandgap voltage increase over the width of the neutral base ($x' \in [x_{e0}, x_{c0}]$) only, but stay constant across the space–charge regions. Choosing the Si-base without Ge contents as reference material, the bandgap voltage differences $\Delta V_{Gp} = \Delta V_G(x' = x_{e0})$ and $\Delta V_{Gx} = \Delta V_G(x' = x_{c0})$, respectively, can be defined. Hence, the intrinsic carrier density within the neutral base with the width $w_{B0} = x_{c0} - x_{e0}$ can be written as

$$n_i^2 = n_{iB,Si}^2 \exp\left(\frac{\Delta V_{Gp} + a_G(x' - x_{e0})}{V_T}\right) \tag{A.3.31}$$

with the slope factor $a_G = (\Delta V_{Gx} - \Delta V_{Gp})/w_{B0}$. Neglecting the (much smaller) dependence of the mobility on field and Ge contents and applying the above relation to (A.3.16) with $p_0 = N_B$ yields for the bias-independent term

$$q\int_0^{x_C} \frac{\mu_{nr} n_{ir}^2}{\mu_n n_i^2} p_C dx' \cong q\int_{x_{e0}}^{x_{c0}} \exp\left(-\frac{\Delta V_{Gp} + a_G(x' - x_{e0})}{V_T}\right) N_B dx' \tag{A.3.32}$$

which gives after evaluating the r.h.s. integral

$$q\int_0^{x_C} \frac{\mu_{nr} n_{ir}^2}{\mu_n n_i^2} p_0 dx' \cong \frac{\frac{V_T}{a_G}\left[\exp\left(-\frac{\Delta V_{Gp}}{V_T}\right) - \exp\left(-\frac{\Delta V_{Gx}}{V_T}\right)\right]}{w_{B0}} \bar{Q}_{p0} = \bar{h}_{g,B0} \bar{Q}_{p0} \tag{A.3.33}$$

with $\bar{Q}_{p0} = qN_B w_{B0}$. As can be seen, the average weighting factor depends exponentially on the bandgap voltages at the *beginning* and the *end* of the neutral base.

For the bias-dependent portion in (A.3.16), one can write at low current densities

$$q\int_0^{x_C} h_g \Delta p dx \cong q\left[\int_{x_e}^{x_{e0}} \frac{N_B}{\exp\left(\frac{\Delta V_{Gp}}{V_T}\right)} dx' + \int_{x_{c0}}^{x_c} \frac{N_B}{\exp\left(\frac{\Delta V_{Gx}}{V_T}\right)} dx' + \int_{x_{jE}}^{x_{jC}} \frac{n}{\exp\left(\tilde{n}\frac{\Delta V_{Gp} + a_G(x'\tilde{n}x_{e0})}{V_T}\right)} dx'\right] \tag{A.3.34}$$

The first two terms represent the depletion components that are only to be evaluated between the SCR boundary (i.e., x_e, x_c) at the given bias point and the respective equilibrium SCR boundaries (i.e., x_{e0}, x_{c0}); it also has been assumed that the bandgap (i.e., Ge mole) change within $(x_{e0}-x_e)$ and $(x_{c0}-x_c)$ is still negligible. The resulting depletion charges are $\bar{Q}_{jE} = qN_B(x_{e0} - x_e)$ and $\bar{Q}_{jC} = qN_B(x_c - x_{c0})$. For the case that the electric field in the base due to Ge grading causes the electrons to travel with saturation drift velocity v_s, i.e. $n = J_T/(qv_s)$ does not depend on x', the resulting base minority charge is then $\overline{Q}_{nB} = J_T w_B/v_s$. With these charge expressions, one obtains after evaluating all terms on the r.h.s. of (A.3.34)

$$q\int_0^{x_C} h_g \Delta p dx \cong \exp\left(-\frac{\Delta V_{Gp}}{V_T}\right) \bar{Q}_{jE} + \exp\left(-\frac{\Delta V_{Gx}}{V_T}\right) \bar{Q}_{jC} + \bar{h}_{g,B0} \bar{Q}_{nB}, \tag{A.3.35}$$

where the last term follows the same evaluation as (A.3.32). The final step is to insert the components in (A.3.33) and (A.3.35) back into (A.3.15) and to normalize the denominator to the base weighting factor, $\overline{h}_{gB0}$ given by Equation A.3.33. The resulting expression then reads

$$J_T = \underbrace{q^2 V_T \frac{\mu_{nB,Si} n_{iB,Si}^2}{\bar{h}_i \bar{h}_\varphi \bar{h}_{g,B0}}}_{c_{10}} \frac{\exp(V_{B'E'}/V_T) - \exp(V_{B'C'}/V_T)}{\bar{Q}_{p0} + h_{jE}\bar{Q}_{jE} + h_{jC}\bar{Q}_{jC} + \bar{Q}_{nB}}. \tag{A.3.36}$$

which has the same form as (A.3.28), but for low current densities and with known analytical expressions of the weighting factors from the above analysis:

$$h_{jE} = \frac{\exp\left(-\frac{\Delta V_{Gp}}{V_T}\right)}{\bar{h}_{g,B0}} = \frac{v\exp(v)}{\exp(v) - 1}, \quad h_{jC} = \frac{\exp\left(-\frac{\Delta V_{Gx}}{V_T}\right)}{\bar{h}_{g,B0}} = \frac{v}{\exp(v) - 1}. \tag{A.3.37}$$

with $v = [\Delta V_{GX} - \Delta V_{GP}]/V_T$ as normalized bandgap difference between beginning and end of the base region (cf. Figure A.3.1b) . The dependence of the weighting factors as a function of v is shown in Figure A.3.5 for a practically relevant range.

From the above results, the forward Early voltage at low injection, can be calculated:

$$V_{Ef} = \frac{\bar{Q}_p}{h_{jC}\bar{C}_{jCi}} \approx \frac{\bar{Q}_{p0}}{\bar{C}_{jCi}} \frac{\exp(v) - 1}{v}. \tag{A.3.38}$$

According to (A.3.37), a 20% difference in Ge across the base region corresponds to an about 40 times increase in Early voltage, which is a significant enhancement factor over a Si-BJT or a SiGe HBT with a Ge box profile ($v = 0$).

In addition to the strong variation in n_i, the mobility varies within the transistor as a function of both doping and bias (via the electric field). The variation caused by the latter is most pronounced in the BC junction and collector region. In general, μ_n and n_i possess an opposite dependence on doping, leading to a partial compensation within h_g. However, the influence of n_i still remains much stronger than that of μ_n. As a consequence, the weighting function h_g always deviates strongly from 1 and has to be considered for all processes.

A.3.5 Further Extensions

All of the considerations so far apply to a 1D transistor structure and quasistatic operation. Extensions in both directions have been investigated and proposed. A solution of the time-dependent continuity equation led to the transient ICCR (TICC) [10], in which the "in-phase" component gives the same expression as the ICCR for the for q.s. transfer current, while the "out-of-phase" solution yields a physical definition of the charging currents flowing through the E and C contact. Hence, the out-of-phase solution defines a physics-based capacitance matrix associated with the E and C terminals, that includes the case of non-quasistatic operation. Extensions of the TICC towards including recombination and non-1D effects were presented in, for example, Ref. [11]. The application of the TICC results in a compact model, however, is quite challenging due to the bias-dependent weighting functions in the integrals defining the charging components.

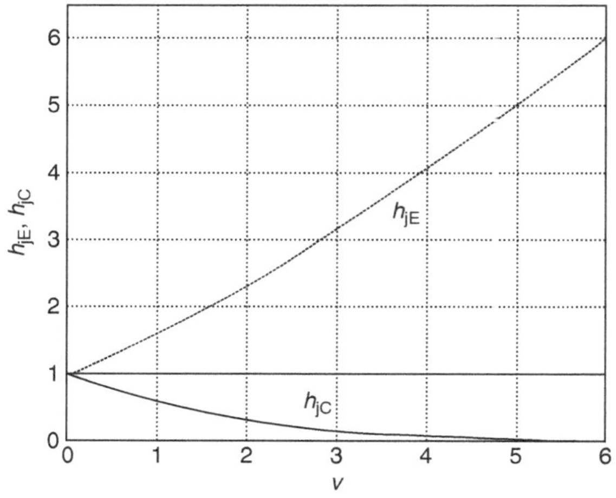

FIGURE A.3.5 Weighting factors of the depletion charges according to Equation A.3.37 as a function of the normalized bandgap difference $v = (\Delta V_{Gx} - \Delta V_{Gp})/V_T$.

As shown in Ref. [7], it is possible to extend the GICCR to two- and three-dimensional transistor structures. The respective derivation is beyond the scope of this chapter, but the result shall be briefly discussed. For instance, the resulting 2D-GICCR reads

$$I_T = c_{10} \frac{\exp\left(V_{\mathrm{B'E'}}/V_{\mathrm{T}}\right) - \left[\exp\left(V_{\mathrm{B'C'}}/V_{\mathrm{T}}\,\frac{2(y_{\mathrm{Bcon}}+b_{\mathrm{Bcon}})}{b_{\mathrm{E}}} + \frac{2b_{\mathrm{Bcon}}}{b_{\mathrm{E}}}\right]}{Q_{\mathrm{p0,T}} + \Delta Q_{\mathrm{p,T}}}, \tag{A.3.39}$$

where the constant c_{10} depends on an enlarged (effective) emitter width

$$b_{\mathrm{E}} = b_{\mathrm{E0}} + 2\int_{b_{\mathrm{E0}}/2}^{y_{\mathrm{Bcon}}} \exp\left(-\frac{\varphi_n(x=0,y)}{V_{\mathrm{T}}}\right)\mathrm{d}y. \tag{A.3.40}$$

b_{E0} is the emitter window width, y_{Bcon} is the edge of the base contact or polysilicon next to the emitter, and b_{Bcon} is the base contact or polysilicon width on mon-silicon. $\Delta Q_{\mathrm{p,T}}$ is defined as in (A.3.29), but now includes, among others, the impact of electron current crowding. This also applies to $Q_{\mathrm{p0,T}}$, which introduces a bias-dependent geometry dependence at higher current densities. In practice, $Q_{\mathrm{p0,T}}$ can be approximated by a constant value to first order.

A.3.6 Summary

A set of integral charge-control relations has been derived and put in perspective to the (classical) literature. It was shown that a "master" equation exists, from which integral charge-control relations of different complexity and accuracy can be derived. The most general form, that is suitable for accurately describing the transfer current in a compact model for HBTs and BJTs, is the GICCR, which includes bandgap differences in the various device regions and also contains the weakest assumptions among the known theories for the transfer current.

The GICCR is a powerful tool to analytically derive the relationship between transfer current, stored charges, and physical as well as structural parameters of a transistor. The GICCR can be very accurate, provided that the respective weigthing factors as a function of device structure and the hole charge as a function of bias are accurately modeled. Notice that the latter is a prerequisite for the description of high-speed applications in any way. Also, since the hole charge has to be continuously differentiable with respect to bias, the transfer current is also automatically continuously differentiable over all bias regions and, hence, is modeled via a single-piece formulation. This is a very desirable feature of the (G)ICCR for compact models.

Applying the "master" equation to compact modeling requires partitioning of the hole charge and analytical approximations for its various components. These measures as well as the determination of charge model parameters and appropriate weighting factor values introduce additional inaccuracies with respect to the results shown here, which are unavoidable though for any compact model equation.

Acknowledgment

The author would like to thank H. Tran for performing simulations and model calculations.

References

1. J.L. Moll and J.M. Ross, The dependence of transistor parameters on the distribution of base layer resistivity, *Proc. IRE*, 44, 1956, S72–S78.
2. H.N. Ghosh, F.H. De La Moneda, and N.R. Dono, Computer-aided transistor design, characterization, and optimization, *Proc. IEEE*, 55, 1967, 1897–1912.
3. H.K. Gummel, A charge-control relation for bipolar transistors", *Bell System Technical J.*, 49, 1970, S115–S120.

4. H. Kroemer, Two-Integral equations pertaining to the electron transport through a bipolar transistor with nonuniform energy gap in the base region, *Solid-State Electron.*, 28, 1985, 1101–1103.
5. M. Schroter, M. Friedrich, and H.-M. Rein, A generalized integral charge-control relation and its application to compact models for silicon based HBTs, *IEEE Trans. Electron Dev.*, 40, 1993, 2036–2046.
6. H.-M. Rein, H. Stübing, and M. Schroter, Verification of the integral charge-control relation for high-speed bipolar transistors at high current densities, *IEEE Trans. Electron Dev.*, 32, 1985, 1070–1076.
7. M. Schroter, A compact physical large-signal model for high-speed bipolar transistors with special regard to high current densities and two-dimensional effects, PhD thesis (in German), Ruhr-University Bochum, Bochum, Germany, 1988.
8. M. Friedrich and H.-M. Rein, Analytical current–voltage relations for compact SiGe HBT models, *IEEE Trans. Electron Dev.*, Vol. ED-46, 2001, 1384–1401.
9. J.J. Ebers and J.L. Moll, Large-signal behavior of junction transistors, *Proc. IRE*, 42, 1761–1772, 1954.
10. H. Klose and A. Wieder, The transient integral charge–control relation—a novel formulation of the currents in a bipolar transistor, *IEEE Trans. Electron Dev.*, ED-34, 1090–1099, 1987.
11. J.S. Hamel and C.R. Selvakumar, The general transient charge–control relation: a new charge–control relation for semiconductor devices, *IEEE Trans. Electron Dev.*, ED-38, 1467–1476, 1991.

A.4
Sample SiGe HBT Compact Model Parameters

Ramana M. Malladi
IBM Microelectronics

This appendix contains a sample set of compact model parameters for a representative $0.32 \times 16.8\ \mu m^2$ first-generation npn SiGe HBT with a peak f_T of 50 GHz, for each of the three dominant higher-order SiGe compact models available in the public domain and in leading circuit simulators: HICUM, MEXTRAM, and VBIC. Each model was carefully calibrated to a comprehensive set of measured dc and ac data.

TABLE A.4.1 HICUM (v 2.1) SiGe HBT Model Parameters

Group	Name	Parameter Description	Value
Transfer current	c10	Constant for ICCR (= Is.qp0)	8.5×10^{-31} A^2 S
	qp0	Zero-bias hole charge in the base	1.4×10^{-13} C
	ich	High-current correction to account for 2D and 3D effects	10 A
	hjei	BE depletion charge weighting factor for HBTs	2
	hjci	BC depletion charge weighting factor for HBTs	0.1
	mcf	Forward nonideality factor for transfer current for HBTs	1
	hfe	Emitter minority charge weighting factor for HBTs	1
	hfc	Collector minority charge weighting factor for HBTs	0.2
Base–emitter currents	ibeis	Internal BE sauration current	7×10^{-20} A
	mbei	Internal BE non-ideality factor	1.005
	ibeps	Peripheral BE saturation current	7×10^{-22} A
	mbep	Peripheral BE non-ideality factor	1
	ireis	Internal BE saturation current (recombination)	1×10^{-17} A
	mrei	Internal BE non-ideality factor (recombination)	2
	ireps	Peripheral BE saturation current (recombination)	1×10^{-20} A
	mrep	Peripheral BE non-ideality factor (recombination)	2
Base–collector currents	ibcis	Internal BC saturation current	1.5×10^{-18} A
	mbci	Internal BC saturation current ideality factor	1.02
	ibcxs	External BC saturation current	2×10^{-19} A
	mbcx	External BC saturation current ideality factor	2
Base–emitter tunnelling current	ibets	BE tunneling saturation current	1×10^{-21} A
	abet	BE tunneling factor	36
Base–collector avalanche current	favl	Avalanche current factor	19.3 V^{-1}
	qavl	Exponent for avalanche current	150×10^{-15} C
Series resistances—base, emitter, and collector	rbi0	Internal base resistance at zero bias	20 Ω
	rbx	Extrinsic base resistance	6 Ω
	fdqr0	Correction factor for modulation by BE and BC SCR	0.2
	fgeo	Geometry factor for current crowding	0.67
	fqi	Ratio of internal to total minority charge	1.0
	fcrbi	Ratio of shunt capacitance (parallel to Rbi) to total internal capacitance	0
Emitter	re	Emitter series resistance	3 Ω
Collector	rcx	Extrinsic collector series resistance	23 Ω
Substrate transistor	iscs	CS diode saturation current	4×10^{-21} A

	msc	CS diode non-ideality factor	1
	itss	Transfer saturation current of substrate transistor	2.5×10^{-19} A
	msf	Forward non-ideality factor of substrate transfer current	1
	msr	Reverse non-ideality factor of substrate transfer current	1
	tsf	Transit time (forward operation)—substrate	2×10^{-12} sec
Substrate network	rsu	Substrate resistance	50 Ω (layout dependent)
	csu	Substrate capacitance	3×10^{-18} F
Self-heating	rth	Thermal resistance	700 K/W
	cth	Thermal capacitance	350 pJ/K
Base–emitter junction capacitance	cjei0	Internal zero-bias BE depletion capacitance	35×10^{-15} F
	vdei	Internal BE built-in voltage	1.0 V
	zei	Internal BE grading coefficient	0.32
	aljei	Maximum internal depletion capacitance divided by cjei0	2.0
	cjep0	Peripheral zero-bias BE depletion capacitance	5×10^{-15} F
	vdep	Peripheral BE built-in voltage	1.0 V
	zep	Peripheral BE grading coefficient	0.32
	aljep	Maximum peripheral depletion capacitance divided by cjep0	2.2
	ceox	Emitter oxide (overlap) capacitance	18×10^{-15} F
Base–collector junction capacitance	cjci0	Internal zero-bias BC depletion capacitance	7×10^{-15} F
	vdci	Internal BC built-in voltage	0.7 V
	zci	Internal BC grading coefficient	0.3
	vptci	Punch-through voltage of internal BC junction	2.5 V
	cjcx0	External zero-bias BC depletion capacitance	30×10^{-15} F
	vdcx	External BC built-in voltage	0.73 V
	zcx	External BC grading coefficient	0.4
	vptcx	Punch-through voltage of external BC junction	100 V
	ccox	BC overlap capacitance	2.5×10^{-15} F
	fbc	Partitioning factor for cjcx and ccox over rbx	0.8
Collector–substrate junction capacitance	cjs0	Zero-bias CS depletion capacitance	40×10^{-15} F
	vds	CS built-in voltage	0.6 V
	zs	CS grading coefficient	0.3
	vpts	Punch-through voltage of CS junction	1×10^{10} V
Diffusion capacitances/transit times—low currents	t0	Low current forward transit time at $V_{cb}=0$ V	2.6×10^{-12} sec
	dt0h	Time constant for base and BC space charge layer width modulation	0.9×10^{-12} sec
	tbvl	Time constant for modeling carrier jam at low V_{ce}	0.7×10^{-12} sec
High currents	tef0	Neutral emitter storage time	40×10^{-15} sec
	gtfe	Exponent for current dependence of neutral emitter storage time	1.0

(*Continued*)

TABLE A.4.1 HICUM (v 2.1) SiGe HBT Model Parameters (*Continued*)

Group	Name	Parameter Description	Value
	thcs	Saturation time constant at high current densities	25×10^{-12} sec
	alhc	Smoothing factor for current dependence of base and collector transit time	0.53
	fthc	Factor for partitioning this into base and collector portion	0.6
	vces	Internal C-E saturation voltage	0.1 V
	rci0	Internal collector resistance at low electric field	20 Ω
	vlim	Voltage separating ohmic (low field) and saturation velocity (high field) regime	0.7 V
	vpt	Collector punch-through voltage	15 V
	tr	Storage time for inverse operation	20×10^{-12} sec
Non-quasistatic effects	alqf	Factor for additional delay time of minority charge	0.125
	alit	Factor for additional delay time of transfer current	0.45
Noise parameters	kf	Flicker noise factor	22×10^{-6}
	af	Flicker noise exponent factor	2.5
	krbi	Noise factor for internal base resistance	1
Temperature effect parameters	tnom	Measurement temperature	25 C
	vgb	Bandgap voltage	1.17 V
	alb	Temperature coefficient of current gain	6×10^{-3}
	alfav	Temperature coefficient of favl	5×10^{-5} K^{-1}
	alqav	Temperature coefficient of qavl	2×10^{-4} K^{-1}
	zetaci	Temperature coefficient for mobility in epi-collector (i.e., for collector resitance)	1.6
	alvs	Relative temperature coefficient of saturation drift velocity	1×10^{-3} K^{-1}
	alces	Relative temperature coefficient of vces	0.4×10^{-3} K^{-1}
	zetarbi	Temperature coefficient for mobility in internal base (i.e., for internal base resistance)	0.6
	zetarbx	Temperature coefficient for mobility in extrinsic base (i.e., for extrinsic base resistance)	0.2
	zetarcx	Temperature coefficient for mobility in extrinsic collector (i.e., for extrinsic collector resistance)	0.2
	zetare	Temperature coefficient for emitter resistance	0
	alt0	First-order temperature coefficient of t0	1×10^{-3} K^{-1}
	kt0	Second-order temperature coefficient of t0	1×10^{-5} K^{-2}

TABLE A.4.2 MEXTRAM 504 SiGe HBT Model Parameters

Group	Name	Parameter Description	Value
Forward and reverse currents	is	Transistor main saturation current	5×10^{-18} A
	ik	Knee current for high-injection effects in the base	4.5×10^{-2} A
	bf	Ideal forward current gain	95
	ibf	Saturation-current of the non-ideal forward base current	2×10^{-17} A
	mlf	Non-ideality factor of the non-ideal forward base current	1.545
	xibi	Sidewall component of ideal base current	0
	bri	Ideal reverse current gain	3.77
	ibr	Saturation current of the non-ideal reverse base current	1.7×10^{-15} A
	vlr	Cross-over voltage of the non-ideal reverse base current	1×10^{-2} V
	xext	Partitioning factor for the extrinsic region	0.19
Early voltage	ver	Reverse Early voltage	4.8 V
	vef	Forward Early voltage	65 V
Weak avalanche	wavl	Epilayer thickness used in weak-avalanche model	2.44×10^{-7} M
	vavl	Voltage determining curvature of avalanche current	0.63 V
	sfh	Current spreading factor of avalanche model (when exavl $=1$)	1.7
Resistances and quasisaturation	re	Emitter resistance	3.0 Ω
	rbc	Constant part of the base resistance	6 Ω
	rbv	Zer-bias value of the bias-dependent base resistance	20 Ω
	rcc	Constant part of the collector resistance	17 Ω
	rcv	Resistance of the un-modulated epilayer	52 Ω
	scrcv	Space charge resistance of the epilayer	54 Ω
	ihc	Critical current for velocity saturation in the epilayer	3.56×10^{-3} A
	axi	Smoothness parameter for the onset of quasi-saturation	0.21
Base–emitter junction capacitance	cje	Zero-bias emitter–base depletion capacitance	42×10^{-15} F
	vde	Emitter–base diffusion voltage	0.9 V
	pe	Emitter–base grading coefficient	0.23
	xcje	Fraction of the emitter–base depletion capacitance that belongs to the side-wall	0
	cbeo	Emitter–base overlap capacitance	18×10^{-15} F
Base–collector junction capacitance	cjc	Zero–bias collector–base depletion capacitance	7×10^{-15} F
	vdc	Collector–base diffusion voltage	0.75 V
	pc	Colector–base grading coefficient	0.28
	xp	Constant part of cjc	1×10^{-3}

(Continued)

TABLE A.4.2 MEXTRAM 504 SiGe HBT Model Parameters (*Continued*)

Group	Name	Parameter Description	Value
	mc	Coefficient for the current modulation of the collector–base depletion capacitance	0.5
	xcjc	Fraction of the collector–base depletion capacitance under the emitter	8.7×10^{-2}
	cbco	Collector–base overlap capacitance	2.5×10^{-15} F
Collector–substrate junction capacitance	cjs	Zero-bias collector–substrate depletion capacitance	45×10^{-15} F
	vds	Collector–substrate diffusion voltage	0.6 V
	ps	Collector–substrate grading coefficient	0.3
	vgs	Bandgap voltage of the substrate	1.17 V
Diffusion capacitances/transit times	mtau	Non-ideality factor for the emitter stored charge	0.388
	taue	Minimum transit time of stored emitter charge	52×10^{-15} sec
	taub	Transit time of stored base charge	1.44×10^{-12} sec
	tepi	Transit time of stored epilayer charge	14.4×10^{-12} sec
	taur	Transit time of reverse extrinsic stored base charge	20×10^{-12} sec
HBT parameters	deg	Bandgap difference over the base	0.03 eV
	xrec	Pre-factor of the recombination part of ideal base current	0
Temperature coeffcients	aqbo	Temperature coefficient of the zero-bias base charge	0.34
	ae	Temperature coefficient of the resistivity of the emitter	0
	ab	Temperature coefficient of the resistivity of the base	1.22
	aepi	Temperature coefficient of the resistivity of the epilayer	1.88
	aex	Temperature coefficient of the resistivity of the extrinsic base	8.7×10^{-7}
	ac	Temperature coefficient of the resistivity of the buried layer	1
	as	Temperature coefficient for Iss and Iks (for a closed buried layer, as=ac and for an open buried layer, as=aepi)	0.76
	dvgbf	Bandgap voltage difference for forward current gain	3.75×10^{-2} V
	dvgbr	Bandgap voltage difference for reverse current gain	4.38×10^{-2} V
	vgb	Bandgap voltage of the base	1.15 V
	vgc	Bandgap voltage of the collector	1.18 V
	vgj	Bandgap voltage: recombination of the emitter–base junction	1.15V
	dvgte	Bandgap voltage difference of emitter stored charge	0.236 V
1/*f* Noise	af	Exponent of the flicker noise	2.5
	kf	Flicker-noise coefficient of the ideal base current	22×10^{-6}
	kfn	Flicker noise coefficient of the non-ideal base current	20×10^{-12}
Substrate transistor	iss	Base-substrate saturation current	2.5×10^{-19} A
	iks	Base-substrate high-injection knee current	50 A
Self-heating network	rth	Thermal resistance	700 K/W
	cth	Thermal capacitance	350 pJ/K

TABLE A.4.3 VBIC SiGe HBT Model Parameters

Group	Name	Parameter Description	Value
Saturation currents and ideality factors	is	Transport saturation current (collector)	4.85×10^{-18} A
	ibei	Ideal base–emitter saturation current	7×10^{-20} A
	iben	Nonideal base–emitter saturation current	1×10^{-15} A
	ibci	Ideal base–collector saturation current	1.5×10^{-18} A
	ibcn	Nonideal base–collector saturation current	1×10^{-34} A
	isp	Parasitic transport saturation current	3×10^{-19} A
	ibcip	Ideal parasitic base–collector saturation current	
	ibcnp	Nonideal parasitic base–collector saturation current	1×10^{-40} A
	ibeip	Ideal parasitic base–emitter saturation current	2.52×10^{-18} A
	ibenp	Nonideal parasitic base–emitter saturation current	1×10^{-28} A
	nf	Forward emission coefficient	1.0003
	nei	Ideal base–emitter emission coefficient	1.026
	nen	Nonideal base–emitter emission coefficient	2.5
	nr	Reverse emission coefficient	1
	nci	Ideal base–collector emission coefficient	1.02
	ncn	Nonideal base–collector emission coefficient	1.00
	nfp	Parasitic forwad emission coefficent	1.00
	ncip	Ideal parasitic base–collector emission coefficient	1.00
	ncnp	Nonideal parasitic base–collector emission coefficient	2
Knee currents	ikf	Forward knee current	4.5×10^{-7} A
	ikr	Reverse knee current	4.8×10^{-3} A
	ikp	Parasitic knee current	10 A
Avalanche breakdown	avc1	Base–collector weak avalanche parameter 1	19.2
	avc2	Base–collector weak avalanche parameter 2	23.6 V
Series resistances	rbi	Intrinsic base resistance	20 Ω
	rbx	Extrinsic base resistance	6 Ω
	rbp	Parasitic base resistance	1 Ω
	re	Emitter resistance	3 Ω
	rcx	Extrinsic collector resistance	23 Ω
	rs	Substrate resistance	50 Ω
Self-heating	rth	Thermal resistance	700 K/W
	cth	Thermal capacitance	350 pJ/W

(*Continued*)

TABLE A.4.3 VBIC SiGe HBT Model Parameters (*Continued*)

Group	Name	Parameter Description	Value
Quasi-saturation parameters	rci	Intrinsic collector resistance	40 Ω
	vo	Epi drift saturation voltage	1×10^{-10} V
	gamm	Epi doping parameter	5×10^{-13} V
	hrcf	High-current RC factor	1×10^{-13}
	qco	Epi-charge parameter	1.4×10^{-15}
Early effect parameters	vef	Forward Early voltage	65 V
	ver	Reverse Early voltage	5.5 V
Base–emitter junction capacitance	cje	Base–emitter zero-bias capacitance	42 fF
	me	Base–emitter grading coefficient	1.0
	pe	Base–emitter built-in potential	0.3 V
	cbeo	Extrinsic base–emitter overlap capacitance	18×10^{-15} F
	fc	Forward bias depletion capacitance limit	0.93
	aje	Base–emitter capacitance switching parameter	−0.1
Base–collector junction capacitance	cjc	Base–collector intrinsic zero-bias capacitance	7×10^{-15} F
	mc	Base–collector grading coefficient	0.3
	pc	Base–collector built-in potential	0.7 V
	cbco	Extrinsic base–collector overlap capacitance	2.5×10^{-15} F
	cjep	Base–collector extrinsic zero bias capacitance	30×10^{-15} F
	ajc	Base–collector capacitance switching parameter	−0.1
Collector–substrate junction capacitance	cjcp	Substrate–collector zero bias capacitance	40×10^{-15} F
	ms	Substrate–collector grading coefficient	0.3
	ps	Substrate–collector built-in potential	0.6 V
	ccso	Fixed collector–substrate capacitance	3e−18 F
	ajs	Substrate–collector capacitance switching parameter	−0.9
Transit times and their bias dependence	tf	Forward transit time	2×10^{-12} sec
	itf	Coefficient of tf dependence of ic	0.32
	vtf	Coefficient of tf dependence of V_{bc}	23.7
	qtf	Variation of tf with base-width modulation	0
	td	Forward excess-phase delay time	0.7×10^{-12} sec
Temperature effect parameters	ea	Activation energy for is	1.17 V
	eaie	Activation energy for ibei	1.17 V
	eaic	Activation energy for ibci and ibeip	1.17 V
	eais	Activation energy for ibcip	1.17 V
	eanc	Activation energy for ibcn/ibenp	1.17 V

	eane	Activation energy for iben	1.17 V
	eans	Activation energy for ibcnp	1.17 V
	xii	Temperature coefficient for ibei, ibci, ibeip, and ibcip	2.0
	xin	Temperature coefficient for iben, ibcn, ibenp, and ibcnp	2.0
	xis	Temperature coefficient for is	1.9
	xre	Temperature coefficient for re	0
	xrb	Temperature coefficient for rbi	0
	xrc	Temperature coefficient for rc	0
	xrs	Temperature coefficient for rs	0
	xvo	Temperature coefficient for vo	0
	tavc	Temperature coefficient for avc2	250×10^{-6} V

References

1. HICUM bipolar transistor model: http://www.iee.et.tu-dresden.de/iee/eb/comp_mod.html.
2. Mextram bipolar transistor model: http://www.semiconductors.phillips.com/acrobat/other/phillips-models/NLUR2000811_7.pdf.
3. VBIC bipolar transistor model: http://www.designers-guide.org/VBIC/references.html.

Index

F

G

H

I

J

K

L

T

U

V

W

X